W9-BDM-158

KAPLAN

GRE* BIOLOGY

By Tim Levin

Simon & Schuster

New York · London · Singapore · Sydney · Toronto

* GRE is a registered trademark of the Educational Testing Service, which is not affiliated with this book.

Kaplan Publishing
Published by Simon & Schuster
1230 Avenue of the Americas
New York, New York 10020

For bulk sales to schools, colleges, and universities, please contact: Order Department, Simon and Schuster, 100 Front Street, Riverside, NJ 08075. Phone: (800) 223-2336. Fax: (800) 943-9831.

The material in this book is up-to-date at the time of publication. However, Educational Testing Service (ETS) may have instituted changes in the test after this book was published. Be sure to carefully read all material you receive regarding the GRE test carefully.

Contributing Editor: Albert Chen
Editor: Ruth Baygell
Cover Design: Cheung Tai
Interior Page Design: Laurel Douglas
Production Editor: Maude Spekes
Desktop Publishing Manager: Michael Shevlin
Editorial Coordinator: Déa Alessandro
Executive Editor: Del Franz

This book would not have been possible without the efforts and insights of the staff of Kaplan, Inc. The author would also like to thank Dr. Deborah Eastman (Biology Department, Southwestern University), Dr. Pauline Carrico (Incyte Genomics), and Ashley Carter and Scott Rifkin (School of Forestry and Environmental Studies, Yale University) for their indispensable contributions in writing portions of the simulated GRE Biology exam; Dr. Jonathan Barasch (College of Physicians and Surgeons, Columbia University) for helping to write some portions of the Organismal Biology section; and Glenn Croston, for his invaluable insight.

Manufactured in the United States of America
Published simultaneously in Canada

June 2002

10 9 8 7 6 5 4 3 2 1

ISBN: 0-7432-3064-7

TABLE OF CONTENTS

ABOUT THE AUTHOR

Tim Levin is a science curriculum developer for Kaplan, and he has worked for the past two years to develop curricular material for Kaplan's MCAT students. Prior to his work with Kaplan, he taught high school and college-level biology for many years. In addition to having taken classes at the University of Vermont College of Medicine and in Columbia University's biotechnology program, he has a master's degree in education from Columbia University. He plans to finish his doctoral degree at Columbia this year.

The material in this book is up-to-date at the time of publication. However, Educational Testing Service may have instituted changes in the test after this book was published. Be sure to carefully read the materials you receive when you register for the test.

If there are any important late-breaking developments—or any changes or corrections to the Kaplan test preparation materials in this book—we will post that information online at kaptest.com/publishing. Check to see if there is any information posted there regarding this book.

Part I

THE GRE BIOLOGY EXAM

The Basics: Test Information and Strategies

ABOUT THE GRE BIOLOGY EXAM

What is the GRE Biology exam?

The GRE Biology exam is a two hour and fifty minute exam designed to test advanced knowledge that a student applying to graduate school in the biological sciences is expected to understand. The test requires knowledge of vocabulary and facts across a variety of biological fields at the equivalent of an upper-level college class. The test consists of approximately 200 multiple-choice questions. There are no essay questions.

What's covered on the exam?

The GRE Biology exam covers three major areas:

Cellular and Molecular Biology	33–34%
Organismal Biology	33–34%
Ecology and Evolution	33–34%

Each area is further broken down into topical areas. The percentages listed below relate to the number of questions you can expect to find for a particular topic. For example, a percentage of 4–5% next to a topic below indicates that 4–5 percent of the 200 questions will deal with that area of biology (about 10 questions).

Cellular and Molecular Biology (33–34%)

A. Cellular Structure and Function (16–17%)
- Biological compounds
- Enzyme activity, receptor binding, and regulation
- Major metabolic pathways and regulation
- Membrane dynamics and cell surfaces
- Organelles: structure, function, synthesis, and targeting
- Cytoskeleton, motility, and shape
- Cell cycle

B. Molecular Biology and Molecular Genetics (16–17%)
- Genetic foundations
- Chromatin and chromosomes
- Genome sequence organization
- Genome maintenance
- Gene expression and regulation in prokaryotes and eukaryotes
- Immunobiology
- Bacteriophages, animal viruses, and plant viruses
- Recombinant DNA technology

Organismal Biology (33–34%)

A. Animal Structure, Function, and Organization (9–10%)
- Exchange with the environment
- Internal transport and exchange
- Support and movement
- Integration and control mechanisms
- Behavior
- Metabolic rates

B. Animal Reproduction and Development (5–6%)
- Reproductive structures
- Meiosis, gametogenesis, and fertilization
- Early development
- Developmental processes
- External control mechanisms

C. Plant Structure, Function, and Organization (6–7%)
- Tissues, tissue systems, and organs
- Water transport, including absorption and transpiration
- Phloem transport and storage
- Mineral nutrition
- Plant energies (respiration and photosynthesis)

D. Plant Reproduction, Growth, and Development (6–7%)
- Reproductive structures
- Meiosis and sporogenesis
- Gametogenesis and fertilization
- Embryogeny and seed development
- Meristems, growth, morphogenesis, and development
- Control mechanisms

E. Diversity of Life (6–7%)
- Monera
- Protista
- Fungi
- Animalia
- Plantae

Ecology and Evolution (33–34%)

A. Ecology (16–17%)
- Environment/organism interaction
- Behavioral ecology
- Population structure and function
- Communities
- Ecosystems

B. Evolution (16–17%)
- Genetic variability
- Evolutionary processes
- Evolutionary consequences

ANATOMY OF THE GRE BIOLOGY EXAM

The multiple-choice questions on the exam all have five answer choices, (A) through (E). The questions at the beginning of the test are more likely to involve factual recall. Here are some examples of these types of questons:

1. Bacterial transduction can be accomplished using

    (A) in vitro fusion of two different cell lines resulting in a hybridoma.
    (B) injection of bacterial DNA into a fertilized mouse egg.
    (C) bacteriophages that infect bacteria in a lysogenic manner.
    (D) implantation of genetically altered bacteria into a host eukaryotic cell.
    (E) inducing transcription of nontranscribed regions of plasmid DNA.

    Answer: (C) Bacteriophages are viruses that infect bacteria. If they infect in a lysogenic manner, which will not kill the bacterium, they will be able to integrate their DNA into the bacterial chromosome. This results in a bacterium that produces viral proteins and also duplicates the viral DNA as it reproduces. This process is called transduction.

2. Actin filaments play a role in all of the following situations EXCEPT in

    (A) release of neurotransmitter from the axon terminal.
    (B) the extension of pseudopods during amoeboid locomotion.
    (C) cytokinesis of animal cells after telophase of mitosis.
    (D) sarcomere shortening within muscle fibers.
    (E) reinforcement of microvilli within the digestive tract.

    Answer: (A) Actin filaments, otherwise known as microfilaments, play a role in all of the situations listed above except in the release of neurotransmitter from the ends of axons, a process mediated by calcium ion influx and the fusion of membrane-bound vesicles with the axon membrane.

The questions toward the end of the exam are grouped in sets and are based on experimental data or situations. These application questions involve more critical thinking and data analysis. Here are some examples of these types of questions:

Questions 3–4

Populations of finches on Espaniola, one of the Galapagos Islands, were studied over nearly 50 years to determine the effects of changing seed size on beak thickness and length. Finches with larger beaks were generally found in areas where seed size was larger, because stronger and thicker beaks serve to puncture the hard outer coating of large seeds more efficiently than smaller beaks. With smaller seeds, smaller-beaked finches were found. The graphs below shows the distribution of seeds on Espaniola from 1950–1990, and the distribution of finch beak size from 1960–1997.

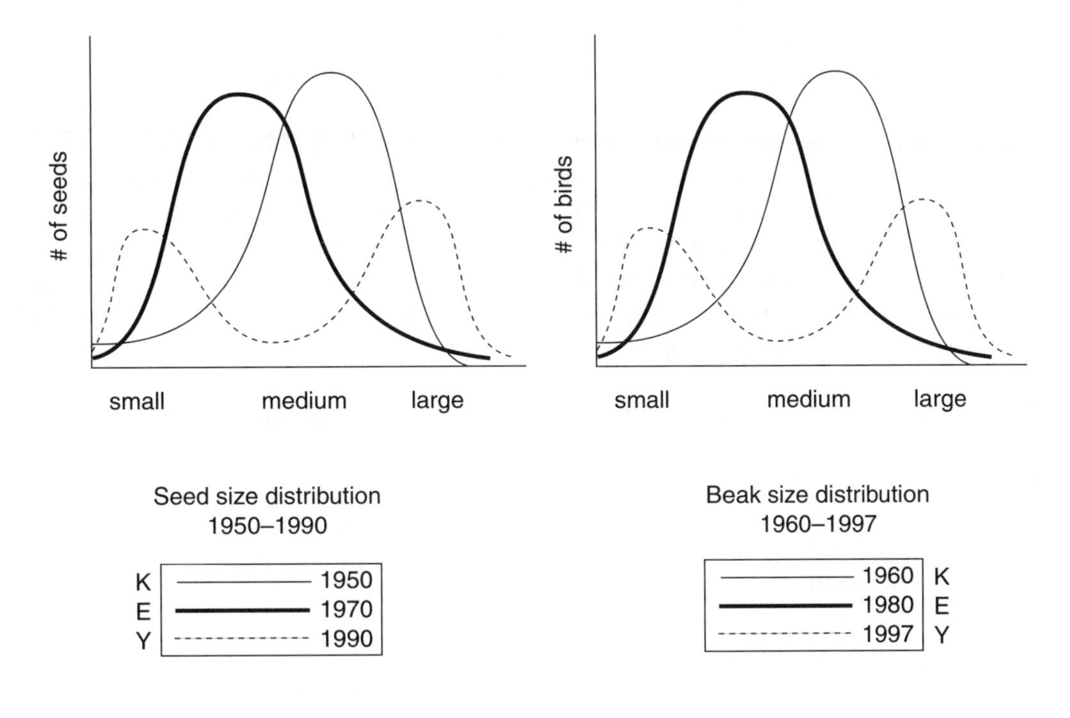

3. The shift in beak size from 1980 to 1997 can be described as
   (A) divergent evolution.
   (B) directional selection.
   (C) diversifying selection.
   (D) convergent evolution.
   (E) Hardy-Weinberg equilibrium.

Answer: (C). From 1980 to 1997, beak size shifts from medium-sized beaks to approximately equal proportions of small and large beaks. This shift suggests that the environment on Espaniola favored individuals with extremes of beak size rather than intermediate individuals. This shift is known as diversifying, or disruptive, selection.

4. The decrease in average seed size from 1950 to 1970 can be best explained by

   (A) a drought on the island that resulted in the allocation of less endosperm in seeds produced by the island's plants.
   (B) the increase in overall beak size in finches during that same period.
   (C) the directional selection of beak size toward smaller beaks during that same period.
   (D) the evolution of smaller plants on the island as a defense against finches with medium beak sizes.
   (E) the evolution of a balanced polymorphism between plants producing smaller seeds and those producing larger seeds.

Answer: (A) Notice in the keys to the graphs that beak size is always changing after seed size changes. In other words, the timing of beak changes follows that of seed changes, which means that beak sizes are not dictating seed sizes, but rather the other way around. Thus, choice (A) makes the most sense here, as a drought would result in the conservation of resources and mean smaller plant sizes and smaller seed sizes over time. In fact, this cycle of droughts and smaller beak sizes alternating with plentiful water and larger beak sizes has been witnessed with many populations of finches on the Galapagos Islands.

You will also see questions on the test that ask you to match statements to a list of answer choices:

Questions 5–7

(A) Mesoderm
(B) Endoderm
(C) Ectoderm
(D) Morula
(E) Blastula

5. This germ layer produces the nervous system.
6. This germ layer produces the digestive system.
7. This germ layer produces the circulatory system.

For question 5, the answer is (C). The ectoderm develops into the epidermis (outer layer of skin), nervous system, and sweat glands. Question 6 is (B), as the endoderm develops into the digestive and respiratory tracts, parts of the liver and pancreas, and the lining of the bladder. And for question 7, (A) is the correct answer: Mesoderm develops into the muscles and skeletal system, circulatory system, excretory system, gonads, and dermis (inner layer of skin).

SCORING THE GRE BIOLOGY EXAM

You will receive one total score plus three subscores in the major topic areas for the GRE Biology Test. Scores for this test usually range from 490 to 790. The range of scores for each administration of the test differs due to standardization of the different editions of the test. So while the 96th percentile for a test administered in April one year might be a total score of 840, the 96th percentile of the November administration might be closer to 810. To clarify these differences, the score report that is sent to you will also tell you how your total score converts to a national percentile rank. Subscores for each of the three sections are reported as two-digit scores ranging from 20–99. Again, like the total scores, these subscores vary with the different administrations of the test.

You are penalized on the test for incorrect answers, and questions left blank are not counted. Your scaled scores on each subsection are computed by subtracting one-fourth the number of incorrect answers from the number of correct answers. It is almost always to your advantage on this test to guess as long as you can rule out one or two answer choices, even if you are not sure of the right answer.

The GRE Biology test is a paper-and-pencil test. Unlike with computer-based exams, you will not know your score immediately after you finish the test. A score report will be sent to you a number of weeks after you have taken the exam. Your score report will look something like this:

MO.	11
YR.	00
TYPE	N
CODE	24
SCORE	840
% BELOW	96
SS1	87
SS2	82
SS3	74
CORRECT	142
INCORRECT	51
OMITS	05

The score report will show your total score (in this student's case, an 840) as well as the percentage rank (96th percentile) and subscores for each of the three subsections (SS1, SS2, and SS3). It will also indicate how many questions were answered correctly and incorrectly out of the approximately 200 questions, as well as the number of questions not answered at all.

TEST-TAKING STRATEGIES

You have 170 minutes to complete about 200 questions, some of which will involve reading short passages and examining data or graphs. You will need to carefully budget your time on the test. Here are some strategies to maximize your score.

1. *Answer the easier questions first.* Easy questions are worth just as many points as hard questions. To maximize your score, you need to answer as many questions correctly as possible, but it doesn't matter if the questions are easy or hard. On your first pass through the test, answer all the easy questions first.

2. *Circle harder questions and come back to them later.* If you come across a tough question, circle it and move on. Don't waste valuable time on hard questions early on in the exam. If you start to answer a question and then find yourself confused, move on and come back to that question later. You're better off spending those extra few minutes answering 3 or 4 easier questions.

3. *Mark up your test booklet.* As a student, you may be used to having teachers tell you not to write in your books. But when taking the GRE Biology exam, it is to your advantage to mark up your test booklet. Label diagrams, cross out incorrect answer choices, and write down key notes. Just remember that no credit is given for any work you do in your answer booklet, so make sure to transfer your final answers to the answer grid.

4. *Be careful with your answer sheet.* Speaking of your answer grid, it's easy to forget to skip a row on your answer grid when you skip a question. Make sure to skip the necessary spaces! Otherwise, you will likely spend valuable time erasing and remarking your answer sheet. Also, make sure to erase fully when you change an answer.

5. *Guess.* Yes, guess! There is a 1/4 point deduction for wrong answers, which cancels out the effect of random guessing. Random guessing, therefore, will not improve your score, but if you can eliminate one or two clearly wrong answer choices, your odds of guessing the right answer improve tremendously.

6. *Pace yourself.* The GRE Biology exam is almost three hours long. It is easy to get distracted and lose focus. Try to keep yourself on task and actively working throughout the exam. Don't get discouraged if you don't know the answers to some questions. You don't need to answer all of the questions correctly to get a good score.

7. *The evening before the exam, relax!* It's tempting to spend the last few hours cramming, but this tends to be counterproductive. Do something low-stress.

MANAGING STRESS

The countdown has begun. Your date with the test is looming on the horizon. Anxiety is on the rise. The butterflies in your stomach have gone ballistic. Your thinking is getting cloudy. Maybe you think you won't be ready. Maybe you already know your stuff, but you're going into panic mode anyway. Don't freak! It's possible to tame that anxiety and stress—before and during the test. Remember, a little stress is good.

Anxiety is a motivation to study. The adrenaline that gets pumped into your bloodstream when you're stressed helps you stay alert and think more clearly. But if you feel that the tension is so great that it's preventing you from using your study time effectively, here are some things you can do to get it under control.

Take Control.

Lack of control is a prime cause of stress. Research shows that if you don't have a sense of control over what's happening in your life, you can easily end up feeling helpless. Try to identify the sources of the stress you feel. Which ones can you do something about? Can you find ways to reduce your stress about any of these sources?

Focus on Your Strengths.

Make a list of areas of strength you have that will help you do well on the test. We all have strengths, and recognizing your own is like having reserves of solid gold at Fort Knox. You'll be able to draw on your reserves as you need them, helping you solve difficult questions, maintain confidence, and keep test stress and anxiety at a distance. And every time you recognize a new area of strength, solve a challenging problem, or score well on a practice test, you'll increase your reserves.

Imagine Yourself Succeeding.

Close your eyes and imagine yourself in a relaxing situation. Breathe easily and naturally. Now, think of a real-life situation in which you scored well on a test or did well on an assignment. Focus on this success. Now turn your thoughts to the GRE Biology test, and keep your thoughts and feelings in line with that successful experience. Don't make comparisons between them; just imagine yourself taking the upcoming test with the same feelings of confidence and relaxed control.

Set Realistic Goals.

Facing your problem areas gives you some distinct advantages. What do you want to accomplish in the time remaining? Make a list of realistic goals. You can't help feeling more confident when you know you're actively improving your chances of earning a higher test score.

Exercise Your Frustrations Away.

Whether it's jogging, biking, pushups, or a pickup basketball game, physical exercise will stimulate your mind and body, and improve your ability to think and concentrate. A surprising number of students fall out of the habit of regular exercise, ironically because they're spending so much time prepping for exams. A little physical exertion will help to keep your mind and body in sync and sleep better at night.

Avoid Drugs.

Using drugs (prescription or recreational) specifically to prepare for and take a big test is definitely self-defeating. (And if they're illegal drugs, you may end up with a bigger problem than the GRE Biology test on your hands.) Mild stimulants, such as coffee or cola can sometimes help as you study, since they keep you alert. On the down side, too much of these can also lead to agitation, restlessness, and insomnia. It all depends on your tolerance for caffeine.

Eat Well.

Good nutrition will help you focus and think clearly. Eat plenty of fruits and vegetables, low-fat protein such as fish, skinless poultry, beans, and legumes, and whole grains such as brown rice, whole wheat bread, and pastas. Don't eat a lot of sugar and high-fat snacks, or salty foods.

Work at Your Own Pace.

Don't be thrown if other test takers seem to be working more furiously than you. Spend your time thinking through your answers in your own style; it will lead to better results. Don't mistake other people's sheer activity as signs of progress and higher scores.

Keep Breathing.

Conscious attention to breathing is an excellent way to keep anxiety down while you take the test. Most people who get into trouble during tests take shallow breaths: They breathe using only their upper chests and shoulder muscles, and may even hold their breath for long periods of time. Conversely, those test takers who breathe deeply in a slow, relaxed manner are likely to be in better control during the session.

Stretch.

If you find yourself getting spaced out or burned out as you're taking the test, stop for a brief moment and stretch. Even though you'll be pausing on the test for a moment, it's a moment well spent. Stretching will help to refresh you and refocus your thoughts.

ADDITIONAL RESOURCES

The GRE subject tests are given each year in November, December, and April. For more information, contact:

Educational Testing Service (ETS)
Rosedale Road
Princeton, New Jersey 08541
Phone: (609) 771-7670
Fax: (609) 683-2040
Email: gre-info@ets.org
Web: www.gre.org

Part II

CELLULAR STRUCTURE AND MOLECULAR BIOLOGY

Introduction to Cellular Structure and Molecular Biology

Biologists are increasingly able to investigate the actual building blocks of life on Earth, the molecular components that build cells, tissues, and organisms. Armed with powerful **molecular biology** techniques, scientists are working to illuminate the inner workings of cells and of the molecules that make up these cells. This section focuses on the structure and function of the biomolecules found in all life, from **proteins** and **sugars** to **lipids** and **nucleic acids**. The GRE will test your knowledge of how these molecules are shaped, how they combine, and what properties they possess. You will also need to be familiar with overall cell structure and the pieces of cells that allow them to move, eat, reproduce, and grow.

A major area covered here is molecular genetics, where the GRE will test you on **DNA's structure and function**, DNA repair, and how DNA can be studied and compared from organism to organism. Various types of genome organization are looked at, from viral and prokaryotic genomes to the more complex eukaryotic ones, and models of **gene control and organization** are discussed at length. Because the field of molecular biology and molecular genetics is expanding so rapidly, this section cannot present every aspect of molecular biology and cell biology, but it gives you an overview of the areas you should be familiar with for the GRE.

CHAPTER ONE

Macromolecules and Bonding

All of the macromolecules that make up living cells are built from small, repeating units of **monomers** (literally meaning "single molecule"), which are linked over and over again to form **polymers** (large molecules of at least several monomers, but generally 100s or 1,000's). There are four main types of biomolecules that you should be familiar with, and you should most definitely know the monomers of each of these larger molecules:

Biomolecule	Monomer(s)	Major Elements
Proteins	Amino acids (20 naturally occurring)	C, H, N, O, (S)
Carbohydrates	Simple sugars (e.g., glucose, fructose)	C, H, O
Lipids	Glycerol and Fatty Acids	C, H, O
Nucleic Acids	Nucleotides (phosphate, sugar, base)	C, H, N, O, P

Each of these will be covered in turn in this chapter.

PROTEINS AND AMINO ACIDS

Proteins are essential for every single biological process. These macromolecules are used as enzymes, for transport and storage, for building of membranes, and for cellular movement. In addition, some proteins act to turn on and off genes and as signaling molecules (messengers) within cells. Still others are used as antibodies in the immune response and as neurotransmitters for cell to cell communication. Proteins also form the vast network of connective tissue that holds cells of more complex organisms together.

Proteins are built from smaller monomers known as **amino acids**, so named for the **amino** (NH_2) group at one end of their structure and the acid (or **carboxyl**, COOH) group at the other end of their structure. As seen below, there is an *R*, or variable, group attached to the central carbon, which differs from amino acid to amino acid—otherwise, amino acids all share the same structure.

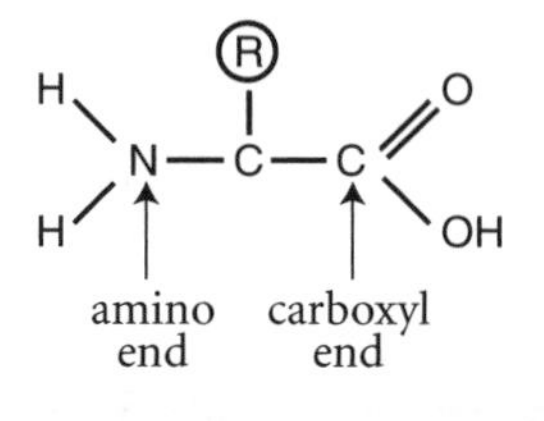

It is these *R* groups that give a protein its complex, 3-D shape, and therefore its function. As amino acids link to each other, these protruding *R* groups interact in a variety of ways: some *R* groups attract each other, some repel, and a few form covalent bonds with other *R* groups. The reason for these attractions or repulsions stems from the properties of each *R* group: some are polar, charged either negatively or positively; some are nonpolar; and others have special functional groups, such as the **sulfhydryl** (-SH) attachment that allows two sulfur-containing amino acids to link sulfur to sulfur creating disulfide bridges across the structure of the protein, essential for its overall 3-D shape.

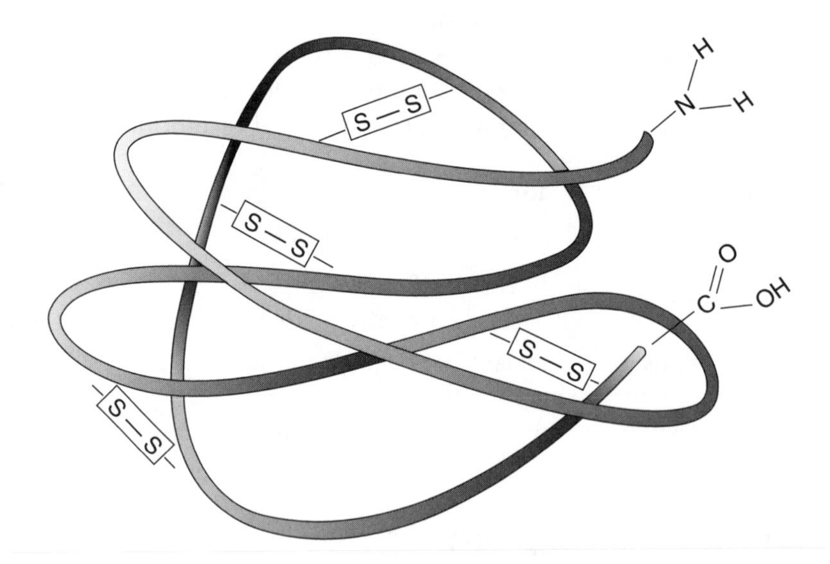

There are 20 naturally occurring amino acids found in proteins that make up living cells. Structures for these amino acids can be found in any college biology text, but you will *not* have to know these structures for the GRE. When amino acids remain monomeric (unbound to other amino acids) at physiologic pH (pH = 7 in most living tissue), they can be found as **zwitterions.** You should be familiar with this term mainly because it illustrates how amino acids can act as natural **pH buffers** for living systems. In the figure below, a hydrogen atom from the carboxyl terminus of the amino acid dissociates and is picked up by the amino terminus. Although the positive and negative charges cancel out overall to give a neutral molecule, the charge separation allows the positively charged amino end to pick up excess

OH^- ions in solution, while the negatively-charged carboxyl end can pick up excess H^+ ions—thus, pH changes are prevented because of the lack of free hydrogen and hydroxide ions in the solution.

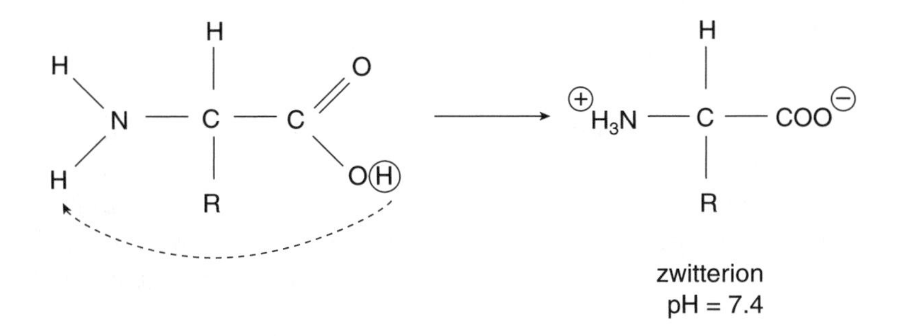

When two amino acids attach, they do so through a reaction known as a **dehydration synthesis**, whereby the carboxyl carbon of one amino acid binds to the amino nitrogen of another in order to form a **dipeptide**, or two attached amino acids. The —OH from the acid group and the —H from the amino group are removed to form water, hence the term dehydration. Amino acids can continue to link in this fashion to form **polypeptides**, or proteins. Most biologically active proteins contain anywhere from several hundred to several thousand linked amino acids.

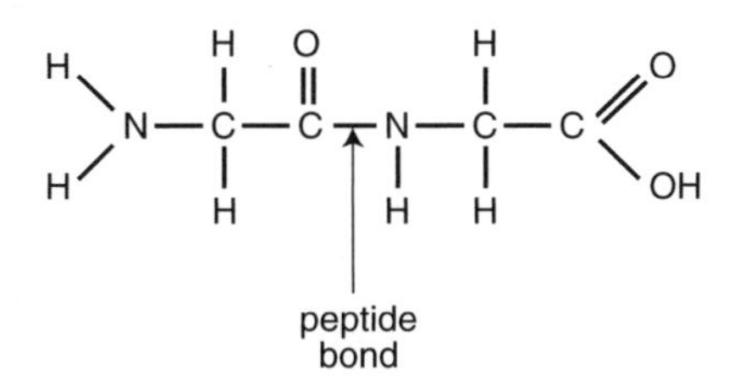

As amino acids attach to one another, they form a long chain that starts to twist and turn according to interactions among the various *R* groups, as discussed earlier. For the GRE, you should be familiar with the four different levels of structure that proteins can have:

- **Primary Structure**: This is simply the linear order of the amino acid chain (e.g., glycine – leucine – glycine – alanine – lysine – etc...); primary structure dictates the secondary and tertiary structures, which give the protein its overall shape and function.

- **Secondary Structure**: Amino acids that are close together in their primary structure will often form either **alpha-helices** (tightly wound coils – like a single, coiled strand of DNA), or **beta-pleated sheets** (straight chains of amino acids held above or below each other in a parallel arrangement due to hydrogen bonding); these helices and sheets add 3-D structure to the overall protein.

- **Tertiary Structure:** Refers to bonding and interactions between amino acid side chains that are further apart from one another than those involved in secondary interactions. This includes disulfide bridges, as mentioned above, and arrangements such as the bunching up of **hydrophilic** amino acids on the outside of a **hydrophobic** core in an aqueous environment. Tertiary structure gives the whole protein its overall 3-D structure essential for function.

- **Quaternary Structure:** Some proteins form *multi-subunit* structures with other proteins in order to build one large protein that exhibits properties which none of the subunits exhibited alone. Examples include hemoglobin, a *tetramer* (four subunits) made of two identical alpha chain proteins and two identical beta chain proteins, and collagen, a *trimer* (three subunits) used to build connective tissue.

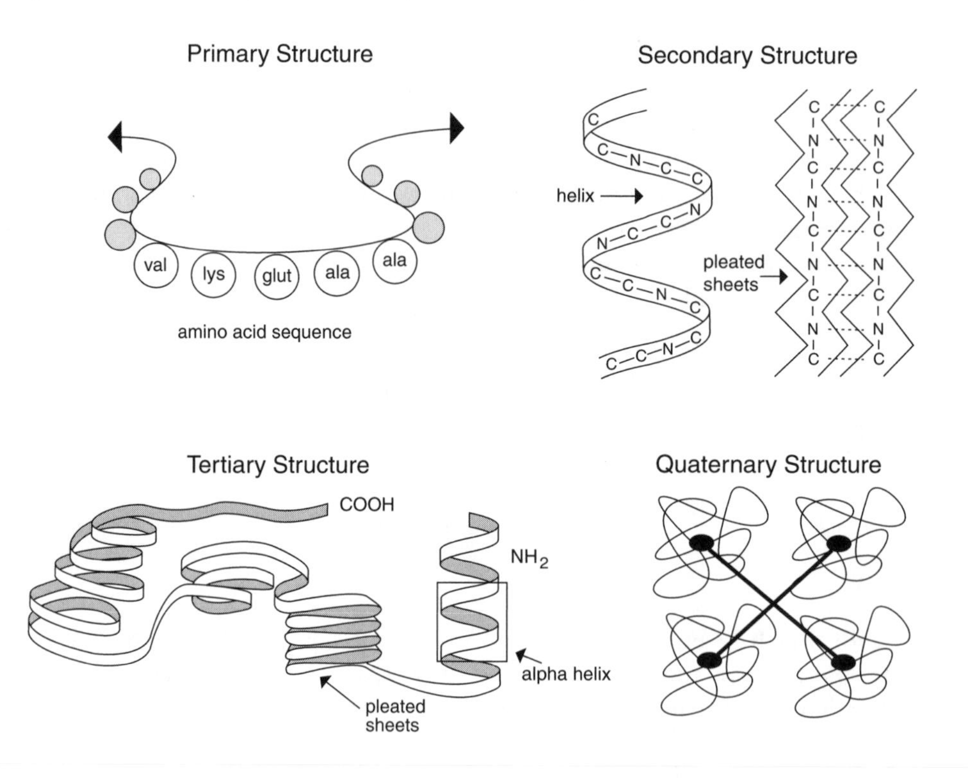

LIPIDS: FATS, OILS, WAXES, STEROIDS

Lipids are a group of nonpolar substances, insoluble in water, which are used for energy storage, membrane formation, and protective coatings. In addition, some lipids function as vitamins or hormones. They come in a variety of forms, some of which combine two or more forms: fatty acids, glycerides, complex lipids, and nonglycerides.

Fatty Acids

These long chains of carbon and hydrogen are components of many lipids. At one end of their long hydrocarbon chain is a carboxylic acid group (—COOH), which helps give fatty *acids* their name. "Fatty" comes from the fact that the long hydrocarbon chains are extremely nonpolar; the term "acid" derives from the —COOH group at one end, which can dissociate, throwing off free hydrogen ions (H^+) to make a solution acidic.

You should remember the following things about fatty acids for the GRE:

- All fatty acids have the following chemical formula: $CH_3(CH_2)_nCOOH$, where n is a number between 12 and 24, always even.

- **Saturated** fatty acids have only single bonds within their hydrocarbon chains, whereas **unsaturated** fatty acids contain one or more double bonds. These double bonds create "kinks" or bends in the fatty acid structure.

WHY ARE UNSATURATED FATTY ACIDS IMPORTANT IN MEMBRANE STRUCTURE?

The kinks in unsaturated fatty acids push other fatty acids near them away and create space within the membrane between lipids. These spaces allow membranes to resist solidifying as temperatures cool, keeping the membranes fluid.

Notice in the figure below how saturated fatty acids differ from unsaturated ones:

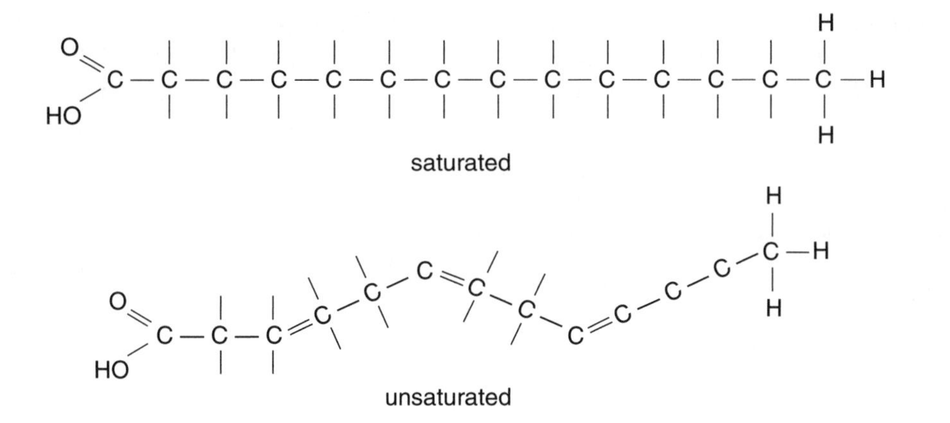

Glycerides

Glycerides are esters made of glycerol plus one to three fatty acid chains. The figure below shows the **esterification** sites encircled with dotted lines. Glycerides are a lipid form also known as fats. Fats with one fatty acid attached to glycerol are called **monoglycerides**; with two fatty acids are **diglycerides**; and with three fatty acids are **triglycerides**.

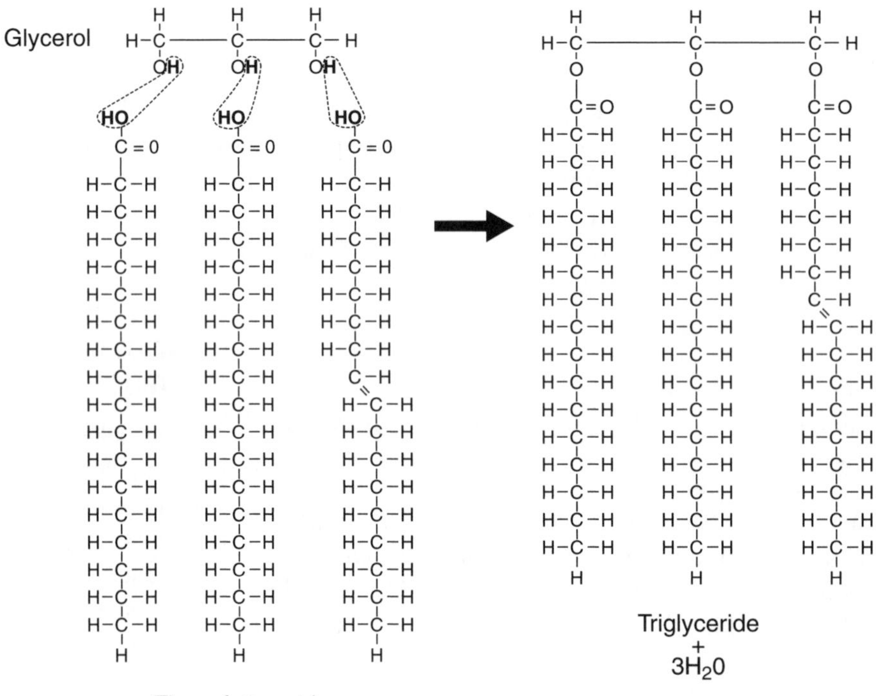

The predominant class of lipid in cell membranes is the **phospholipids**, made of a glycerol + two fatty acid chains + a phosphate group attached to a variable "R" group. The phosphate group makes ones end of this molecule polar and hydrophilic, while the fatty acids create nonpolar "tails" that hang off the opposite end of the molecule. This "amphipathic" molecule—meaning, having two poles—allows a lipid bilayer to form around cells. This lipid bi-layer has two layers of phospholipids, in which the polar heads of one layer face the extracellular space, and the polar heads of the other layer face the intracellular space. Non-polar triglycerides are used to store energy in adipose tissue. Remember that triglycerides are overall very hydrophobic molecules.

A closer look at these phospholipids shows their chemical structure:

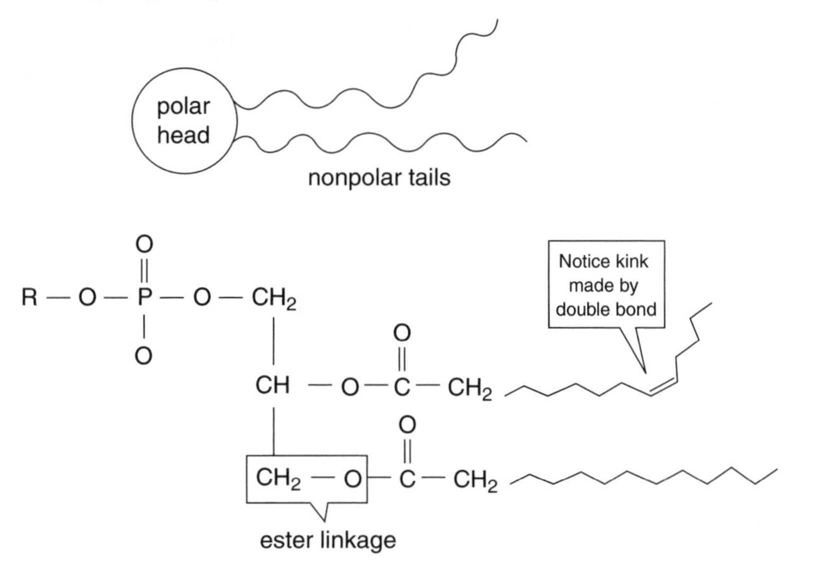

Phospholipids can also be used in the process of **emulsification**, in which amphipathic lipids surround a fat droplet and break it into smaller particles. The fatty acid tails surround the nonpolar substance and cause it to break into many smaller pieces. In this manner, fatty acids and phospholipids *act as detergents*, similar to how soap molecules extrude dirt from skin.

Complex Lipids

Lipids are known to form lipid complexes, or aggregates, which are various types of lipids grouped together with proteins or other molecules. These complexes can include triglycerides, phospholipids, steroids such as cholesterol, and proteins. A cell's plasma membrane, as seen above, is a good example of a lipid complex, containing many different types of lipids and proteins.

Lipid complexes are also useful for transporting various substances around the body, enclosing them within a lipid and protein barrier so that they can move through the bloodstream more easily. These complexes include **chylomicrons**, which surround and transport fats (triglycerides) from the intestines to other tissues, and HDL (high-density lipoproteins) and LDL (low-density lipoproteins), which carry cholesterol from the intestines to various sites around the body.

Non-Glycerides

This group of lipids is not shaped like the triglycerides, which have glycerol and fatty acids as building blocks. Rather, this group includes the **steroids**, such as cholesterol, and **sphingolipids**. The steroids, characterized by a common core of four fused carbon rings, differ from one another only in the groups that hang off these fused rings. These variable groups give the steroids their unique properties.

Common Steroid Core

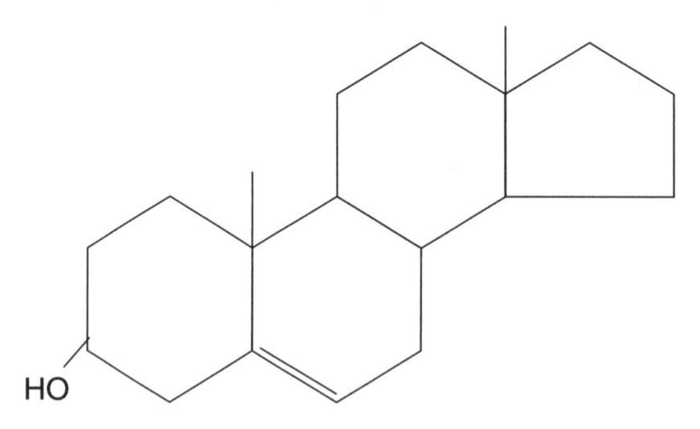

Cholesterol is a steroid commonly found within the cell membranes of mammals, an essential component for maintaining membrane fluidity. As temperatures fall, the stiff, fused-ring structure of cholesterol keeps the phospholipids in the membranes from collapsing in on one another, aided by the kinks in the fatty acids within those phospholipids. In addition, as temperatures rise, cholesterol's rigid structure adds stability and firmness between the otherwise very fluid phospholipids.

Other steroids include the sex hormones estrogen and testosterone, as well as a variety of other hormones. As will be discussed later, because steroids are lipids, they are able to pass right through the lipid bilayer of cell membranes and directly affect DNA in the cells, whereas non-steroid hormones must effect changes via cell-surface receptors and second messengers.

CARBOHYDRATES

Carbohydrates are derived from simple sugars, **monosaccharides** that contain carbon, hydrogen, and oxygen in a specific ratio. Their general formula is $(CH_2O)_n$, where n represents the number of carbons in the sugar ring or chain. The most common monosaccharide is glucose, used as an energy source in all living organisms and having the chemical formula $C_6H_{12}O_6$. Simple sugars exist either in a linear form or in a cyclic form, caused when the aldehyde or ketone group bonds to the terminal hydroxyl group in order to form a ring with oxygen in it.

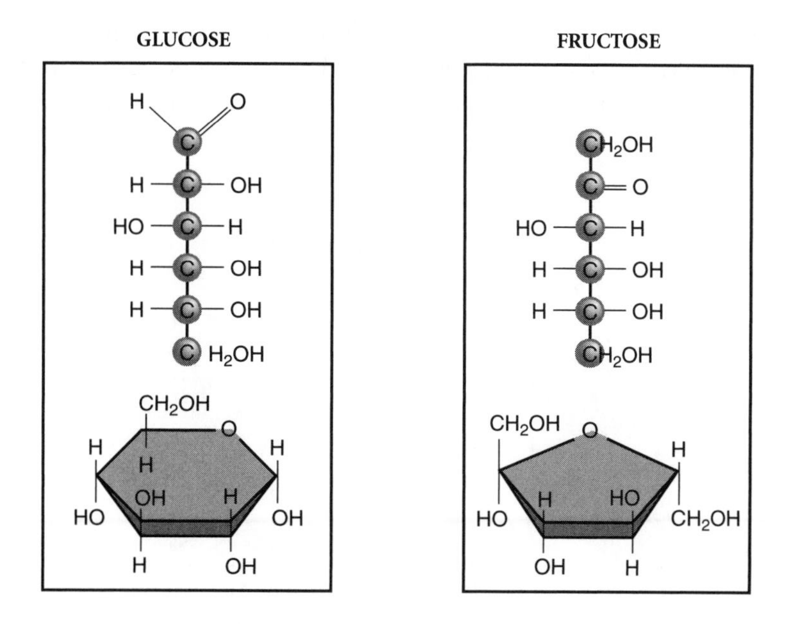

The simple sugar fructose, you will notice from the diagram above, has the same chemical formula as glucose, yet a different structure. These two sugars are known as structural isomers. Carbohydrates are joined to one another via **glycosidic bonds**, as pictured below. These bonds bridge from one sugar to another across an atom of oxygen, and come in two varieties: an **alpha linkage**, which results in kinks so that long chains of simple sugars bonded via alpha linkages take on a twisty and branching configuration; or, a **beta linkage**, which allows for the formation of straight chains of sugars.

You may see the terms "**oligosaccharide**" and "**polysaccharide**" on test day: "oligo-" means multiple but short, so an oligosaccharide is a carbohydrate made of 2–10 simple sugars attached to each other in a row. A polysaccharide is generally more than 10 sugars.

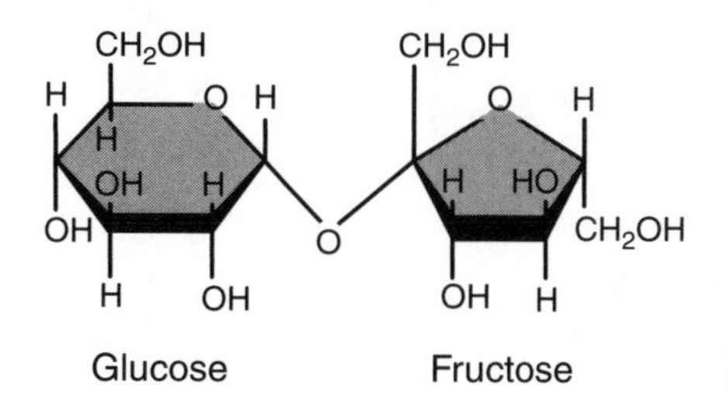

Alpha linkages are found in sugar storage molecules such as **glycogen** or **starch** (chains of glucose like "candy necklaces" that the organism can break down as energy needs require), while beta linkages are found in structural carbohydrates such as **cellulose** (which builds plant cell walls).

Another carbohydrate to be aware of for the GRE is the amino-sugar, **chitin**, which builds the hard exoskeletons of insects and crustaceans such as lobsters and shrimp. Chitin is a polysaccharide of amino sugars, and is very similar to cellulose in its structure because it links monosaccharides via beta glycosidic linkages.

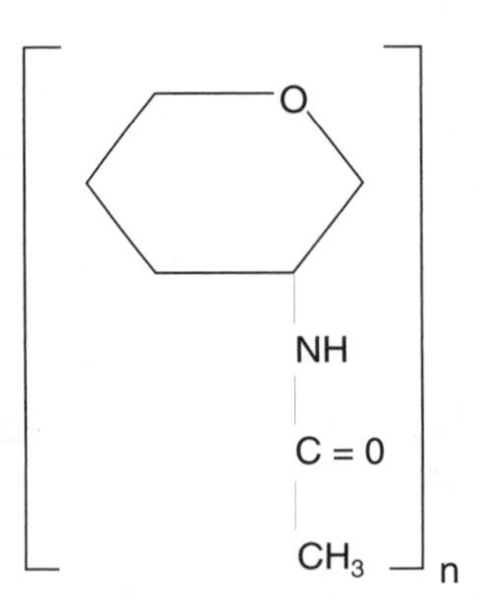

NUCLEIC ACIDS

Although later chapters will more fully detail DNA and RNA structure and function, it is beneficial to review the basic characteristics of these biomolecules in this section.

In addition to being the main energy molecule of cells, ATP can be used as an RNA nucleotide–the base A (adenine) attached to a ribose sugar.

Nucleic acids are information storage molecules, as well as molecules which form the structures that build proteins for all cells. Their basic unit, or monomer, is the **nucleotide**, a complex molecule with three distinct parts: a phosphate group, a sugar group, and a **nitrogenous base**. Individually, some of these nucleotides also perform crucial functions in cells. Below, **ATP** (adenosine triphosphate) is pictured.

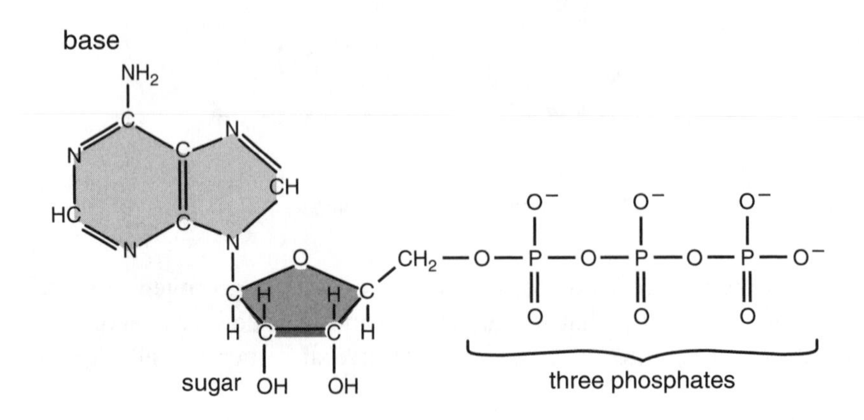

DNA is a double-stranded molecule that is held together by **hydrogen bonds** between complementary base pairs in an **antiparallel fashion:** guanine bonds with cytosine (G-C base pair) and adenine bonds with thymine (A-T base pair). There are more than 10^9 nucleotides in one human cell! DNA nucleotides always use deoxyribose as their sugar, and their nitrogenous bases are one of four possibilities: A, C, G, or T.

Note how the sugar in the diagram above has –OH groups hanging off both the 2' and 3' carbons. This is a **ribose** sugar, found in RNA, whereas **deoxyribose** sugar replaces one of those –OHs with an –H (hence, the term *deoxy*). RNA also differs from DNA in that it is a single-stranded molecule, with bases A, C, G, and U (**uracil**).

ENZYMES AND THEIR BINDING AND ACTIVATION

Enzymes

Most enzymes are proteins with specific 3-D structures that allow them to bind to very particular molecules (called **substrate** molecules) and increase the rate of reactions between these molecules. In many cases, enzymes bind to larger molecules and break them into smaller ones. Enzymes can synthesize or break down molecules at the rate of thousands or millions per second—they are extremely fast-acting! They allow reactions to occur that either would not take place or would take place far too slowly under normal conditions to be useful.

You should be familiar with the term **catalyst** for the GRE, which refers to any chemical agent that accelerates a reaction without being permanently changed in the reaction. Enzymes are biological catalysts, which can be used over and over again.

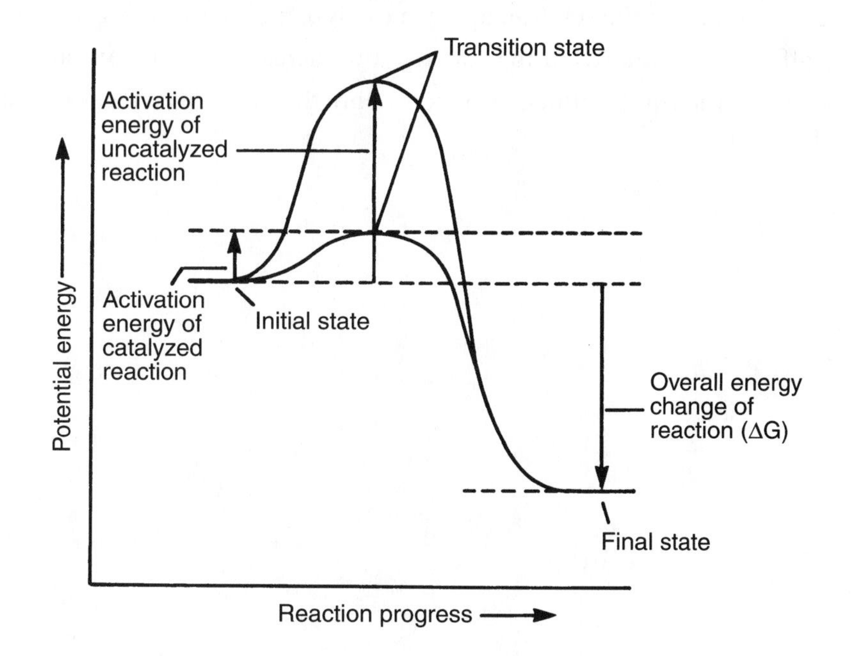

The figure above compares an uncatalyzed reaction with an enzymatically catalyzed reaction. The **activation energy**, or the energy which is required to start up the reaction, is much lower in the catalyzed reaction, yet the overall **free energy** change (ΔG) is the same for both reactions. The laws of thermodynamics can be used to predict if a reaction will occur or not. If products have less free energy (G) than the reactants, the reaction has an overall negative change in G ($\Delta G < 0$) and will occur spontaneously. If products have more free energy than reactants, the reaction is an uphill one, needing a great deal of supplied energy to occur. Free energy is a measure of the **potential energy** of the molecules in a reaction. Those with higher potential energy, or higher G, to start out are more likely to react and lower their G through the reaction than vice-versa. What that means is that reactions having a $\Delta G < 0$ are deemed "favorable." Keep in mind that most biosynthetic reactions have $\Delta G > 0$ and will not occur spontaneously without the help of both enzymes and ATP.

However, although thermodynamics and ΔG alone may predict that a reaction is favorable or can occur spontaneously, the **kinetics**, or rate, of the reaction may be so slow that these reactions are not feasible for living systems. Sure, a hamburger will break down eventually if exposed to enough acid in your stomach, but without digestive enzymes to help speed up this breakdown, the hamburger might take months to break down sufficiently to be useful to you. Thus, although hamburger breakdown may be spontaneous, the limiting factor in the reaction is the reaction rate, which is dependent on energy being provided in order to start this breakdown. This energy is the activation energy, and enzymes provide a foundation on which molecules can react so that the energy needed to start a reaction is not as great as it would have to be without the enzyme's presence.

So, to recap, enzymes:

- Lower the activation energy of a reaction
- Do not get used up in the reaction and can catalyze millions of reactions per second
- Do not affect the overall ΔG of the reaction, but increase the reaction rate
- Do not change the equilibrium of reactions, only the rate at which equilibrium is reached

Binding and Activation

Enzymes are very specific for the molecules they bind to and the reactions they catalyze. Each enzyme has a name that usually indicates exactly what it does, and the name often ends in "*-ase*." For example, lactase enzyme breaks the complex sugar lactose into the simple sugars glucose and galactose. The enzyme pyruvate decarboxylase removes a carbon from the 3-carbon molecule pyruvate. Thinking about enzyme names in this way may be helpful on the GRE.

Reaction rates increase as more and more enzyme is added to a particular environment. If the enzyme concentration is kept constant, the reaction rate will plateau at a maximum speed as substrate concentration increases, since the enzymes can only work so fast. This maximum reaction rate is termed $\mathbf{V_{max}}$ and is illustrated in the graph below. This point occurs when the enzymes become **saturated** with substrate. Enzymes become saturated at high substrate concentrations because substrate must bind to enzymes at a particular place: the **active site.**

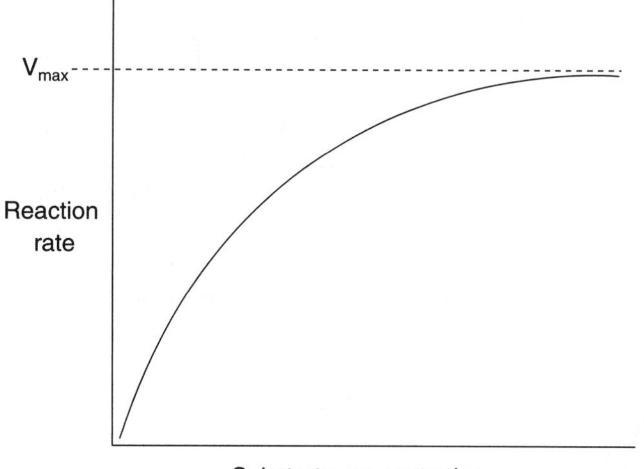

All enzymes possess an active site, a three-dimensional pocket within their structure in which substrate molecules can be held in a certain orientation to facilitate a reaction. The two models of enzyme-substrate interaction are shown below. In the **Lock-and-Key** model, the spatial structure of an enzyme's active site is exactly complementary to the spatial structure of the substrate, so that the enzyme and substrate fit together as lock and key. In the **Induced Fit** model, the active site has flexibility that allows the 3-D shape of the enzyme to shift in order to accommodate the incoming substrate molecule.

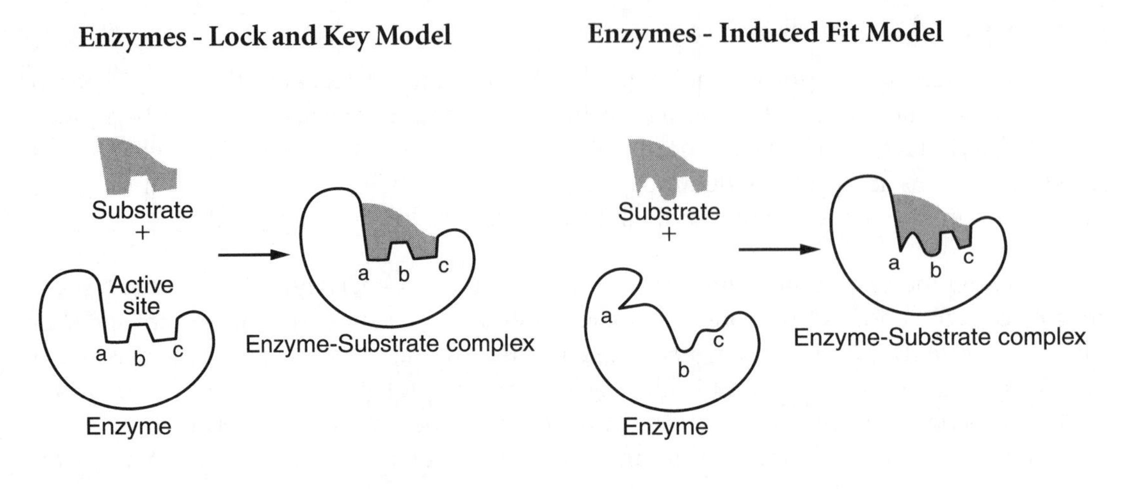

WHY ARE VITAMINS IMPORTANT IN YOUR DIET?

Vitamins are important **cofactors** for enzymes. Without them, some enzymes simply don't work. Deficiencies in thiamine (Vitamin B1), a cofactor of enzymes called alpha-ketolases, cause accumulations of pyruvate and lactate and the reduction of Krebs Cycle products. Thiamine is also needed for acetylcholine release in neurons. Thiamine deficiencies lead to disorders of cellular metabolism and nerve signal conduction.

During an enzymatic reaction, the substrate is held in the active site, which has two distinct domains: a catalytic site and a binding site. While the binding site consists of amino acids to which the substrate forms temporary bonds, the catalytic site is where the reaction actually occurs. The substrate molecule will go through a brief **transition state**, where it is neither the original substrate molecule nor the product. When the product forms, the enzyme and substrate separate.

This rate of conversion is influenced not only by the substrate concentration, but also by such factors as temperature and pH. All enzymes have temperature and pH optimums, conditions under which these enzymes can perform most efficiently. For humans, these optimums are around 98.6 degrees Fahrenheit, body temperature, and a pH of 7.4, the pH of blood plasma. On either side of these optimums, reaction rates fall quickly.

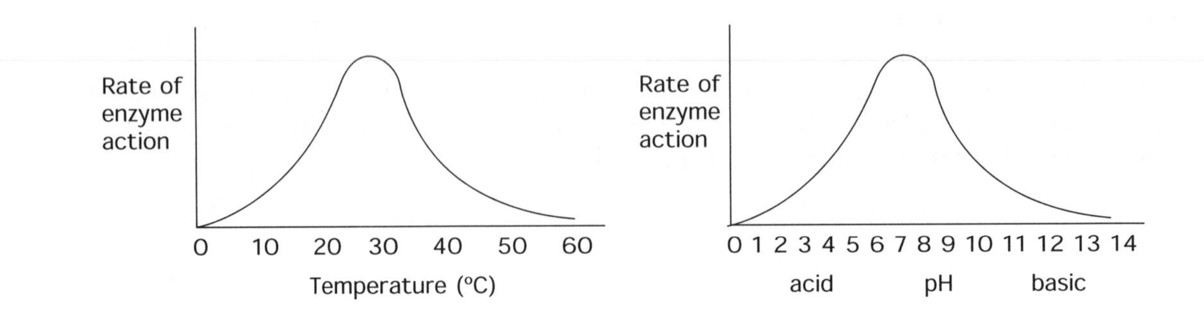

Regulation of Enzymes

Cells must regulate enzyme action in order to keep these rapid reactions under control. Most enzymes are inactive most of the time, and enzyme pathways are regulated in complex fashions to ensure efficiency and safety. There are several kinds of inhibition that cells use to control enzymes:

Feedback Inhibition occurs when the end-product of an enzyme-catalyzed reaction works to block the original enzymes that started the reaction in the first place. In the diagram below, this would occur if product C could bind to the active site of enzyme 1, thereby preventing any more of compound A from becoming compound B.

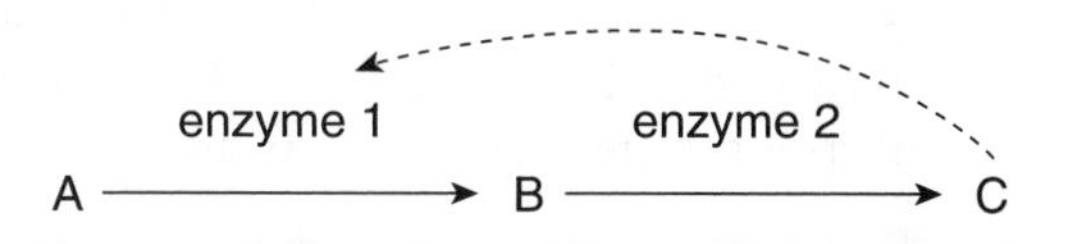

A commonly cited example of this process is how ATP can block the enzyme that produces ATP from ADP. This enzyme adds a free-floating, inorganic phosphate (P_i) to the ADP to build ATP. As ATP gets used up, however, this inhibition is removed because there are fewer ATP molecules to sit in the active sites of the ADP → ATP converting enzyme.

Competitive Inhibition reduces the productivity of enzymes because certain molecules are competing with the substrate for the enzyme's active sites. Although competitive inhibitors can usually be overcome by the addition of more substrate molecules, if the competitor is present, the reaction will take longer to complete. V_{max} will not change since the enzyme itself can still facilitate the same number of reactions per unit of time, if enough substrate is present.

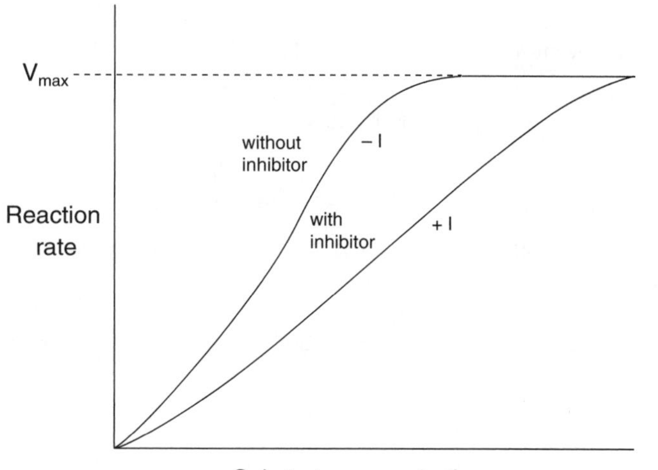

A classic example of competitive inhibition is the class of antibacterial drugs called *sulfa drugs*. These antibiotics were the first effective agents used against bacterial infections in humans. Discovered in the 1930's, sulfa drugs act to block bacterial synthesis of **folic acid**, an important cofactor for enzymes. They act as competitive inhibitors of the structurally related compound, *p-amino benzoic acid* (PABA), which is incorporated into folic acid as folic acid is made in bacterial cells.

The drugs are competitive inhibitors because they sit in the active site of an enzyme used by bacteria to synthesize folic acid and the sulfa gets incorporated into the folic acid instead of the usual PABA, making the folic acid nonfunctional. Below are the compounds used by bacteria to build folic acid, and you will see that PABA is one of the key ingredients. Notice how similar to PABA's structure are the structures of two common sulfa drugs, sulfanilamide and sulfisoxazole. Because of enzyme specificity, competitive inhibitors must have a chemical structure very similar to the compound they are substituting.

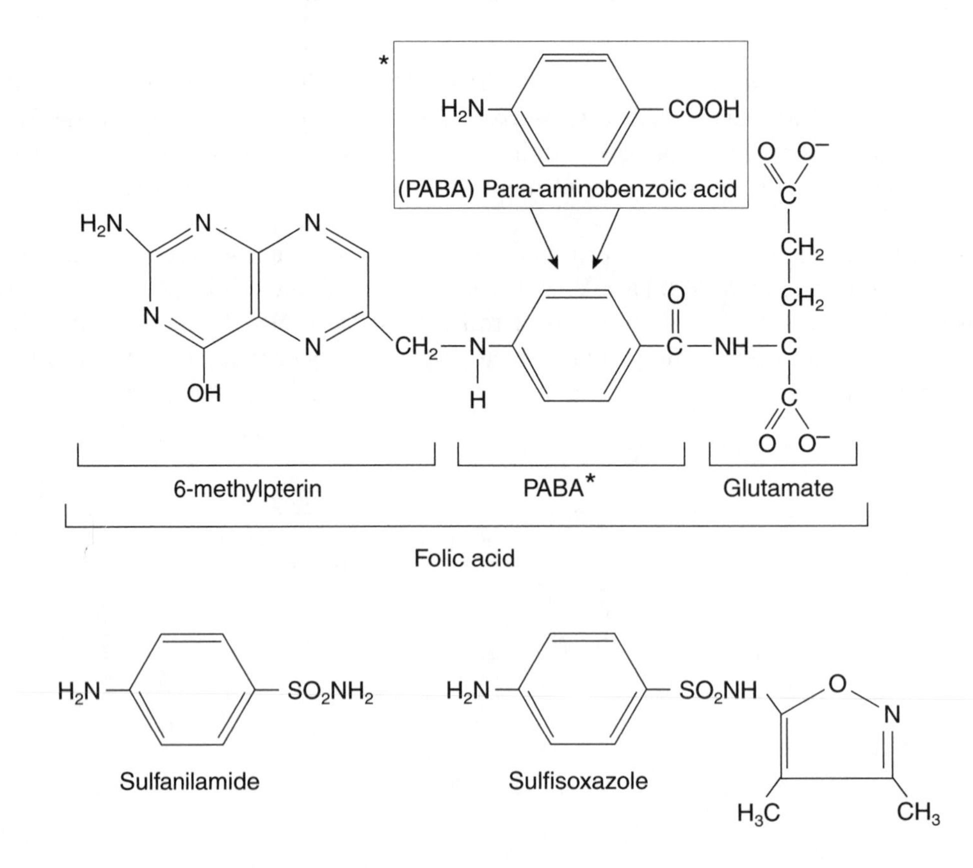

If bacteria cannot make folic acid, nucleotides of DNA are not produced and cell division cannot occur because DNA is not replicated. Bacteria are extremely sensitive to sulfa drugs, since they must synthesize their own folic acid, while humans can get the acid from dietary sources. This is what makes sulfa an ideal antibiotic, despite its frequent side-effects.

Irreversible Inhibitors are competitive inhibitors that chemically and covalently bind to an enzyme's active site. This renders the enzyme permanently inactive, so that increasing substrate concentration does nothing to increase the reaction rate. An example of an inhibitor that works in this fashion is the nerve gas *sarin*, which irreversibly binds to the enzyme acetylcholinesterase, responsible for breaking down the neurotransmitter acetylcholine (ACh) at the neuromuscular junction. Without this enzyme, you cannot break down the ACh present in the synapse between nerve cells, and muscles become rigid due to long-lasting stimulation. Breathing eventually stops as respiratory muscles become paralyzed.

Pseudoirreversible Inhibitors are competitive inhibitors that have extremely high affinities for enzyme active sites. While they do not irreversibly bind, they are very hard to displace once sitting in the binding site. The cancer drug, methotrexate, is an example here. Sitting in the active site of an enzyme responsible for converting dietary folic acid into the folic acid form needed for DNA synthesis and replication, methotrexate effectively stops rapidly dividing cells from replicating.

Noncompetitive Inhibitors do not compete with substrate for the active site, but act elsewhere on the enzyme's structure, changing the enzyme's overall 3-D shape. These compounds decrease an enzyme's efficiency because the active site on the enzyme is altered when compounds bind to the enzyme in other places.

Finally, substances can bind to sites other than the active site and, in doing so, increase or decrease the efficiency of the enzyme. They can control enzymes **allosterically**. Most enzymes having allosteric sites are ones with multiple subunits, such as hemoglobin. In hemoglobin, oxygen atoms binding to any one of the four subunits induce a conformational change in the other subunits so that they are more effective at grabbing oxygen around them—this is an example of a positive allosteric effect.

CHAPTER TWO

Cellular Metabolic Pathways

Cells combine enzymatic regulation with the biomolecules discussed above (as well as many others) into essential pathways, through which they are able to create from scratch certain necessary molecules, build energy, communicate with other cells, and break down nutrients, wastes, and toxins. These pathways include glycolysis, fermentation, and cellular respiration, as well as photosynthesis and other biosynthetic reactions.

RESPIRATION, FERMENTATION, AND PHOTOSYNTHESIS

Cellular metabolism is the sum total of all chemical reactions that take place in a cell. These reactions can be generally categorized as either anabolic or catabolic. Anabolic processes are energy-requiring, involving the biosynthesis of complex organic compounds from simpler molecules. Catabolic processes release energy as they break down complex organic compounds into smaller molecules. The metabolic reactions of cells are coupled so that energy released from catabolic reactions can be harnessed to fuel anabolic reactions.

Transfer of Energy

The Flow of Energy

The ultimate energy source for living organisms is the sun. Autotrophic organisms, such as green plants, convert sunlight into energy stored in the bonds of organic compounds (chiefly glucose) during the anabolic process of **photosynthesis**. Autotrophs do not need an exogenous supply of organic compounds. Heterotrophic organisms obtain their energy catabolically, via the breakdown of organic nutrients that must be ingested. Below is an energy flow diagram for biological systems; note that some energy is dissipated as heat at every stage.

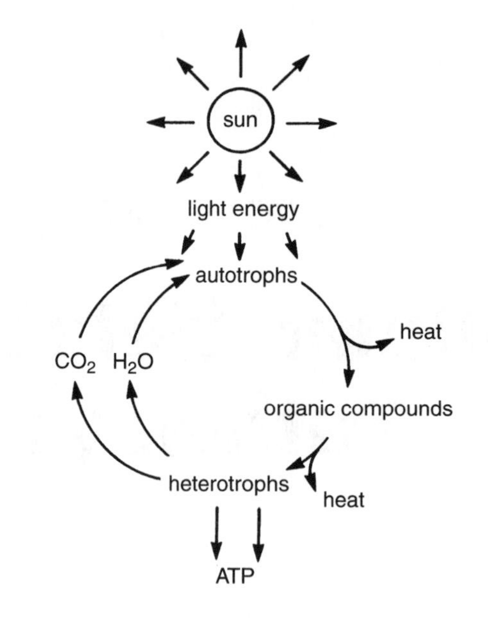

Glucose plays an essential role in the energetics of cell metabolism. The production of glucose ($C_6H_{12}O_6$) by autotrophs involves the breaking of C–O and O–H bonds in CO_2 and H_2O, and the forming of C–H, C–O, C–C, and O–H bonds in glucose. The net reaction of photosynthesis:

$$6CO_2 + 6H_2O + \text{Energy} \longrightarrow \underset{\text{glucose}}{C_6H_{12}O_6} + 6O_2$$

All organisms metabolize glucose and other organic molecules to release the stored bond energies. The net reaction of glucose catabolism, which is essentially the reversal of photosynthesis, is:

$$\underset{\text{glucose}}{C_6H_{12}O_6} + 6O_2 \longrightarrow 6CO_2 + 6H_2O + \text{Energy}$$

Energy Carriers

During metabolism, the cell uses various molecular carriers, such as ATP and the coenzymes NAD^+, $NADP^+$, and FAD, to shuttle electrons between reactions.

ATP: ATP, or adenosine triphosphate, is the cell's main energy currency. Through its formation and degradation, cells have a quick way of releasing and storing energy. ATP is synthesized during glucose catabolism. As you learned in chapter 1, ATP is composed of the nitrogenous base adenine, the sugar ribose, and three phosphate groups. The energy of ATP is stored in the covalent bonds attaching these phosphate groups, often referred to as high-energy bonds.

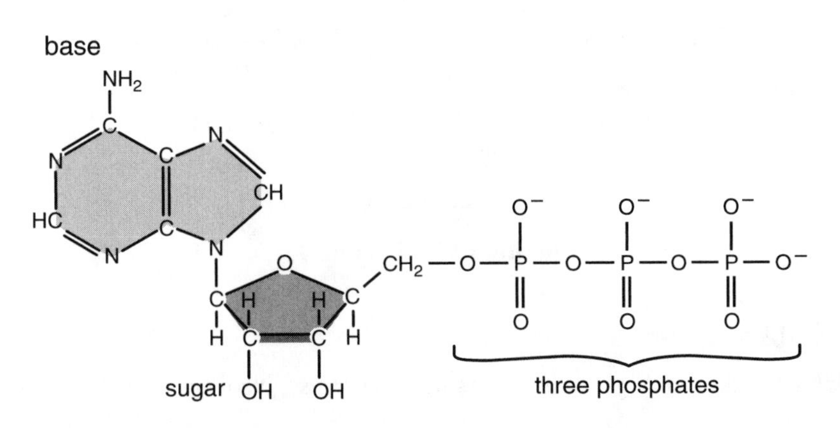

Hydrolysis of ATP to ADP (adenosine diphosphate) and P_i (inorganic phosphate) releases stored bond energy that the cell can use in metabolic processes. Approximately 7 kcal of energy are released per mole of ATP. This provides energy for **endergonic** (endothermic) reactions such as muscle contraction, motility, and the active transport of substances across plasma membranes. ATP may also be hydrolyzed into AMP (adenosine monophosphate) and PP_i (pyrophosphate):

$$\text{ATP} \longrightarrow \text{ADP} + P_i + 7 \text{ kcal/mole}$$

$$\text{ATP} \longrightarrow \text{AMP} + PP_i + 7 \text{ kcal/mole}$$

Alternatively, ADP and P_i combine to form ATP; in this way, the cell regenerates its ATP supply. This process requires energy (supplied by the degradation of glucose):

$$\text{ADP} + P_i + 7 \text{ kcal/mole} \longrightarrow \text{ATP}$$

NAD^+, $NADP^+$, and FAD: A second mechanism by which the cell stores chemical energy is in the form of high potential electrons. Electrons are transferred as hydride ions (written as $H{:}^-$) or as pairs of hydrogen atoms. During glucose oxidation, hydrogen atoms are removed. Most of these are accepted by the carrier coenzymes NAD^+ (nicotinamide adenine dinucleotide), FAD (flavin adenine dinucleotide), and $NADP^+$ (nicotinamide adenine dinucleotide phosphate). These molecules transport the high-energy electrons of the hydrogen atoms to a series of carrier molecules on the inner mitochondrical membrane that are collectively known as the **electron transport chain**.

Oxidation refers to the loss of an electron. NAD^+, $NADP^+$, and FAD are referred to as oxidizing agents because they cause other molecules to lose electrons and undergo **oxidation**. In the process, they themselves undergo **reduction**; that is, they gain electrons. For example, when NAD^+ accepts electrons in the form of a hydride ion it is reduced to NADH, while the donating molecule is oxidized. Likewise, when FAD accepts electrons in the form of hydrogen atoms it is reduced to $FADH_2$. In their reduced forms, NADH, NADPH, and $FADH_2$ all behave as reducing agents. NADH transfers its electrons to another electron acceptor (e.g., the first carrier of the electron transport chain), thereby reducing it, and in the process NADH is oxidized back to NAD^+. Thus, these coenzymes temporarily store and release energy in the form of electrons through their successive oxidations and reductions.

oxidized form		reduced form
NAD^+	$\xrightarrow{H:^-}$	NADH
FAD	$\xrightarrow{2H}$	$FADH_2$

Reduction of NAD^+ and FAD

Glucose Catabolism

The degradative oxidation of glucose occurs in two stages, glycolysis and cellular respiration.

Glycolysis

The first stage of glucose catabolism is **glycolysis.** Glycolysis is a series of reactions that lead to the oxidative breakdown of glucose into two molecules of **pyruvate** (the ionized form of pyruvic acid), the production of ATP, and the reduction of NAD^+ into NADH. All of these reactions occur in the cytoplasm and are mediated by specific enzymes. The glycolytic pathway is outlined below.

Step	Compound	Energy
Step 1	Glucose	ATP → ADP
Step 2	Glucose 6-phosphate	
Step 3	Fructose 6-phosphate	ATP → ADP
Step 4	Fructose 1,6-diphosphate	
Step 5	Glyceraldehyde 3-phosphate* (PGAL) ⇄ Dihydroxyacetone phosphate	
Step 6	1,3-Diphosphoglycerate	ADP → ATP
Step 7	3-Phosphoglycerate	
Step 8	2-Phosphoglycerate	
Step 9	Phosphoenolpyruvate	ADP → ATP
	Pyruvate	

* NOTE: Steps 5 – 9 occur twice per molecule of glucose (see text).

Glycolytic Pathway: Note that at step 4, fructose 1,6-diphosphate is split into 2 three-carbon molecules: dihydroxyacetone phosphate and glyceraldehyde 3-phosphate (PGAL). Dihydroxyacetone phosphate is isomerized into PGAL so that it can be used in subsequent reactions. Thus, 2 molecules of PGAL are formed per molecule of glucose, and all of the subsequent steps occur twice for each glucose molecule.

From one molecule of glucose (a six-carbon molecule), 2 molecules of pyruvate (a three-carbon molecule) are obtained. During this sequence of reactions, 2 ATP are used (in steps 1 and 3) and 4 ATP are generated (2 in step 6, and 2 in step 9). Thus, there is a net production of 2 ATP per glucose molecule. This type of phosphorylation is called **substrate level phosphorylation**, since ATP synthesis is directly coupled with the degradation of glucose without the participation of an intermediate molecule such as NAD^+. One NADH is produced per PGAL, for a total of 2 NADH per glucose.

Clinical Correlate:

Deficiency of pyruvate kinase, the final enzyme in the glycolytic pathway (Step 9), leads to a form of anemia (hemolytic anemia). Since red blood cells don't have mitochondria, they depend on glycolysis as their only source of ATP. ATP, in turn, is required to maintain the integrity of the cell membrane. Deficiency of pyruvate kinase leads to a shortage of ATP, and therefore an inability of the red blood cell to maintain its shape. The spleen tends to sequester and destroy these deformed red blood cells, causing hemolytic anemia.

The net reaction for glycolysis is:

$$\text{Glucose} + 2\text{ADP} + 2\text{P}_i + 2\text{NAD}^+ \longrightarrow 2\text{Pyruvate} + 2\text{ATP} + 2\text{NADH} + 2\text{H}^+ + 2\text{H}_2\text{O}$$

This series of reactions occurs in both prokaryotic and eukaryotic cells. However, at this stage, much of the initial energy stored in the glucose molecule has not been released, and is still present in the chemical bonds of pyruvate. Depending on the capabilities of the organism, pyruvate degradation can proceed in one of two directions. Under anaerobic conditions (in the absence of oxygen), pyruvate is reduced during the process of **fermentation**. Under aerobic conditions (in the presence of oxygen), pyruvate is further oxidized during **cell respiration** in the mitochondria.

Fermentation: NAD^+ must be regenerated for glycolysis to continue in the absence of O_2. This is accomplished by reducing pyruvate into ethanol or lactic acid. Fermentation refers to all of the reactions involved in this process—glycolysis and the additional steps leading to the formation of ethanol or lactic acid. Fermentation produces only 2 ATP per glucose molecule.

Alcohol fermentation commonly occurs only in yeast and some bacteria. The pyruvate produced in glycolysis is decarboxylated to become **acetaldehyde**, which is then reduced by the NADH generated in step 5 of glycolysis to yield **ethanol**. In this way, NAD^+ is regenerated and glycolysis can continue.

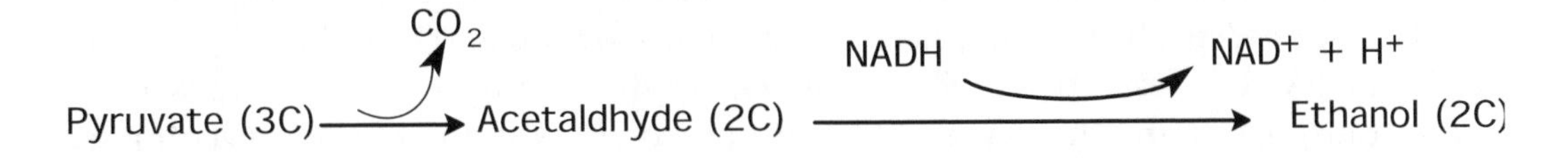

Lactic acid fermentation occurs in certain fungi and bacteria, and in human muscle cells during strenuous activity. When the oxygen supply to muscle cells lags behind the rate of glucose catabolism, the pyruvate generated is reduced to lactic acid. As in alcohol fermentation, the NAD^+ used in step 5 of glycolysis is regenerated when pyruvate is reduced. In humans, lactic acid may accumulate in the muscles during exercise, causing a decrease in blood pH that leads to muscle fatigue. Once the oxygen supply has been replenished, the lactic acid is oxidized back to pyruvate and enters cellular respiration. The amount of oxygen needed for this conversion is known as **oxygen debt.**

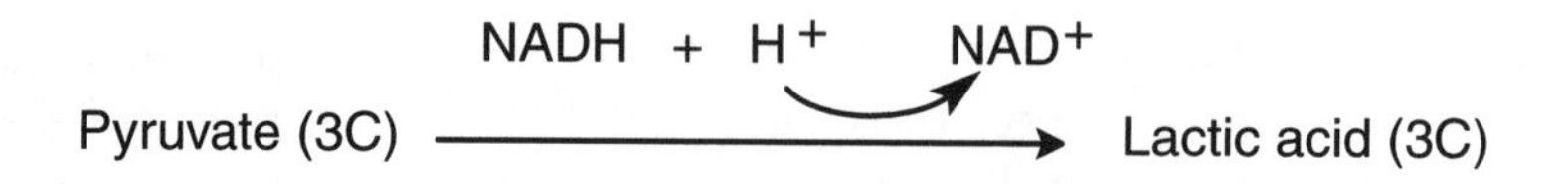

Cellular Respiration

Cellular respiration is the most efficient catabolic pathway used by organisms to harvest the energy stored in glucose. Whereas glycolysis yields only 2 ATP per molecule of glucose, cellular respiration can yield 36–38 ATP. Cellular respiration is an aerobic process; oxygen acts as the final acceptor of electrons that are passed from carrier to carrier during the final stage of glucose oxidation. The metabolic reactions of cell respiration occur in the eukaryotic mitochondrion and are catalyzed by reaction-specific enzymes.

Cellular respiration can be divided into three stages: pyruvate decarboxylation, the citric acid cycle, and the electron transport chain.

Pyruvate Decarboxylation: The pyruvate formed during glycolysis is transported from the cytoplasm into the mitochondrial matrix where it is decarboxylated; i.e., it loses a CO_2, and the acetyl group that remains is transferred to coenzyme A to form **acetyl CoA.** In the process, NAD^+ is reduced to NADH.

$$\text{Pyruvate (3C)} + \text{Coenzyme A} \xrightarrow{NAD^+ \;\rightarrow\; NADH + H^+} \text{Acetyl CoA (2C)}$$

Citric Acid Cycle: The citric acid cycle is also known as the Krebs cycle or the **tricarboxylic acid cycle** (TCA cycle). The cycle begins when the two-carbon acetyl group from acetyl CoA combines with oxaloacetate, a four-carbon molecule, to form the six-carbon **citrate.** Through a complicated series of reactions, 2 CO_2 are released, and **oxaloacetate** is regenerated for use in another turn of the cycle.

For each turn of the citric acid cycle, 1 ATP is produced by substrate level phosphorylation via a GTP intermediate. In addition, electrons are transferred to NAD^+ and FAD, generating NADH and $FADH_2$, respectively. These coenzymes then transport the electrons to the **electron transport chain,** where more ATP is produced via **oxidative phosphorylation** (see below). Studying the cycle, we can do some bookkeeping; keep in mind that for each molecule of glucose, 2 pyruvates are decarboxylated and channeled into the citric acid cycle.

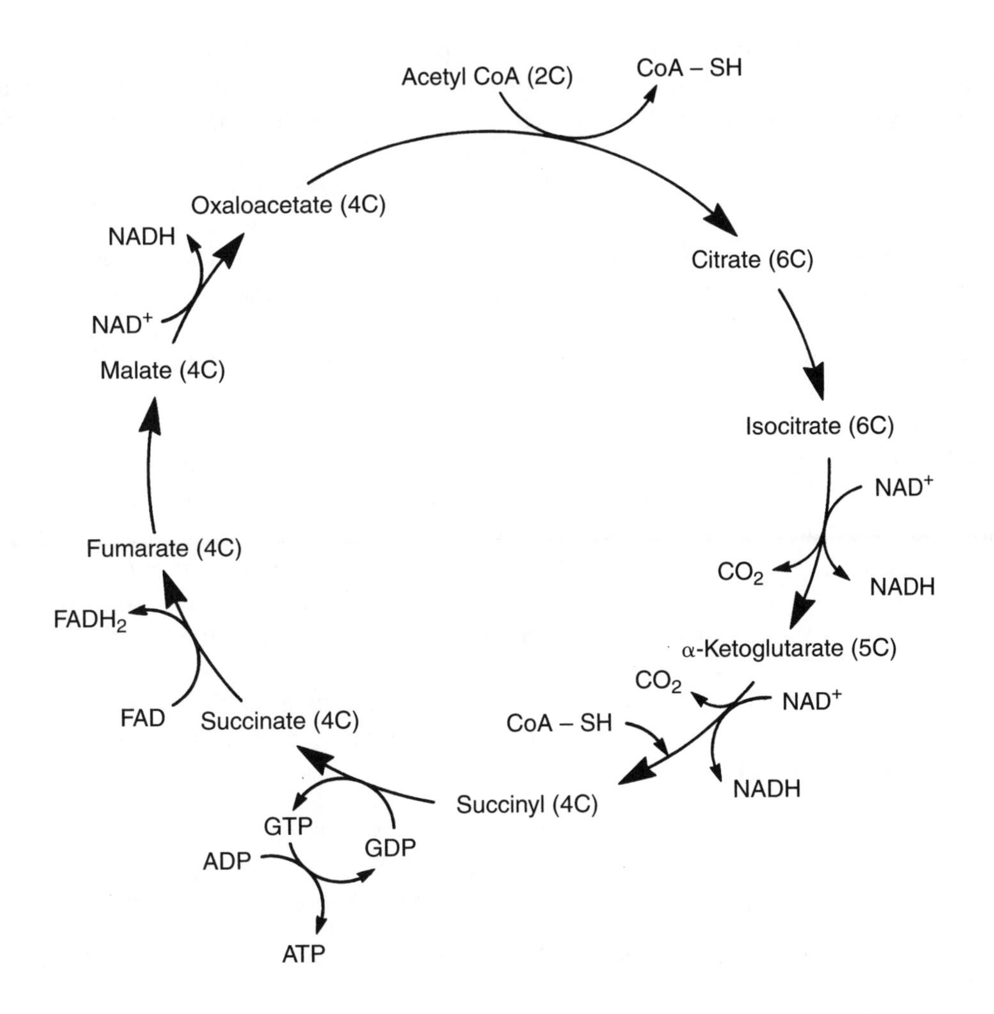

The *net reaction* of the citric acid cycle per glucose molecule is:

$$2\text{Acetyl CoA} + 6NAD^+ + 2FAD + 2GDP + 2P_i + 4H_2O \longrightarrow 4CO_2 + 6NADH + 2FADH_2 + 2ATP + 4H^+ + 2CoA$$

Electron Transport Chain: The electron transport chain (ETC) is a complex carrier mechanism located on the inside of the inner mitochondrial membrane. During oxidative phosphorylation, ATP is produced when high energy potential electrons are transferred from NADH and $FADH_2$ to oxygen by a series of carrier molecules located in the inner mitochondrial membrane. As the electrons are transferred from carrier to carrier, free energy is released, which is then used to form ATP. Most of the molecules of the ETC are **cytochromes**, electron carriers that resemble hemoglobin in the structure of their active site. The functional unit contains a central iron atom, which is capable of undergoing a reversible redox reaction; that is, it can be alternatively reduced and oxidized.

Everything the human body does to deliver inhaled oxygen to tissues (discussed in later chapters) comes down to the role oxygen plays as the final electron acceptor in the electron transport chain. Without oxygen, ATP production is not adequate to sustain human life. Similarly, the CO_2 generated in the citric acid cycle is the same carbon dioxide we exhale.

FMN (flavin mononucleotide) is the first molecule of the ETC. It is reduced when it accepts electrons from NADH, thereby oxidizing NADH to NAD^+. Sequential redox reactions continue to occur as the electrons are transferred from one carrier to the next; each carrier is reduced as it accepts an electron, and is then oxidized when it passes it on to the next carrier.

The last carrier of the ETC, cytochrome a_3, passes its electron to the final electron acceptor, O_2. In addition to the electrons, O_2 picks up a pair of hydrogen ions from the surrounding medium, forming water.

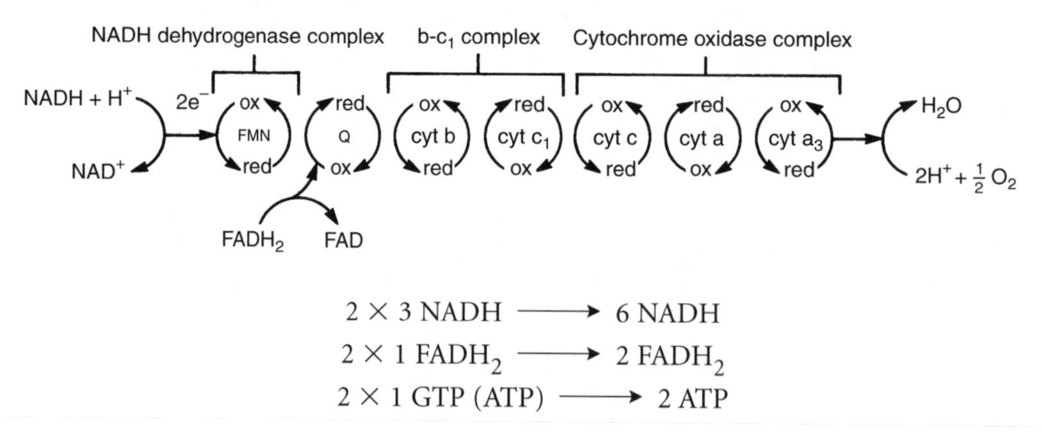

Without oxygen, the ETC becomes backlogged with electrons. As a result, NAD^+ cannot be regenerated and glycolysis cannot continue unless lactic acid fermentation occurs. Likewise, ATP synthesis comes to a halt if respiratory poisons such as **cyanide** or **dinitrophenol** enter the cell. Cyanide blocks the transfer of electrons from cytochrome a_3 to O_2. Dinitrophenol uncouples the electron transport chain from the proton gradient established across the inner mitochondrial membrane.

ATP generation and the proton pump. The electron carriers are categorized into three large protein complexes: NADH dehydrogenase, the b–c_1 complex, and cytochrome oxidase. There are energy losses as the electrons are transferred from one complex to the next; this energy is then used to synthesize 1 ATP per complex. Thus, an electron passing through the entire ETC supplies enough energy to generate 3 ATP. NADH delivers its electrons to the NADH dehydrogenase complex, so that for each NADH, 3 ATP are produced. However, $FADH_2$ bypasses the NADH dehydrogenase complex and delivers its electrons directly to carrier Q (**ubiquinone**), which lies between the NADH dehydrogenase and b–c_1 complexes. Therefore, for each $FADH_2$, there are only two energy drops, and only 2 ATP are produced.

The operating mechanism in this type of ATP production involves coupling the oxidation of NADH to the phosphorylation of ADP. The coupling agent for these two processes is a proton gradient across the inner mitochondrial membrane, maintained by the ETC. As NADH passes its electrons to the ETC, hydrogen ions (H^+) are pumped out of the **matrix**, across the inner mitochondrial membrane, and into the **intermembrane space** at each of the three protein complexes. The continuous translocation of H^+ creates a positively charged acidic environment in the intermembrane space. This electrochemical gradient generates a proton-motive force, which drives H^+ back across the inner membrane and into the matrix. However, to pass through the membrane (which is impermeable to ions), the H^+ must flow through specialized channels provided by enzyme complexes called ATP synthetases. As the H^+ pass through the ATP synthetases, enough energy is released to allow for the phosphorylation of ADP to ATP. The coupling of the oxidation of NADH with the phosphorylation of ADP is called **oxidative phosphorylation.**

Review of Glucose Catabolism. It is important to understand how all of the events described above are interrelated. Below is a eukaryotic cell with a mitochondrion magnified for detail.

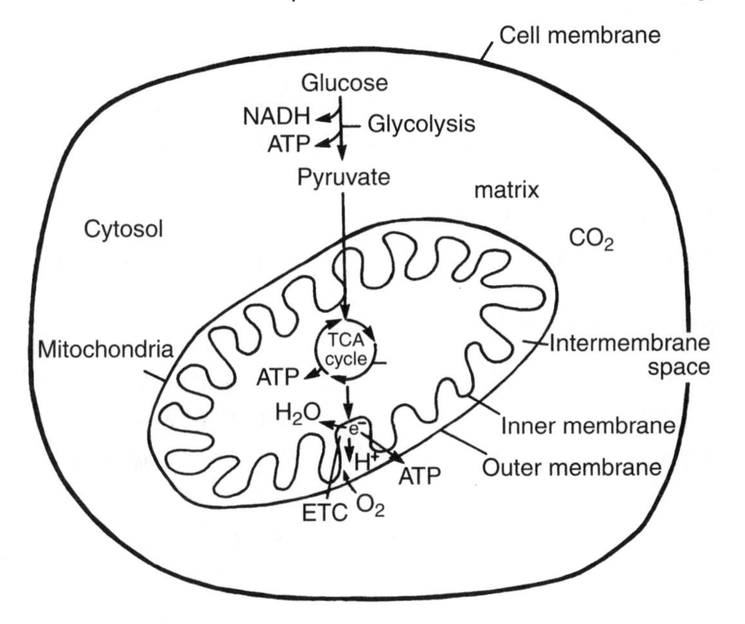

To calculate the net amount of ATP produced per molecule of glucose we need to tally the number of ATP produced by substrate level phosphorylation and the number of ATP produced by oxidative phosphorylation.

- **Substrate Level Phosphorylation**

  Degradation of one glucose molecule yields a net of 2 ATP from glycolysis and 1 ATP for each turn of the citric acid cycle with two turns per glucose. Thus, a total of 4 ATP are produced by substrate level phosphorylation.

- **Oxidative Phosphorylation**

  Two pyruvate decarboxylations yield 1 NADH each for a total of 2 NADH per glucose. Each turn of the citric acid cycle yields 3 NADH and 1 $FADH_2$, for a total of 6 NADH and 2 $FADH_2$ per glucose molecule. Each $FADH_2$ generates 2 ATP, as previously discussed. Each NADH generates 3 ATP except for the 2 NADH that were reduced during glycolysis; these NADH cannot cross the inner mitochondrial membrane, and must transfer their electrons to an intermediate carrier molecule, which delivers the electrons to the second carrier protein complex, Q. Therefore, these NADH generate only 2 ATP per glucose. So the 2 NADH of glycolysis yield 4 ATP, the other 8 NADH yield 24 ATP, and the 2 $FADH_2$ produce 4 ATP, for a total of 32 ATP by oxidative phosphorylation.

The total amount of ATP produced during eukaryotic glucose catabolism is 4 via substrate level phosphorylation plus 32 via oxidative phosphorylation, for a total of 36 ATP. (For prokaryotes the yield is 38 ATP, because the 2 NADH of glycolysis don't have any mitochondrial membranes to cross and therefore don't lose energy.) See the table below for a summary of eukaryotic ATP production.

Glycolysis	
2 ATP invested (steps 1 and 3)	–2 ATP
4 ATP generated (steps 6 and 9)	+4 ATP (substrate)
2 NADH × 2 ATP/NADH (step 5)	+4 ATP (oxidative)
Pyruvate Decarboxylation	
2 NADH × 3 ATP/NADH	+6 ATP (oxidative)
Citric Acid Cycle	
6 NADH × 3 ATP/NADH	+18 ATP (oxidative)
2 $FADH_2$ × 2 ATP/$FADH_2$	+4 ATP (oxidative)
2 GTP × 1 ATP/GTP	+2 ATP (substrate)
Total	+36 ATP

Alternative Energy Sources

When glucose supplies run low, the body utilizes other energy sources. These sources are used by the body in the following preferential order: other carbohydrates, fats, and proteins. These substances are first converted to either glucose or glucose intermediates, which can then be degraded in the glycolytic pathway and TCA cycle.

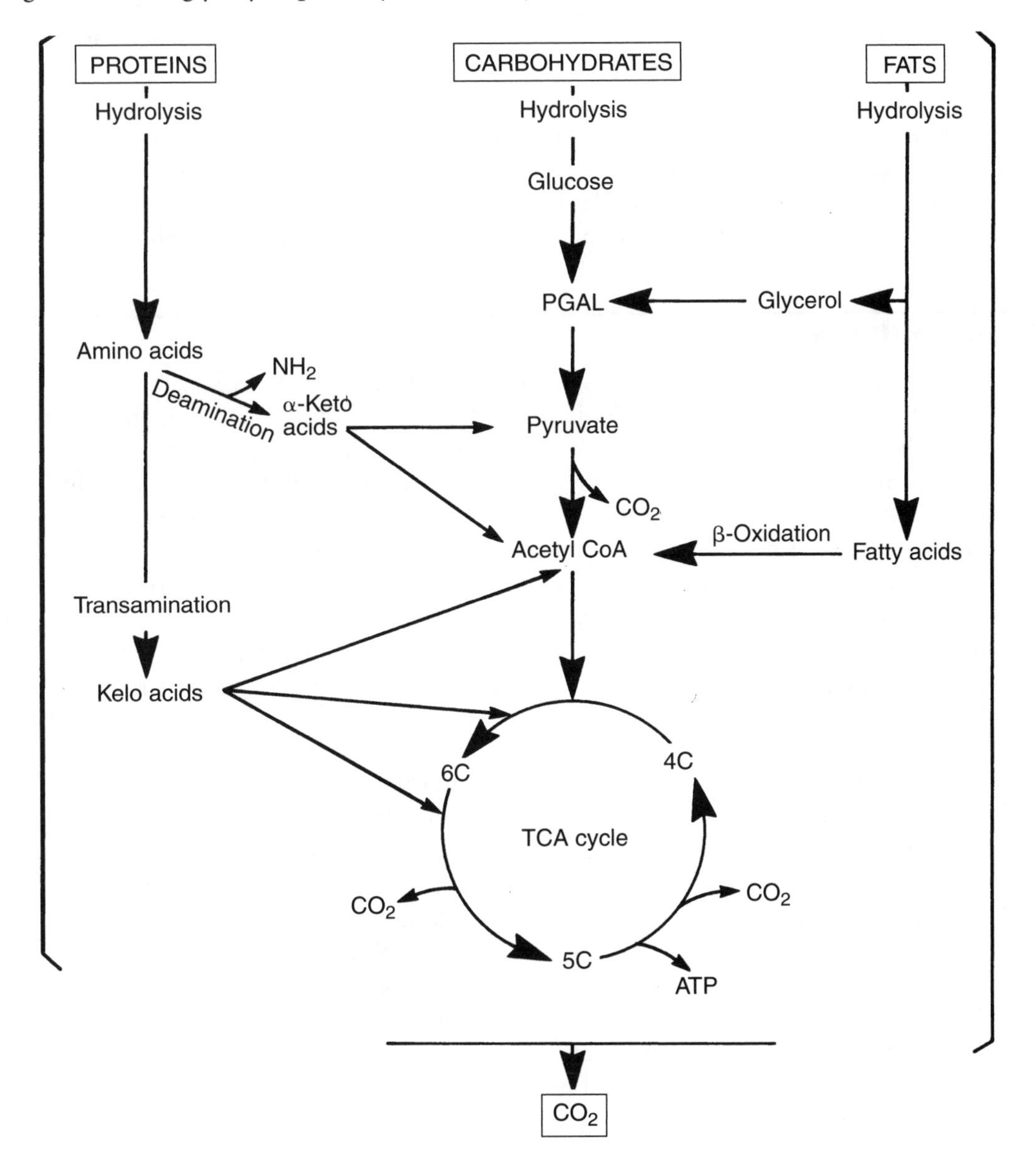

Disaccharides are hydrolyzed into monosaccharides, most of which can be converted into glucose or glycolytic intermediates. Glycogen stored in the liver can be converted, when needed, into glucose-6-phosphate, a glycolytic intermediate.

Fat molecules are stored in adipose tissue in the form of triglyceride. When needed, they are hydrolyzed by lipases to fatty acids and glycerol, and are carried by the blood to other tissues for oxidation. Glycerol can be converted into PGAL, a glycolytic intermediate. A fatty acid must first be "activated" in the cytoplasm; this process requires 2 ATP. Once activated, the fatty acid is transported into the mitochondrion and taken through a series of beta-oxidation cycles that convert it into two-carbon fragments, which are then converted into acetyl CoA. Acetyl CoA then enters the TCA cycle. With each round of β-oxidation of a saturated fatty acid, 1 NADH and 1 $FADH_2$ are generated.

Of all the high-energy compounds used in cellular respiration, fats yield the greatest number of ATP per gram. This makes them extremely efficient energy storage molecules. Thus, while the amount of glycogen stored in humans is enough to meet the short-term energy needs of about a day, the stored fat reserves can meet the long-term energy needs for about a month.

Photosynthesis

To survive, all organisms need energy. ATP is an energy intermediary used to drive biosynthesis and other processes. ATP is generated in mitochondria using the chemical energy of glucose and other nutrients. Where does the chemical energy of glucose come from? The energy foundation of almost all ecosystems is photosynthesis. Plants are autotrophs, or self-feeders, that generate their own chemical energy from the energy of the sun through photosynthesis. The chemical energy that plants get from the sun is used to produce glucose. This glucose can then be burned in plant mitochondria to make ATP, which is used to drive all of the energy-requiring processes in the plant, including the production of proteins, lipids, carbohydrates, and nucleic acids. Similarly, animals can eat plants to extract the energy for their own metabolic needs. In this way, photosynthesis is the energy foundation of most living systems.

Photosynthesis occurs in plants in the chloroplast, an organelle that is specific to plants. In prokaryotes that perform photosynthesis, there are no chloroplasts; instead, photosynthesis occurs in association with the plasma membrane or infoldings of the membrane. Chloroplasts are found mainly in the cells of the mesophyll, the green tissue in the interior of the leaf. The leaf contains pores on its surface called stomata that allow carbon dioxide in and oxygen out to facilitate photosynthesis in the leaf. The chloroplast has an inner and outer membrane; within the inner membrane there is a fluid called the **stroma**. In addition, the interior of the chloroplast contains a series of membranes called the **thylakoid** membranes that form stacks called **grana**. The structure of the chloroplast is similar in many ways to that of the mitochondrion, with an inner and outer membrane in which the inner membrane is impermeable to most ions.

Photosynthesis can be summarized by this equation:

$$6CO_2 + 12H_2O + \text{light} \rightarrow C_6H_{12}O_6 + 6O_2 + 6H_2O$$

Photosynthesis involves the reduction of CO_2 to a carbohydrate. It can be characterized as the reverse of respiration, in that reduction occurs instead of oxidation. Photosynthesis has

two main parts, the **light reactions** and the **Calvin-Benson cycle** (also called the Calvin cycle). The light reactions occur in the interior of the thylakoid while the Calvin-Benson cycle occurs in the stroma.

Light Reactions

The first part of photosynthesis is made up of light reactions, in which light energy is used to generate ATP, oxygen, and the reducing molecule NADPH. The molecule that captures light energy to start photosynthesis is a pigment called chlorophyll in the thylakoid membranes of the chloroplast. Chlorophylls absorb most wavelengths of visible light, with the exception of green. Since chlorophyll does not absorb green, it reflects green light, making plants appear green. Chlorophyll is used by two complex systems called photosystems I and II in the thylakoid membrane. Each photosystem is a complex assembly of protein and pigments in the membrane. When photons strike chlorophyll, electrons are excited and transferred through the photosystems to a **reaction center**. When electrons reach the reaction center, the reaction center gives up excited electrons that enter an electron transport chain where they are used to generate chemical energy as either reduced NADPH or ATP.

Two different processes occur in the photosystems, **cyclic photophosphorylation** and **noncyclic photophosphorylation**. Both are used to generate ATP, but in different ways. The ATP in turn is used to generate glucose in the dark reactions. Cyclic photophosphorylation occurs in photosystem I to produce ATP. In the cyclic method, electrons move from the reaction center, through an electron transport chain, then back to the same reaction center again (see figure). The reaction center in photosystem I includes a chlorophyll called P700 because its maximal light absorbance occurs at 700 nm. This process does not produce oxygen and does not produce NADPH.

Cyclic Photophosphorylation

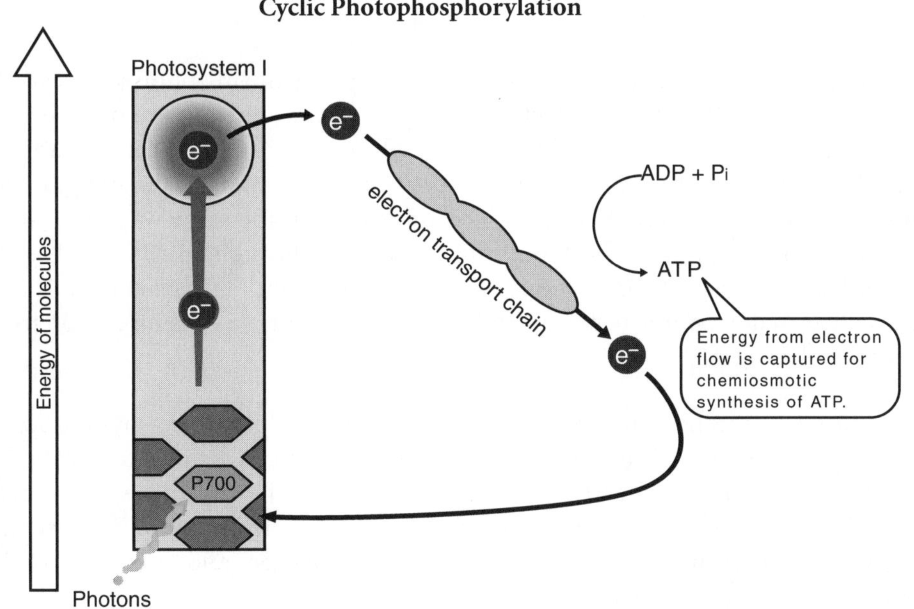

Noncyclic photophosphorylation starts in photosystem II (see figure). In noncyclic photophosphorylation, chlorophyll pigment absorbs light and passes excited electrons to a reaction center, a process equivalent to cyclic photophosphorylation. The photosystem II reaction center contains a P680 chlorophyll, distinct from photosystem I. From the photosystem II reaction center, the electrons are passed to an electron transport chain. In this case however the electrons are not returned to the reaction center at the end of the electron transport chain but are passed to photosystem I. Photosystem II replaces the electrons it lost by getting them from water, producing oxygen in the process. The electrons that enter photosystem I in this case are used to produce NADPH.

Noncyclic Photophosphorylation: Photosynthesis Light Reactions

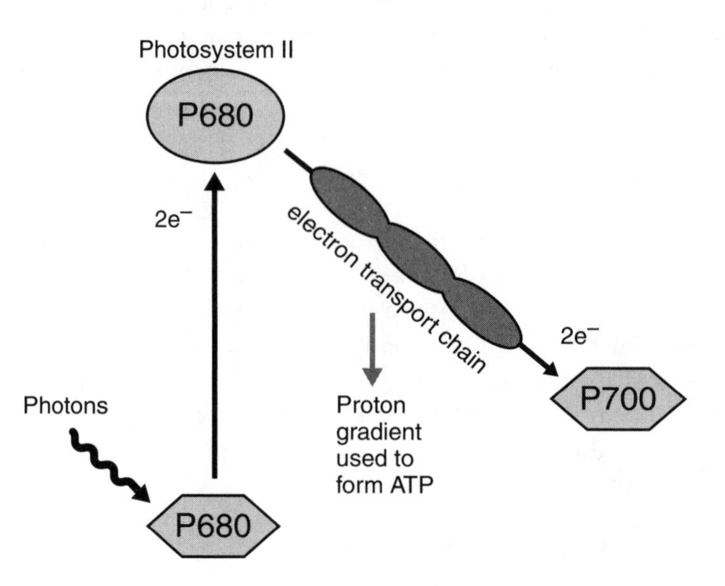

So far, we have not addressed the mechanism used to produce ATP during photosynthesis. As the electrons work their way through the electron transport chains, protons are pumped out of the stroma and into the interior of the thylakoid membranes, creating a proton gradient. This proton gradient generated by an electron transport chain is similar to the pH gradient created in mitochondria during aerobic respiration and is used in the same way to produce ATP. Protons flow down this gradient back out into the stroma through an ATP synthase to produce ATP, similar once again to mitochondria. The NADPH and ATP produced during the light reactions are used to complete photosynthesis in the Calvin cycle, using carbon from carbon dioxide to make sugars.

The oxygen produced in the light reactions is released from the plant as a byproduct of photosynthesis. Starting about 1.5 billion years ago, photosynthesis helped to create the oxygen rich atmosphere found on earth today, which allowed the evolution of animals requiring the efficient energy metabolism provided by aerobic respiration. The oxygen produced through photosynthesis today maintains Earth's oxygen and is a key to the continued functioning of the biosphere.

Calvin-Benson Cycle

The first portion of photosynthesis, in which ATP, oxygen and NADPH are produced is sometimes called the light reactions since light energy is required to energize electrons and drive the reactions forward. In the remaining part of photosynthesis, the energy captured in the light reactions as ATP and NADPH is used to drive carbohydrate synthesis. This cycle, also known as the **Calvin-Benson cycle**, fixes CO_2 into carbohydrates, reducing the fixed carbon to carbohydrates through the addition of electrons. The NADPH provides the reducing power for the reduction of CO_2 to carbohydrate, and air provides the carbon dioxide. CO_2 first combines with, or is fixed to, ribulose bisphosphate (RuBP), a five-carbon sugar with two phosphate groups. The enzyme that catalyzes this reaction is called **rubisco** and is the most abundant enzyme on earth. The resulting six-carbon compound is promptly split, resulting in the formation of two molecules of 3-phosphoglycerate, a three-carbon compound. The 3-phosphoglycerate is then phosphorylated by ATP and reduced by NADPH, which leads to the formation of phosphoglyceraldehyde (PGAL). This can then be utilized as a starting point for the synthesis of glucose, starch, proteins, and fats.

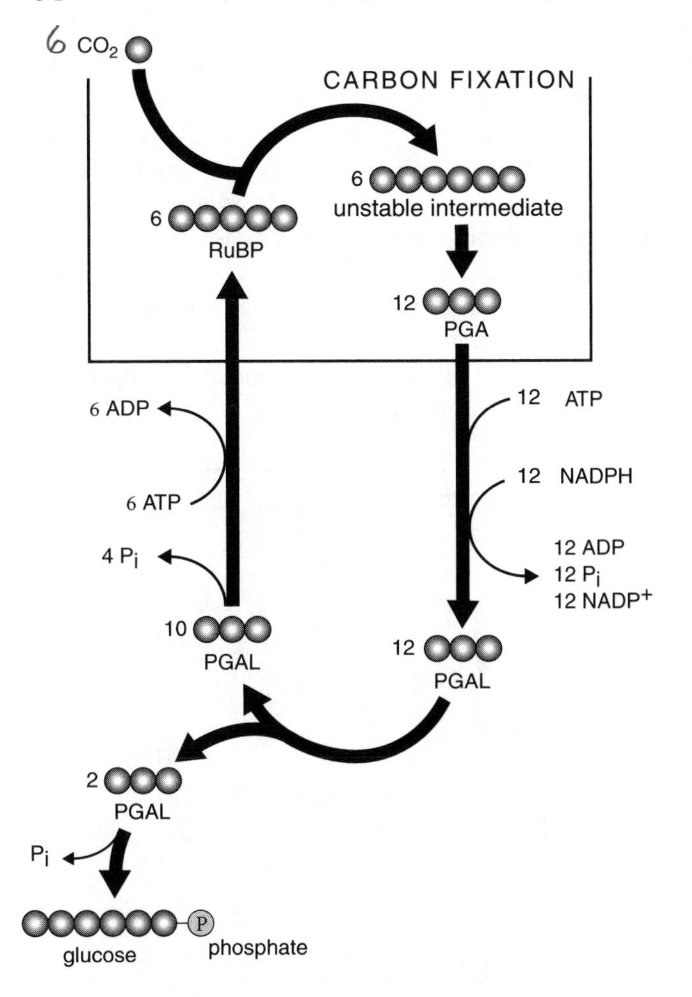

Rubisco is not completely specific for CO_2 however, and will also catalyze a reaction with oxygen and ribulose bisphosphate in a process called **photorespiration**. This reaction produces a two-carbon molecule called glycolate, that passes to peroxisomes where it is oxidized to CO_2. This pathway reverses some of the energy captured in photosynthesis, undoing carbon fixation without producing energy. It is not clear if there are any advantages to this, but the consequences can be predicted. By reversing a percentage of carbon fixation, photorespiration reduces the overall efficiency of photosynthesis.

Some plants have evolved changes in photosynthesis that bypass photorespiration to increase the overall efficiency of photosynthesis. This altered process is called **C_4 photosynthesis.** C_4 plants have an enzyme called PEP carboxylase in their leaves, which catalyzes carbon dioxide fixation with phosphoenolpyruvate. The product of this reaction is oxaloacetate, a 4-carbon molecule (thus the name C_4 photosynthesis). Since a different enzyme is used for carbon fixation and PEP carboxylase does not react with oxygen, there is minimal photorespiration in C_4 plants, making photosynthesis in C_4 plants more efficient.

C_4 plants perform the Calvin-Benson cycle as do C_3 plants. In C_4 plants, however, carbon fixation and the Calvin-Benson cycle occur in different cells. C_4 plants have different leaf structure than C_3 plants to accommodate the different biochemistry they use (see figure). **Mesophyll** cells in C_4 plants line the surface of the leaves and carry out carbon fixation to produce oxaloacetate. The oxaloacetate diffuses from the mesophyll cells into neighboring cells in the interior of the leaf called **bundle sheath cells.** In bundle sheath cells, CO_2 is removed from oxaloacetate, producing pyruvate as well as CO_2. The pyruvate is transported back to the mesophyll cells, and the CO_2 is used in the bundle sheath cells to enter the Calvin-Benson cycle. The carbon dioxide enters the Calvin-Benson cycle with catalysis by rubisco, as in C_3 plants, so that C_4 plants actually perform carbon fixation twice. The movement of oxaloacetate from mesophyll to bundle sheath cells tends to pump CO_2 into bundle sheath cells, maintaining the high CO_2 concentration required to avoid photorespiration.

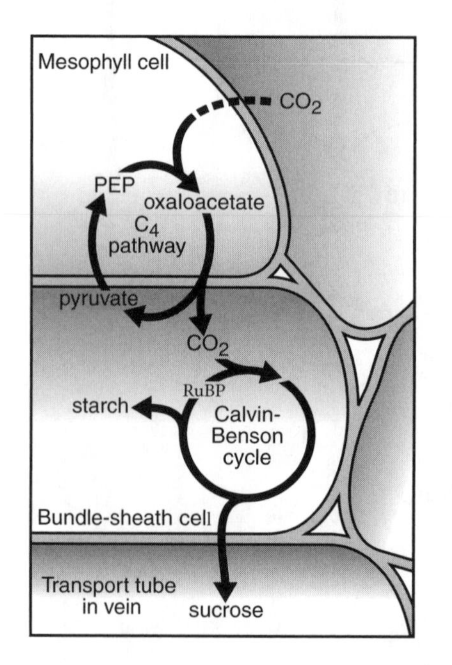

C_4 plants tend to live in dry, hot environments; examples of C_4 plants are corn and crab grass. The carbon fixation step catalyzed by rubisco does not occur in hot conditions with low CO_2 because rubisco has a lower affinity for CO_2 than PEP carboxylase does. In hot environments, plants, close the stomata on their leaves and limit the supply of CO_2. In C_3 plants these conditions favor photorespiration, but C_4 plants carry on carbon fixation under these conditions with PEP carboxylase. C_3 plants are more efficient at cooler temperatures, however. In ecosystems in which C_3 and C_4 plants compete, the plant which dominates may depend on the temperature of the environment.

Another class of plants called **CAM** plants has also adapted to perform photosynthesis in hot, dry conditions. CAM plants keep their stomata closed during the day to conserve water, and like C_4 plants, keep up carbon fixation using PEP carboxylase. CAM plants, however, perform carbon fixation during the night when the stomata are open and store the CO_2 for use in the Calvin-Benson cycle. The products of carbon fixation are stored for use during the day.

HORMONES AND SECOND MESSENGERS

General Principles of Signaling

Two main ways that animal cells communicate with one another are (a) via signaling molecules that are secreted by the cells, and (b) through molecules that rest on the cells' surfaces and remain attached to the cell even as they signal other cells. The target cell receives information through receptors on its surface, which are generally membrane-spanning proteins with binding domains on the outside of the plasma membrane.

Although some signaling molecules may act far away from the cell that secreted them, many bind only to receptors on cells in the immediate vicinity. **Paracrine** signaling refers to the process of signaling only nearby cells, with signaling molecules quickly pulled out of the extracellular matrix. **Synaptic** signaling occurs only in nerve cells, over extremely short distances, as electric signals reach axon terminals and cause the release of chemicals called neurotransmitters. These chemical messengers bind to nearby nerve cells and cause an electrical signal to be propagated or continued. Synaptic signaling will be discussed in detail later on. It allows for rapid communication over long distances without a diminished effect, which can happen when sending individual molecules over long distances, such as through the bloodstream. **Endocrine** signaling refers to the secretion of chemical messengers into the bloodstream (or other liquid medium, such as is the case with plants) for widespread distribution throughout the entire organism.

Cells communicate with each other using a wide-variety of molecules, some which can pass through the cell membranes of the target cells and act directly within the cell's cytoplasm and others which bind to plasma membrane receptors and act through "second messengers." All of these molecules can be considered **hormones**, circulating signals that are released by specialized cells and travel throughout the bloodstream.

Steroid Hormones

Steroid hormones, such as the sex hormones testosterone and estrogen, are lipids with cholesterol-based structures and can pass through the cell membrane to act directly on the DNA in the cell nucleus. Often, steroid hormones must bind to intracellular receptor proteins in order to cross the nuclear membrane or regulate transcription. The steroid-receptor complex is generally considered to be a **transcription factor**, since it is able to bind to "enhancer" regions on the DNA, turning on the transcription of certain genes.

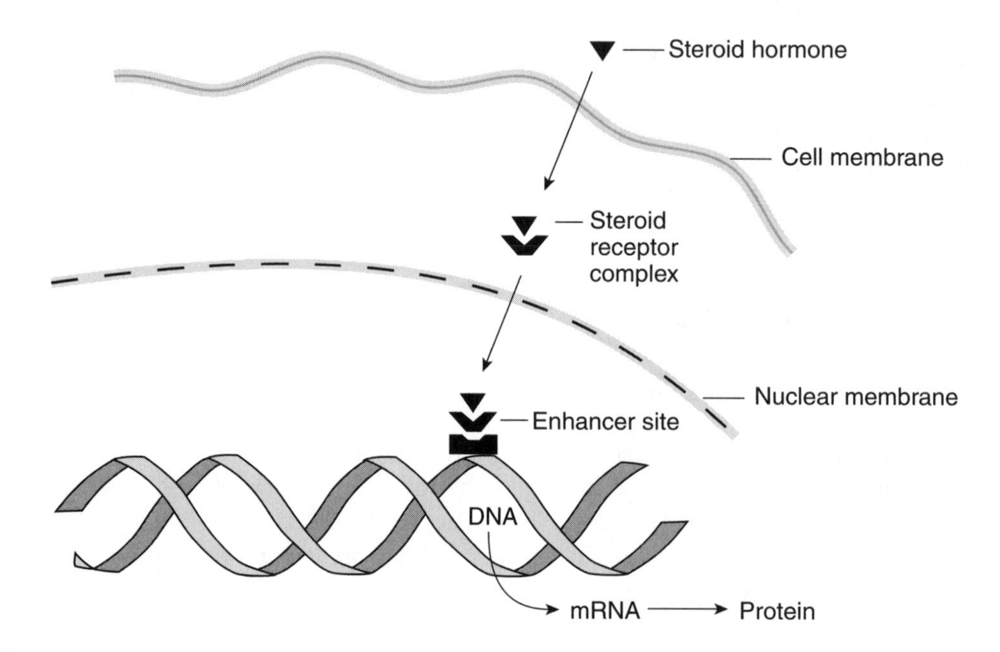

The basis for the production of steroid hormones, as well as compounds like Vitamin D and the thyroid hormones, is cholesterol. Steroid hormones travel through the bloodstream bound to carrier molecules, from which they dissociate when entering a cell. *Because these hormones are lipids and do not dissolve in water, their effects can last for hours or days* in the bloodstream after being released, a much longer period of time than water-soluble, nonsteroid hormones can last.

Another molecule worth mentioning here is not a steroid, though it can easily pass through a cell's lipid bilayer because it is so small. **Nitric oxide** (NO), recently recognized as an important signaling molecule, passes into the cell's cytoplasm, acting on the enzyme guanylyl cyclase, which in turn produces **cyclic-GMP**. Cyclic-GMP is an important molecule that, in the case of NO, causes smooth muscle cells in blood vessels to relax when stimulated by acetylcholine. Keep in mind, too, that different cells can respond to the same hormone or neurotransmitter in different ways. For example, although acetylcholine can cause smooth muscle cells to relax, it causes skeletal muscle cells to contract. It is postulated that these differences are caused by the variety of receptor complexes that can respond to the same hormone.

Nonsteroid Hormones

Many compounds that are unable to cross the plasma membrane and enter the cytoplasm still act as powerful signaling molecules. Examples include peptides, such as **atrial natriuretic peptide**, released by the heart to influence water absorption in the kidneys, **calcitonin**, involved in calcium regulation, and **glucagon**, involved in blood sugar regulation. More hormones will be discussed specifically in section III.

These nonsteroid hormones act via signal-transduction pathways, whereby binding of the hormone to a cell-surface receptor induces a conformational change in the receptor protein, setting off an intracellular cascade to alter the cell's behavior in some way. **Cell-surface receptors** come in three different forms: Ion-Channel-linked, G-Protein-linked, and Enzyme-linked. All of these receptors can bind to the hormone or chemical signal (also referred to as the "**ligand**") very accurately and very tightly and turn on intracellular signals.

Ion-channel-linked receptors are also called *ligand-gated channels.* These membrane-spanning proteins undergo a conformational change when a ligand binds to them so that a "tunnel" is opened through the membrane to allow the passage of a specific molecule. These ligands can be neurotransmitters or peptide hormones, and the molecules that pass through are often ions, such as sodium (Na^+) or potassium (K^+), which can alter the charge across the membrane (more in section III). The ion channels, or pores, are opened only for a short time, after which the ligand dissociates from the receptor and the receptor is available once again for a new ligand to bind.

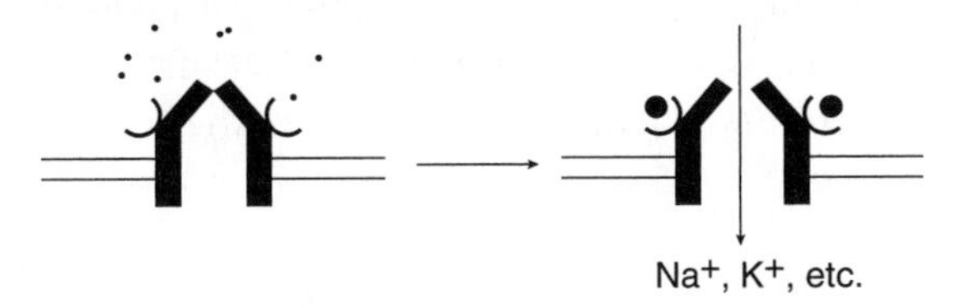

G-protein-linked channels cause G-proteins to dissociate from the cytoplasmic side of the receptor protein and bind to a nearby enzyme. This enzyme continues the signaling cascade by inducing changes in other intracellular molecules; in addition, it can also cause other membrane channels to open in areas some distance from the originating receptor.

Most G-proteins activate what are known as "second messengers," small intracellular molecules like **cyclic AMP** (cAMP), calcium, and phosphates, which in turn activate key enzymes or transcription factors involved in essential reactions. Signaling cascades involving G-proteins can be very complex and involve many different enzymes and conversions, which prevents the reactions from running out of control.

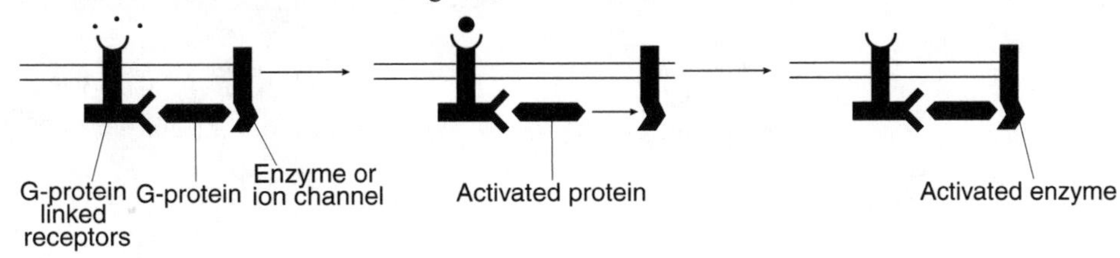

Enzyme-linked receptors can act directly as enzymes, catalyzing a reaction inside the cell, or they can be associated with enzymes that they activate within the cell. Most enzyme-linked receptors turn on a special class of enzymes called protein **kinases**, which add free-floating phosphate groups to proteins, regulating their activity. Protein **phosphorylation** is an essential means of intracellular signaling and control, as proteins become activated or deactivated simply by the addition or removal of phosphates. Protein kinases often target other protein kinases, initiating a cascade of kinase activity. Protein phosphatases reverse the action of protein kinases.

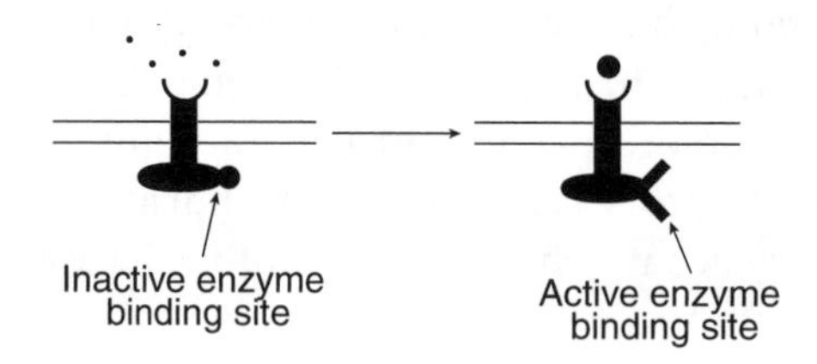

Signal Integration

G-protein-linked and enzyme-linked receptors use complex relays of signal proteins to amplify and/or regulate their signal transduction. In some cases, a measure of safety requires that two different receptors on the cell surface become activated in order to turn on a particular intracellular protein. In other cases, signals at different receptors lead to the phosphorylation of different proteins that activate together only when *both* proteins have been phosphorylated. This "signal integration" leads to a measure of *control over reactions* and the ability to use multiple inputs to cause a certain effect that can vary in degree.

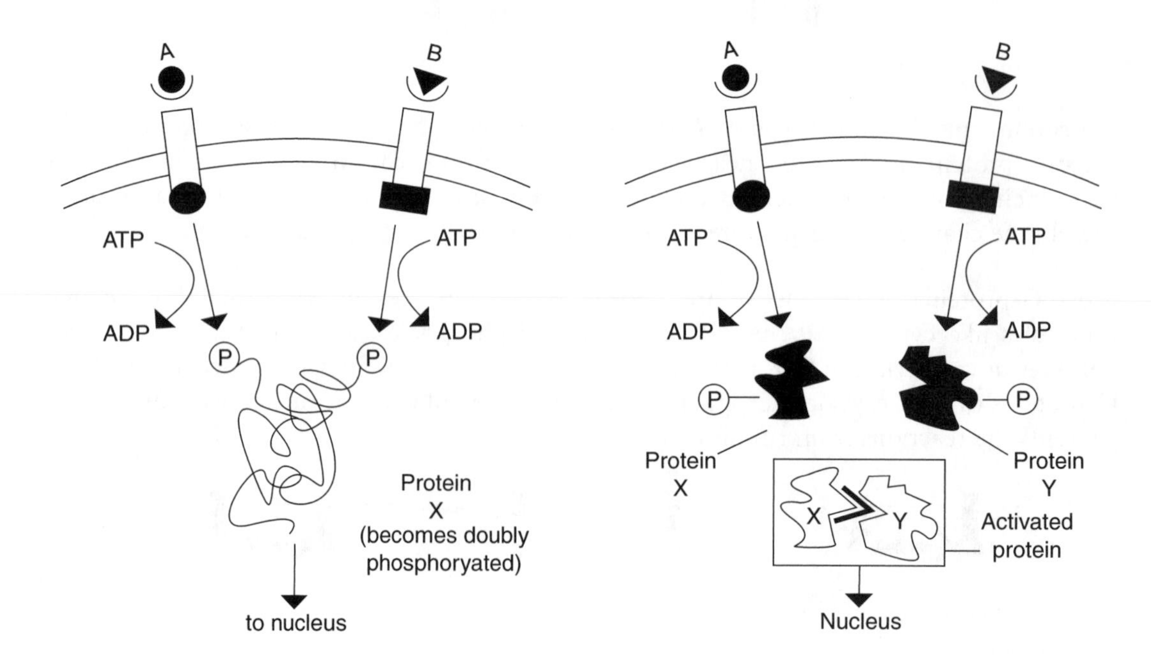

CHAPTER THREE

Cell Membranes and Organelles

Although Robert Hooke is widely credited with the first discovery of cells back in the mid-1600s, it was not until 200 years later that cells were recognized as the basic units of life's structure and function. The formation of **the cell theory** was soon to follow, an overarching explanation for observations that had been made over several centuries. The cell theory has three major components:

i. Cells are the basic units of structure and function in all living things.
ii. Cells are capable of carrying out all necessary activities for life.
iii. All new cells come from preexisting cells.

Cells are separated from their external environment by a semipermeable lipid bilayer and occasionally by a semipermeable cell wall in addition to this membrane. This separation is crucial for the cell to be able carry out reactions necessary for life in a space insulated from external factors and compounds. Eukaryotic cells have the benefit of using specialized **organelles** within each cell to compartmentalize these separate reactions even further. Yet, cells must also integrate themselves with their external environment, exchanging key materials with the outside world and with other cells nearby, and they often need to anchor themselves in some fashion to this outside world so that they are not swept away. For this exchange and anchoring, many complex proteins form membrane channels, junctions, and other cell-to-cell connections.

This chapter investigates the structure and function of the cell membrane and cell surface proteins, as well as organelle structure and function.

MEMBRANE STRUCTURE AND CELL SURFACE PROTEINS

The cell (or **plasma**) **membrane** makes up as much as 50 percent of the mass of any given cell, and it is responsible for carefully regulating the passage of materials into and out of the cell. According to the **fluid-mosaic model**, this semipermeable membrane consists of a double layer of phospholipids, which possess both a **hydrophilic** phosphate ("head") region and a **hydrophobic** fatty acid ("tail") region. Lipid molecules such as these that assemble together spontaneously in solution always do so in such as way that their hydrophobic tails are on the inside of a bilayer and their hydrophilic heads are on the outside. The diagram below of a typical phospholipid shows the representation that is most often seen in textbooks alongside the actual chemical structure.

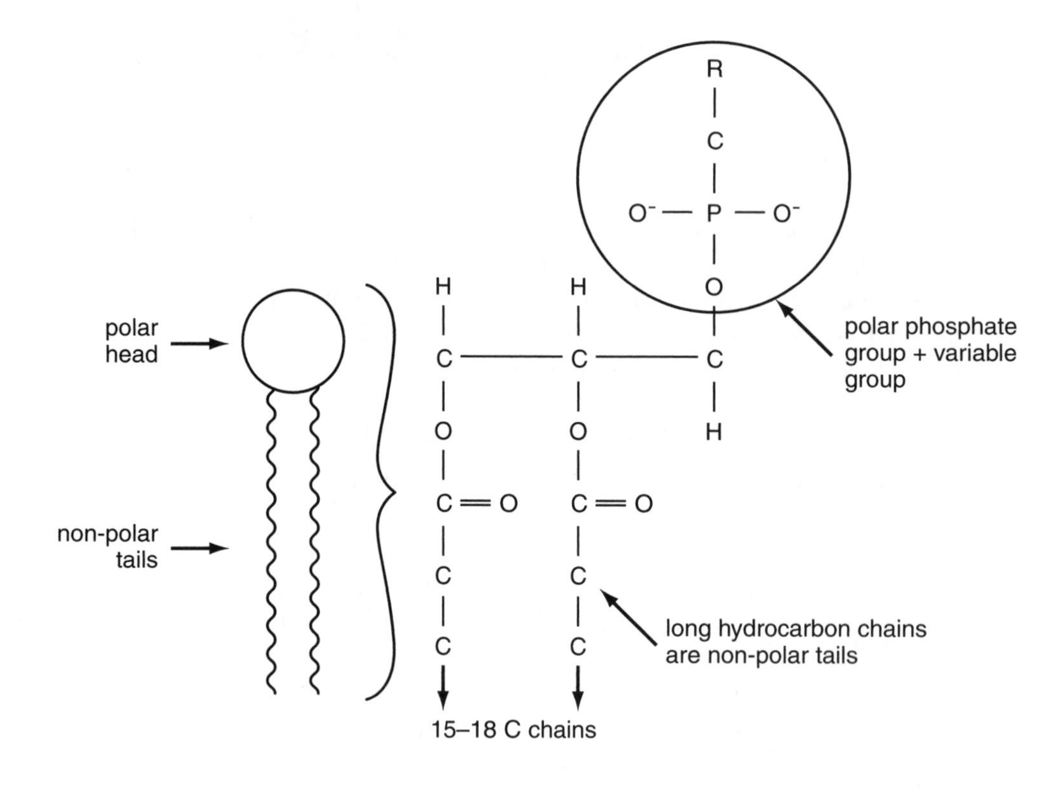

The variable "R" group on the head end of the molecule comes in several different forms depending on the phospholipid present. There are four major phospholipid types in mammalian cells: phosphotidylcholine, sphingomyelin, phosphotidylserine, and phosphotidylethanolamine. All have an overall neutral charge, except for the lipid with a serine R-group, which has a negative charge.

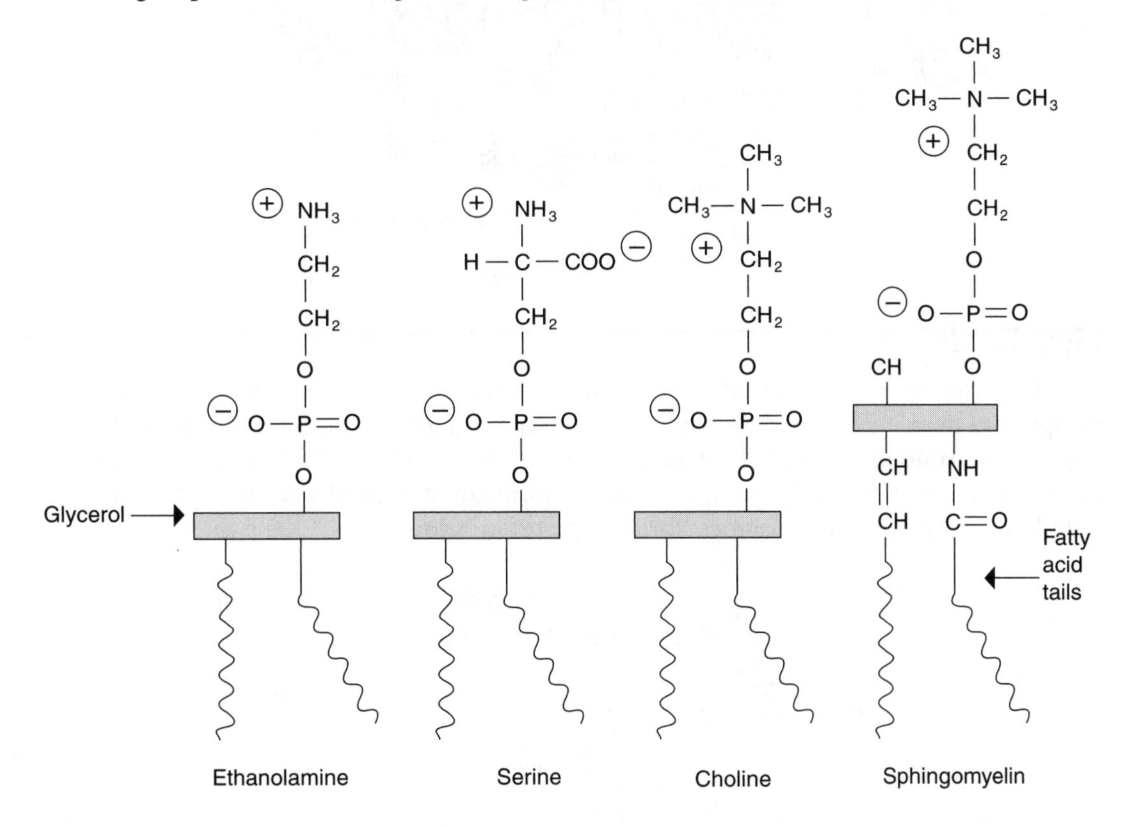

It is clear that the variety of **phospholipids** present in the membrane allows specificity for reactions, since it is known that many membrane proteins must have certain kinds of lipids next to them in order to function properly. Keep in mind that the external membrane layer may be composed of different percentages of certain phospholipids than the internal layer, which again adds to reaction specificity.

Other lipids are also present in cell membranes aside from the four major ones mentioned above. These include **molecules** such as phosphotidylinositol, a lipid essential in cell signaling, used to convert nonsteroid hormone messages into "second messages" sent into the cell cytoplasm. In addition, cholesterol plays a crucial role described in chapter 1. An overall view of the cell membrane is presented below:

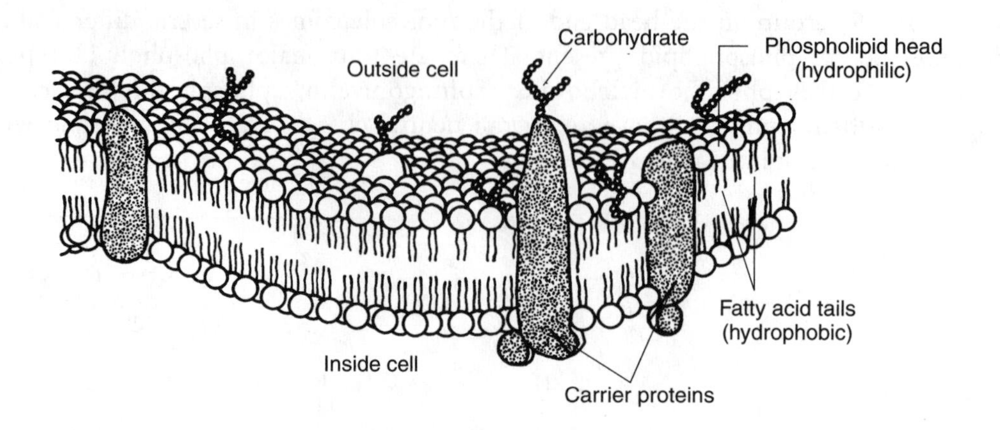

Glycolipids

As will be discussed in detail later, the Golgi apparatus often adds complex carbohydrate groups onto proteins destined for the cell membrane and for export. But it can also add these sugar groups onto lipids headed for the membrane, resulting in the formation of **glycolipids.** The most common type of glycolipid is the **ganglioside**, structurally related to but far more complex than the simpler galactocerebroside pictured below:

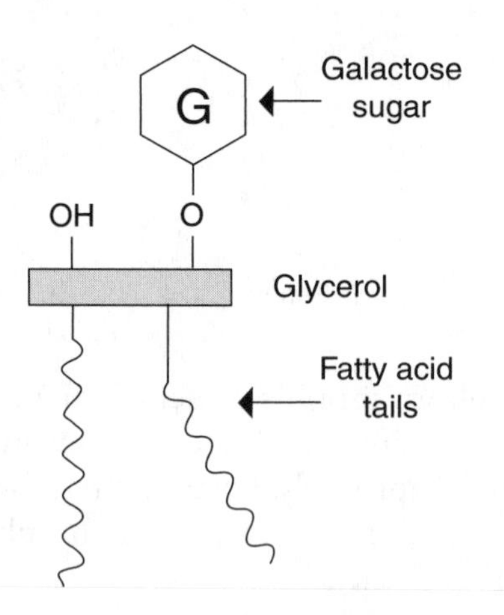

While you will not need to know the actual structures of glycolipids or of the phospholipids pictured above, seeing their chemistry helps to clarify how they fit into the cell membrane. While the role of glycolipids remains unclear, they are thought to aid in the *transmission of electrical impulses* along cell membranes as well as in *cell-to-cell recognition.*

Membrane Proteins

Membrane proteins, like the membrane phospholipids, usually have carbohydrate groups attached to them so that the outside surface of the plasma membrane is extremely sugar rich.

Membrane-spanning proteins have regions that are hydrophobic as well as regions that are hydrophilic, with the nonpolar (hydrophobic) regions passing through the nonpolar interior of the membrane and the polar areas sticking both into the cytoplasm and out into the extracellular space. Other proteins can be located completely intracellularly or extracellularly, anchored to the cell membrane by a variety of special lipids. These proteins are made in the cytosol and bind into the cell membrane only because they subsequently have a lipid molecule attached onto their structure. Below are some of the ways in which proteins regularly associate with the lipid bilayer:

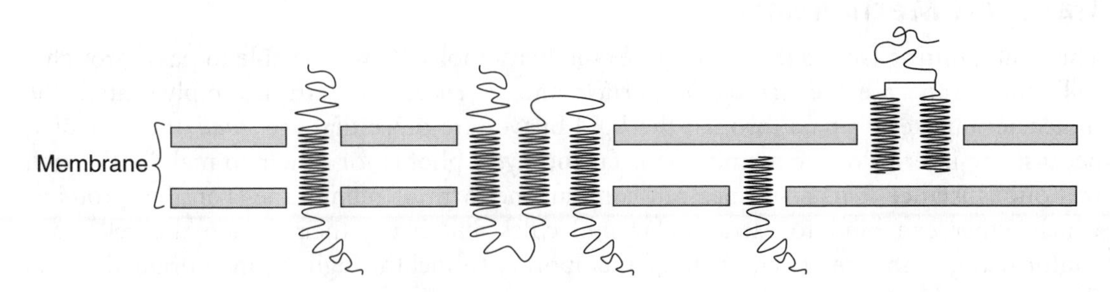

In most membrane-spanning proteins, the most favorable configuration for the amino acid region that crosses within the hydrophobic membrane interior is that of an **alpha-helix.** This screw-type formation minimizes contact of the polar peptide bonds between the amino acids with the nonpolar fatty acid tails of the phospholipids. This leaves only the nonpolar amino acid side chains to stick out from the helix and contact the membrane's hydrophobic interior.

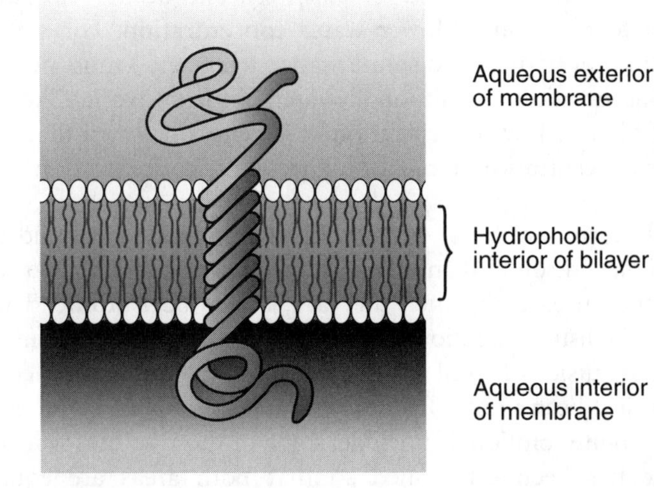

Most membrane-spanning segments are approximately 20–30 amino acids in length and contain mainly amino acids with hydrophobic side chains. Remember that some proteins, as seen in the figure at the top of the page, can pass through the membrane multiple times with multiple transmembrane domains threaded through the bilayer. These "multi-pass" proteins have several alternating regions of hydrophobic and hydrophilic amino acids.

Transmembrane proteins are involved not only in carrying materials across the membrane, but also in cell recognition, cell adhesion, cell signaling, and enzymatic reactions. Recall that most proteins sticking up from the surface of the membrane are covered in carbohydrates on the extracellular surface. The term used to describe the protein- and carbohydrate-rich coating on the cell surface is **glycocalyx**. Keep in mind that these sugars reside exclusively on the exterior of the membrane.

Transport Mechanisms

The main limiting factors that determine whether a molecule will be able to pass through a cell's membrane are the *size of the particle* and *its charge* (polarity). Simply stated, the molecules quickest to pass through the lipid bilayer are those that are *small* and *nonpolar*, because the interior of the membrane is far too hydrophobic for others to make it through without assistance. This assistance can come in the form of membrane-spanning proteins, which either can bind to extracellular molecules and bring them inside the cell via a conformational change or can open up a temporary tunnel through the membrane lipids so that the molecules can pass through.

Simple diffusion refers to the movement of particles down their concentration gradient from a region of higher concentration to a region of lower concentration. This form of transport takes place directly through the cell membrane lipid bilayer without using any form of energy or membrane proteins in order to move particles. Again, small nonpolar molecules move most freely by simple diffusion. Examples include water, carbon dioxide, and oxygen. The simple diffusion of water is referred to as osmosis and occurs from a region of higher water concentration to a region of lower water concentration. For water to be in high concentration, the amount of dissolved solute (salts, sugars, etc.) must be low, and vice versa for water in low concentration. So, although water diffusion works like any other passive diffusion in term of high $\rightarrow$ low concentration, it is generally stated that water moves from an area of low *solute* concentration to one of higher *solute* concentration.

Solutions low in solute concentration relative to other solutions are said to be **hypotonic**, whereas solutions higher in solute than others are **hypertonic**. When two solutions have the same solute concentration as each other, they are said to be **isotonic**. The cell membrane effectively separates two distinct solutions: one is the extracellular environment and the other is the cytoplasm. If the outside of a cell is hypertonic, higher in solute concentration, than the inside (e.g., a cell has just been moved from fresh water to salt water), water will move out of the cell into the high solute solution. If possible, some of those solutes will also move into the cell until a balance has been established so that both areas are equivalent in solute concentration. In hypotonic solutions, cells generally take on water, sometimes until they burst.

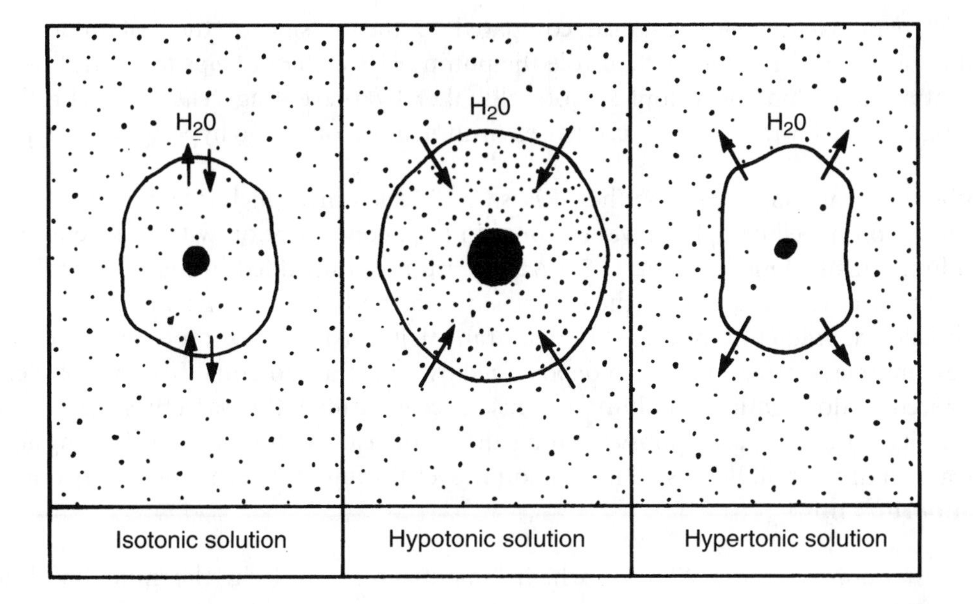

Facilitated diffusion, otherwise known as passive transport, involves the use of channel or carrier proteins imbedded in the membrane to allow molecules to diffuse down a gradient. The structures of the proteins involved in this type of transport are very similar and amino acid sequences are highly conserved across many species. In some cases, these proteins act as pores for ions; in other cases, they may open and close in response to external signals. Keep in mind that, because cells naturally have a negatively charged cytoplasm, the opening of ion channels favors the movement of positively charged ions into the cytoplasm. This combination of solute concentration and an electrical gradient is called the **electrochemical gradient** and is the key determinant of what moves into and out of membranes when passive transport channels are open.

In isotonic solutions, water and solutes can and do move across the membrane, yet movement inward is always balanced by reciprocal movement outward. This means that overall concentrations of water and solute on opposite sides of the membrane do not change.

Active transport: Some membrane proteins use the energy of ATP to change the protein's conformation in the membrane so that molecules can be brought into and out of the cell against their concentration gradients. These **ATPase pumps** are found in every membrane of cells and they are extremely important in the maintenance of unequal concentrations of certain ions across the lipid bilayer—something that we will later see is *essential for processes like nerve signal conduction.*

A commonly cited example of active transport is the Na^+-K^+ ATPase membrane pump, whose conformational change uses the energy of ATP breakdown to pull 2 K^+ (potassium) ions into a cell while kicking out 3 Na^+ (sodium) ions at the same time. This transport of molecules in opposite directions is known as **antiport**, and it can be contrasted with pumps that pull two different molecules in the same direction (**symport**). Because the pump, which is present in all cell membranes, pumps out three positive charges for every two it brings in, the inside of the cell remains negatively charged compared to the outside of the cell under normal conditions. Yet, the more important role the pump plays is that it helps to control the solute concentration within the cytoplasm of cells, thereby preventing cells from shrinking or swelling too much when the extracellular environment becomes too hypertonic or hypotonic.

Another ion that you may see on the GRE is Ca^{++} (calcium), which is kept in extremely low concentration in cell cytoplasm, yet is stored in high concentration within the endoplasmic reticulum. This is done by using **Ca^{++}-ATPase** pumps, imbedded in the ER membrane, to actively transport calcium from the cytoplasm into the ER lumen. This naturally sets up a strong calcium gradient across the ER membrane that is used, for example, by muscle cells to regulate muscle contraction. When depolarized by an action potential from a nerve cell, the specialized endoplasmic reticulum of muscle cells (called the sarcoplasmic reticulum) releases its store of calcium ions, flooding the cytoplasm with Ca^{++} and leading to rapid contraction of the cell. Because only one ion moves through these channels, they are known as **uniport** pumps.

As a last example, those ATPases which "manufacture" ATP in the mitochondria and chloroplast as part of the **electron transport chain** are simply ATPase membrane transport proteins working in a reverse manner from how they usually work. Rather than ATP hydrolysis driving changes in protein structure so that ions can pass through the membrane, it seems that these pumps are driven by the flow of H^+ ions moving through them.

Endo- and exocytosis are two mechanisms of transport that can move large molecules and even entire cells through the cell membrane. In order accomplish this, the cell membrane actually invaginates, or pinches inward, to form a pocket in which the material to be transported can fall. In the case of endocytosis, this invagination pinches off completely, forming a vesicle that contains the transported material and can move freely within the cytoplasm. In exocytosis, a vesicle containing material to be expelled simply merges with the lipid bilayer and the material is pushed off into the extracellular space.

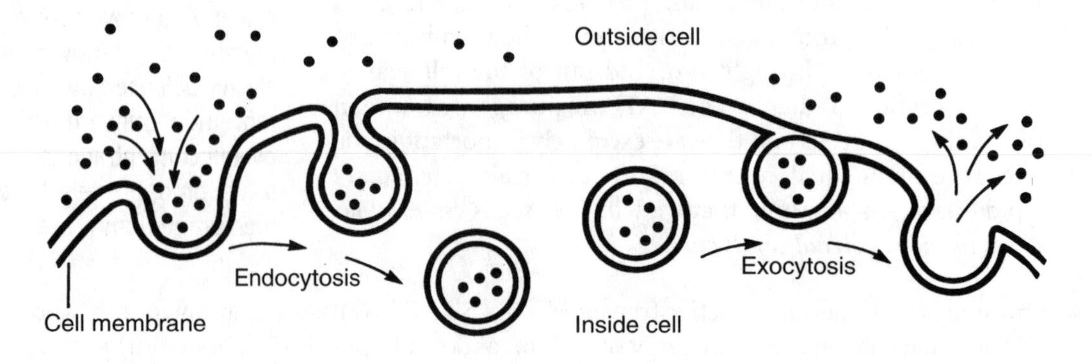

CELL JUNCTIONS AND THE CELL WALL

In order for cells to form complex tissues, secure cell-to-cell bonds must hold them together. These cell junctions come in a variety of forms and serve many different purposes: occluding junctions, otherwise known as **tight junctions**, seal spaces between cells; anchoring junctions, which include **desmosomes**, connect one cell's cytoplasm to another through anchoring proteins; and, communicating junctions, including **gap junctions** and **plasmodesmata** in plants, allow cells to directly exchange cytoplasmic material via channels that cross both cells' membranes.

We will touch on occluding and anchoring junctions briefly here, but most questions on the GRE involving cell junctions will involve the third type: communicating.

Occluding (Tight) Junctions

These connections are thought to be formed by proteins which tightly wind between the adjacent plasma membranes of neighboring cells, binding the cells together at those points so tightly that *nothing can diffuse between cells or past the junction.* In such a way, tight junctions form a total barrier to transport and diffusion where they exist.

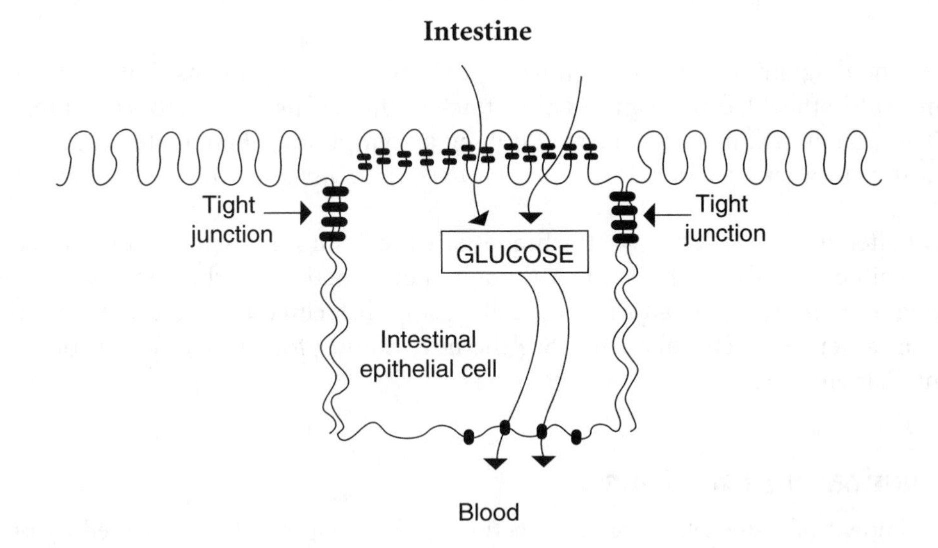

Where they are most useful, as shown in the diagram above, is in places like the *intestines,* where specialized cells absorb nutrients from one side of the cell (the intestinal side) and transport them through the cell and out the other end (into the bloodstream). Certain transporters (e.g., sodium ion-driven glucose transporters) exist on the intestinal side of the cell but not on the bloodstream side, and glucose that enters the bloodstream is prevented from diffusing back into the intestinal tract by tight junctions.

Anchoring Junctions

Found between cells subjected to fair amounts of stress, either from shearing forces or contacting forces, these junctions connect one cell's cytoplasm to another's via a series of proteins. Regardless of classification, anchoring junctions not only allow cells to adhere to neighbors, but also may allow them to contract themselves into large tubelike tissues as the fibers holding the junctions together contract across the cells' cytoplasms.

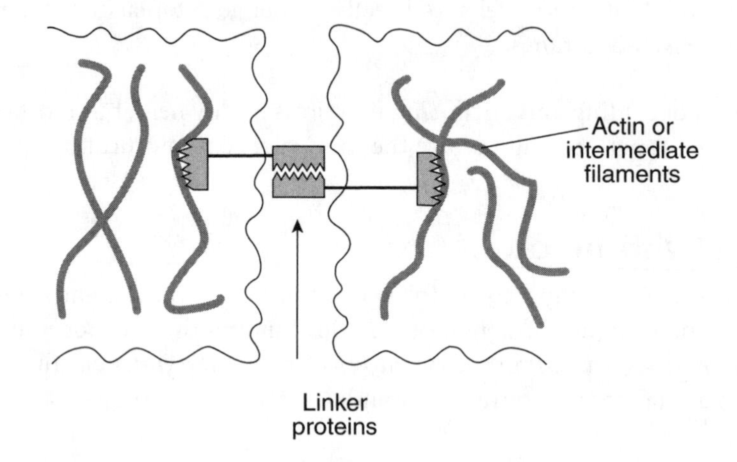

Notice in the diagram above that anchoring junctions involve proteins that attach to actin filaments within the cell cytoplasm and also attach to "linker" proteins across the intercellular space. Thus, the junction is not an actual linking of cytoplasm, where material can be freely exchanged, but rather a *physical joining so that the cells do not shear away from each other.*

The most often mentioned of these attachments are the desmosomes, found in heart cells and between epithelial cells in the skin. Although they function in the same way as other anchoring junctions, these attach two cells using **intermediate filaments** within the cytoplasm rather than actin filaments. See the next chapter for more information on these different filament types.

Communicating Junctions

The best known of these cell-to-cell connections are the gap junctions, formed by proteins called **connexins**, which build tubes or pores between two adjacent cells' cytoplasms. It is through these pores that ions and other material can pass from one cell to the other. In cells that rapidly transmit chemical or electrical signals across tissues, gap junctions are everywhere. Because chemical and electrical transmission is mediated through the movement of ions and other messengers, gap junctions *allow for undisrupted and very fast signal transmission* across wide areas of tissue. In heart cells, gap junctions allow for rhythmic contractions of large sections of the heart all at once and for waves of muscle contraction such as found in the esophagus. These junctions also allow for coordination of *rapid and complex movements* such as a fish's tail-flip to escape an incoming predator. The diagram below shows a generic gap junction with many **connexin** proteins that form a **connexon**:

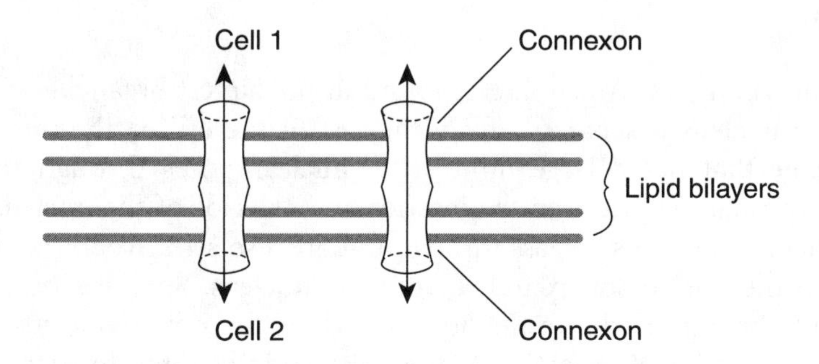

Plasmodesmata are plant cells' equivalent of gap junctions, particularly useful for the free flow of nuclei from one cell to another. These junctions in plants are not nearly as complex in structure as those in animals, but serve essentially the same purpose. Plant viruses, however, often exploit plasmodesmata because the openings allow the virus particles to spread rapidly from one section of the plant to others.

ORGANELLE STRUCTURE AND FUNCTION

Eukaryotic Cells and Organelles

All multicellular organisms (such as you, or a tree, or a mushroom) and all protists such as amoebas and paramecia are eukaryotic. The eukaryotes include the protists, fungi, animals and plants. Eukaryotic cells are enclosed within a lipid bilayer cell membrane, as are prokaryotic cells. Unlike prokaryotes, eukaryotic cells contain membrane-bound organelles (see figure). An organelle is a structure within the cell with a specific function that is separated from the rest of the cell by a membrane. The presence of membrane-bound organelles in eukaryotes allows eukaryotic cells to compartmentalize activities in different parts of the cell, making them more efficient. Compartments within a cell can allow the cell to carry out activities such as ATP production and consumption within the same cell and control each independently.

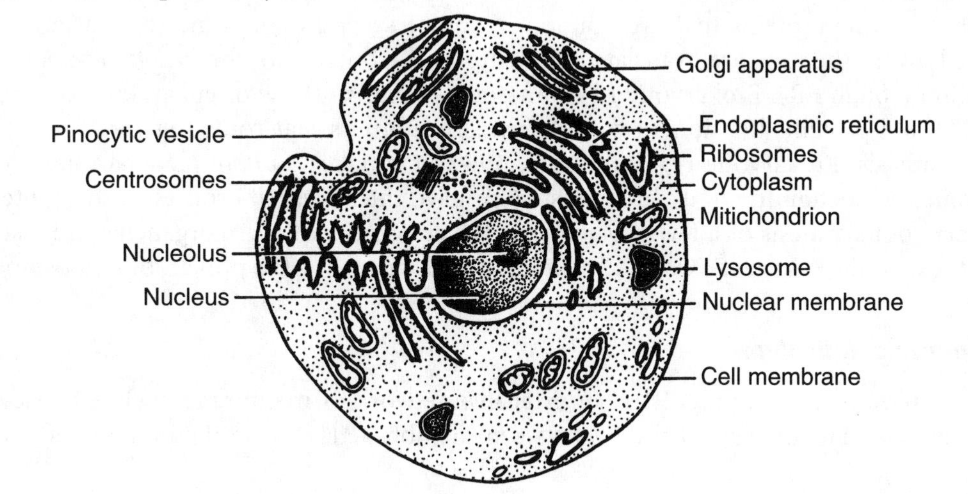

Nucleus

The genetic material, the DNA genome, is found in the largest organelle of the animal cell, the nucleus. The nucleus is separated from the rest of the cell by the **nuclear envelope**, a double-membrane that has a large number of nuclear pores through the envelope for communication of material between the interior and exterior of the nucleus. The pores are large enough to allow proteins to pass through but are also selective in the proteins that are transported into the nucleus or excluded from the nucleus. Special sequences in proteins signal a protein to be imported into the nucleus. While the prokaryotic genome is generally found in a single circular piece of DNA, the eukaryotic genome in each cell is split into chromosomes. Chromosomes contains the DNA genome complexed with structural proteins called histones that help to package the large strands of DNA in each chromosome within the limited space of the nucleus. Genes in the DNA genome are read (transcribed) to make RNA, which is processed in the nucleus before it is exported to the cytoplasm, where the RNA is read in turn (translated) to make proteins. The basic information flow of the cell is DNA to RNA to protein. The DNA genome is replicated in the nucleus when the cell divides. Other metabolic activities such as energy production are excluded from the nucleus. The structure and function of the eukaryotic genome will be presented later in more detail.

A dense structure within the nucleus in which ribosomal RNA (rRNA) synthesis occurs is known as the nucleolus. The nucleolus is not surrounded by a membrane, but is the site of assembly of ribosomal subunits from RNA and protein components. After assembly, the ribosomal subunits are exported from the nucleus to the cytoplasm to carry out protein synthesis.

Ribosomes

Ribosomes are not organelles but are large complex structures in the cytoplasm that are involved in protein production (translation) and are synthesized in the nucleolus. They consist of two subunits, one large and one small. Each ribosomal subunit is composed of ribosomal RNA (rRNA) and many proteins. Free ribosomes are found in the cytoplasm, while bound ribosomes line the outer membrane of the endoplasmic reticulum (ER). Proteins that are destined for the cytoplasm are synthesized by ribosomes free in the cytoplasm, while proteins that are bound for one of several membranes or that are to be secreted from the cell are translated on ribosomes bound to the rough endoplasmic reticulum (rough ER). Prokaryotic ribosomes are similar to those of eukaryotes, composed of rRNA and proteins that form two different size subunits that come together to perform DNA synthesis. Prokaryotic ribosomes are, however, smaller and simpler than eukaryotic ribosomes. Mitochondria and chloroplasts also have their own ribosomes in their interior and carry out synthesis of a few proteins, but the ribosomes of these organelles are distinct from those of the eukaryotic cytoplasm and more closely resemble prokaryotic ribosomes.

Endoplasmic Reticulum

The endoplasmic reticulum (ER) is an extensive network of membrane-enclosed spaces in the cytoplasm. The interior of the ER between membrane layers is called the lumen and at points in the ER the lumen is continuous with the nuclear envelope. If a region of the ER has ribosomes lining its outer surface, it is termed rough endoplasmic reticulum (rough ER); without ribosomes, it is known as smooth endoplasmic reticulum. Smooth ER is involved in lipid synthesis and the detoxification of drugs and poisons and has the appearance of a network of tubes, while rough ER is involved in protein synthesis and is a series of stacked plates. Proteins that are secreted, found in the cell membrane, the ER, or the Golgi are made by ribosomes on the rough ER. Proteins synthesized on the rough ER cross into the lumen of the rough ER during synthesis. A hydrophobic sequence of amino acids at the amino terminus of proteins as they are synthesized determines whether the protein will be sorted into the secretory pathway starting at the rough ER or synthesized in the cytoplasm. Proteins that are secreted will have only one hydrophobic signal sequence, the signal peptide, and will be inserted into the ER lumen when they are synthesized, then released from the cell later. Proteins that are destined to be membrane bound have hydrophobic transmembrane domains that are threaded through the rough ER membrane as the protein is synthesized. When the protein reaches the correct membrane destination along the secretory pathway, additional signals in the protein sequence and structure will cause the protein to stay localized at the correct location.

Small regions of ER membrane bud off to form small round membrane-bound vesicles that contain newly synthesized proteins. These cytoplasmic vesicles are then transported to the Golgi apparatus, the next stop along the secretory pathway.

Golgi Apparatus

The Golgi is a stack of membrane-enclosed sacs, usually located in the cell between the ER and the plasma membrane (see figure). The stacks closest to the ER are called the cis Golgi and the stacks farthest from the ER, closer to the plasma membrane, are called the trans Golgi. Vesicles containing newly synthesized proteins bud off of the ER and fuse with the cis Golgi. In the Golgi, these proteins are modified and then repackaged for delivery to other destinations in the cell. For example, the Golgi carries out post-translational modification of proteins through glycosylation, the process of adding sugar groups to proteins to form glycoproteins. Many proteins destined for the plasma membrane have carbohydrate groups added to the surface of the protein facing the exterior of the cell.

After processing in the cis Golgi, proteins are packaged in vesicles that move to the next layer in the stack, where they fuse and release their contents. Proteins proceed in this manner from one stack to the next until they reach the trans Golgi. In the trans Golgi, proteins are sorted into vesicles based on signals in different proteins that indicate their final destination. The nature of the signal varies, but includes the protein primary sequence, its structure, and post-translational modifications. Once packaged into vesicles, the vesicles move on to their final destination. The final destination for a protein may include the lysosome, the plasma membrane or the exterior of the cell. Some proteins are retained in the Golgi or the ER. Proteins that are destined for the plasma membrane as transmembrane proteins are inserted in the membrane in the ER as they are synthesized, and maintain their orientation in the membrane as they move from ER to Golgi to vesicle to the plasma membrane. Proteins that are secreted from the cell are inserted in the ER lumen during protein synthesis, and remain in the lumen of the ER to the Golgi, where they form secretory vesicles. The last step in secretion is the fusion of the secretory vesicle with the plasma membrane, releasing the contents of the vesicle to the cellular exterior.

Lysosomes

Lysosomes contain hydrolytic enzymes involved in intracellular digestion that break down proteins, carbohydrates and nucleic acids. For white blood cells, the lysosome may degrade bacteria or damaged cells. For a protist, lysosomes may provide food for the cell. They also aid in renewing a cell's own components by breaking them down and releasing their molecular building blocks into the cytosol for reuse. A cell in injured or dying tissue may rupture the lysosome membrane and release its hydrolytic enzymes to digest its own cellular contents.

The lysosome maintains a slightly acidic pH of 5 in its interior, a pH at which lysosomal enzymes are maximally active. The contents of the lysosome are isolated from the cytoplasm by the lysosomal membrane, keeping the pH distinct from the neutral pH of the cytoplasm. The distinct pH optimum and compartmentalization of lysosomal enzymes prevents them from degrading the rest of the cellular contents.

Peroxisomes

Peroxisomes contain oxidative enzymes that catalyze reactions in which hydrogen peroxide is produced and degraded. Peroxisomes break fats down into small molecules that can be used for fuel; they are also used in the liver to detoxify compounds, such as alcohol, that may be harmful to the body. The peroxides produced in the peroxisome would be hazardous to the cell if present in the cytoplasm, since these molecules are highly reactive and could covalently alter macromolecules such as DNA. Compartmentalization of these activities within the peroxisome reduces this risk.

Mitochondria

Mitochondria are the source of most energy in the eukaryotic cell as the site of aerobic respiration. Mitochondria are bound by an outer and an inner phospholipid bilayer membrane (see figure). The outer membrane has many pores and acts as a sieve, allowing molecules through on the basis of their size. The area between the inner and outer membranes is known as the intermembrane space. The inner membrane has many convolutions called **cristae**, as well as a high protein content that includes the proteins of the electron transport chain. The area bounded by the inner membrane is known as the mitochondrial **matrix**, and is the site of many of the reactions in cell respiration, including electron transport, the Krebs cycle, and ATP production.

Mitochondria are somewhat unusual in that they are semiautonomous within the cell. They contain their own circular DNA and ribosomes, which enable them to produce some of their own proteins. The genome and ribosomes of mitochondria resemble those of prokaryotes more than eukaryotes. In addition, they are able to self-replicate through binary fission. Mitochondria are believed to have developed from early prokaryotic cells that began a symbiotic relationship with the ancestors of eukaryotes, with the mitochondria providing energy and the host cell providing nutrients and protection from the exterior environment. This theory of the origin of mitochondria, and the modern eukaryotic cell, is called the endosymbiotic hypothesis.

Specialized Plant Organelles

Plants lack centrioles, but also have some organelles that are not found in animal cells.

Chloroplasts

Chloroplasts are found only in plant cells and some protists. With the help of one of their primary components, chlorophyll, they function as the site of photosynthesis, using the energy of the sun to produce glucose. Chloroplasts have two membranes, an inner and an outer membrane. Additional membrane sacs called **thylakoids** inside the chloroplast are derived from the inner membrane and form stacks called **grana**. The fluid inside the chloroplast surrounding the grana is the **stroma**. The thylakoid membranes contain the chlorophyll of the cell.

Like mitochondria, chloroplasts contain their own DNA and ribosomes and exhibit the same semi-autonomy. They are also believed to have evolved via symbiosis of a photosynthetic early prokaryote that invaded the precursor of the eukaryotic cell. In this arrangement, the chloroplast precursor cell provided food and received protection. Photosynthetic prokaryotes today carry out photosynthesis in a manner similar to the chloroplast.

Vacuoles are membrane-enclosed sacs within the cell. Many types of cells have vacuoles, but plant vacuoles are particularly large, taking up 90% of the cell volume in some cases. Plants use the vacuole to store waste products, and the pressure of liquid and solutes in the vacuole helps the plant to maintain stiffness and structure as well.

All plant cells have a cellulose cell wall that distinguishes them from animal cells, which lack a cell wall. The cell wall of plants is also distinct from the peptidoglycan cell wall of bacteria and the chitin cell wall of fungi. The cell wall provides structure and strength to plants.

Comparison of Cell Properties:

Property	Prokaryote	Eukaryote
DNA genome	Small, circular, no histones	Large segmented chromosomes with histones, packaging
Nucleus	None	Yes
Membrane-bound organelles	None	ER, Golgi, Mitochondria, chloroplasts, etc.
ATP production	Plasma membrane	Mitochondria
Cell division/reproduction	Binary fission	Mitosis/meiosis/sexual reproduction
Cell wall	Yes, in most	Plants and fungi: yes, animals: no
Flagella	Yes: but different structure	Yes: with microtubules
RNA processing	Simple, no splicing	5'-cap, poly-A tail, mRNA splicing
Transcription and translation	Together in cytosol	Separated
Cytoskeleton	None	Yes

CHAPTER FOUR

The Cytoskeleton and Cell Shape

Cells are dynamic, moving entities, which not only change their shape and exhibit complex movements, but also transport organelles and various proteins within their own cytoplasm from place to place. All of this would not be possible without the cytoskeleton, comprised of a massive network of three types of fibers: actin, myosin, and intermediate filaments. Each of these protein types is used for specific functions, and each is described in this chapter.

ACTIN AND MYOSIN

Actin molecules exist in the cytoplasm either as a globular monomer (called G-actin) or as a long filament (F-actin) made of these monomers linked one after another. It is through the rapidly changing length of actin filaments that the cell regulates such complex movements as phagocytosis and pseudopod extension. When actin monomers build together, they form a *double-stranded helical protein* filament that is constantly being built on or degraded.

Actin is the most abundant protein in the cell cytoplasm, making up perhaps 20 percent of all protein therein. It is highly concentrated just inside the plasma membrane where attachments to the membrane regulate cell movement. Rapidly growing ends of actin filaments, in fact, can push parts of the cell membrane in certain directions while shrinking filaments pull other membrane regions toward this growing edge. In such a way, the cell can move using an ameboid motion. In addition, actin filaments are responsible for the pinching off of a cell in cytokinesis after mitosis has split the duplicated chromosomes to opposite ends of the dividing cell. In this case, actin filaments attached to the membrane act like the drawstring on a bag when the filaments are degraded and shortened.

The mechanism of actin growth and disassembly involves ATP. As actin monomers are added to growing filaments, they trap ATP within their structure. In order to disassemble, enzymes hydrolyze this ATP and the monomer falls off. The rate of ATP hydrolysis controls the rates of actin filament growth and destruction. Remember for the GRE that actin filaments are also known as microfilaments—don't confuse this with microtubules on test day!

The **myosins** are motor proteins, which enable actin and other filaments to pull along one another in a shearing or contractile motion. Myosin fibers have small arm-like extensions which can attach to actin filaments and pull them "hand-over-hand," much like a tug-o'-war. Myosin will be discussed in more detail in section III when we address the contraction of skeletal muscles.

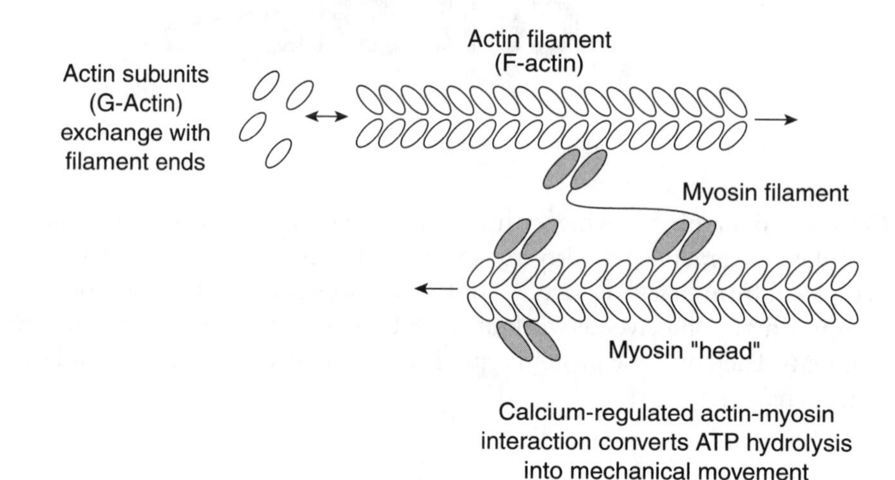

Calcium-regulated actin-myosin interaction converts ATP hydrolysis into mechanical movement

MICROTUBULES AND INTERMEDIATE FILAMENTS

Microtubules are *cellular conveyor belts* that rapidly shrink and grow out of a region near the center of the cell (called the centrosome). It is here that one end of the microtubule is anchored and stabilized. Microtubules are used to rapidly transport vesicles, organelles, and even chromosomes across the cell by anchoring these cargo items on specialized proteins that slide along the microtubules from place to place. They may appear to grow and shrink rapidly as their polymer filaments add or lose monomers. The microtubule is made of a protein called tubulin, which organizes itself into a series of rings as seen below. Tubulin dimers fall off ends of the microtubules as GTP (not ATP, as with actin formation) is hydrolyzed.

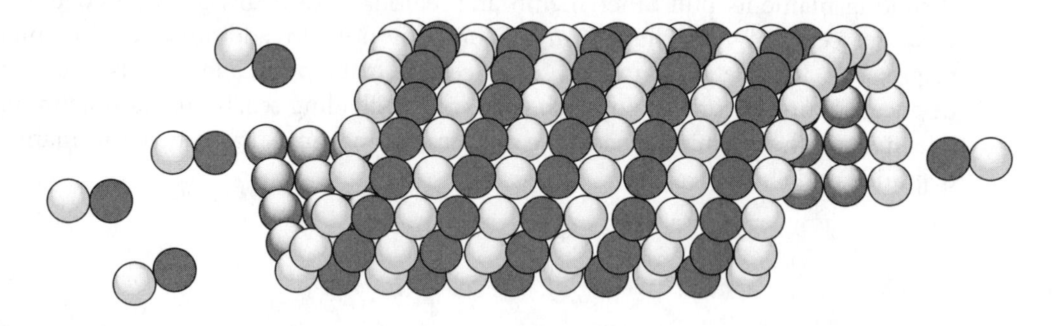

As a microtubule stabilizes in length, which may be very temporary, microtubule-associated proteins (MAPs) attach to the tubulin on one end and to vesicles, organelles, or granules on the other end. These MAPs then rapidly shuttle the cargo down the length of the microtubule. You may have heard of **kinesins**, which are motor molecules shuttling cargo toward the outer perimeter of the cell, and **dyneins**, which pull toward the microtubule-organizing center (centrosome).

Dynein is also found in flagella and cilia, which are long and thin extensions of the cell cytoplasm used by many cells for movement. Keep in mind that the cilia and flagella of prokaryotes is structurally unrelated to the cilia and flagella of eukaryotes.

Cilium Cross-Section

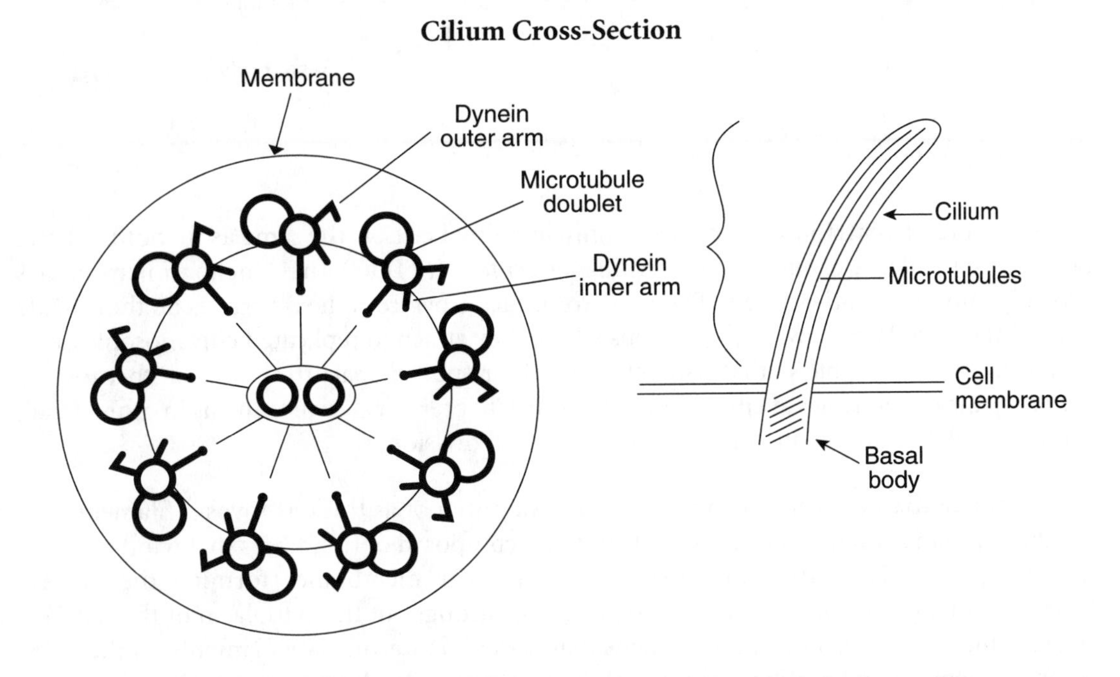

As you can see above, cilia and flagella are comprised of long, stabilized microtubules arranged in a **"9+2" structure** (nine pairs of microtubules surrounding two central microtubules for added stability). These nine doublets slide past each other as dynein proteins grab neighboring tubules and pull them. This rapid sliding generates the force needed for the cilia or flagella to quickly beat back and forth and cause movement.

Cilia and flagella are both anchored into the cell membrane by arrangements of microtubule triplets, which are called **basal bodies.** Because the microtubules in cilia and flagella must be rebuilt often, tubulin dimers use these basal bodies as the foundation to make new microtubules which are used to maintain cilia and flagella.

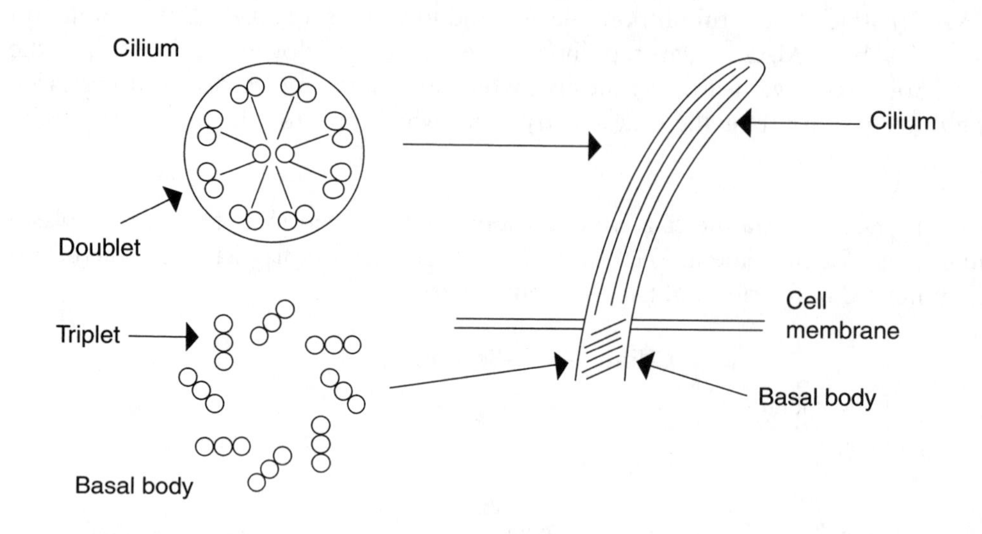

The structure of **centrioles**, which occur only in animal cells, is the same as that of the basal bodies. Centrioles are *microtubule-organizing centers* (MTOC) that anchor microtubules growing into the mitotic spindle. These microtubules grow from the centrioles on the outside edge of the cytoplasm, toward the nucleus where they attach to replicated chromosomes, and then quickly disassemble so that sister chromatids are pulled apart from one another toward the two pairs of centrioles at either end of the cell. Chapter 5 will cover mitosis in more detail, but microtubules are essential for the process of cell division.

Intermediate filaments are so named because they are not as thick as myosin filaments but are thicker than actin filaments. These strands are composed of thin fibers that wind together into long coils. They are found beneath the nuclear membrane (forming the nuclear "lamina") to give the membrane stability, and also throughout the cytoplasm of the cell. You may see the names of some of the proteins that make up intermediate filaments on the GRE, so be aware of them: **keratins** (found in skin, hair, and nails), **laminins** (which make up the nuclear lamina), and vimentin.

CHAPTER FIVE

The Cell Cycle

The continuity of life depends upon cells being able to go through a life cycle in which they grow and develop, then copy their DNA and divide. In single-celled organisms such as the protists and the bacteria, cell division can lead to the birth of a new organism. In multicelled organisms (fungi, plants, and animals) cell division is used not only to replace or repair worn-out and damaged tissues, but also to create gametes, or sex cells, which can combine to produce new organisms. While bacteria and protists have no need for two different cell division processes, more complex, multicellular life must differentiate between "normal, everyday cell division" and cell division that creates sex cells—hence, the development of meiosis for gamete production to accompany mitosis for "normal, everyday cell division."

The process of meiosis will be focused on in chapter 17, where reproduction is explained. This chapter will deal primarily with the key aspects of mitosis and what happens when cell division is not properly regulated.

HOW CAN SCIENTISTS TELL THAT PROTEINS IN TWO DIFFERENT SPECIES HAVE A COMMON EVOLUTIONARY ORIGIN?

They look for significant overlap in the sequence of amino acids between the two proteins. More than anything else, this will indicate common evolutionary origin, because similar amino acid sequence means similar DNA codes for both proteins. Similar DNA codes are strong evidence for common evolutionary origin.

MITOSIS AND CELL CYCLE REGULATION

Research has shown that the proteins regulating the life cycle of cells are extremely well-conserved across almost all eukaryotes. The proteins that small organisms like yeast use for controlling the timing of their cell division are nearly the same proteins that humans use. There is no typical length of time it takes a cell to complete its cell cycle. Some cells, such as those in certain insects, can grow, copy their DNA, and divide into two new cells in under 10 minutes. Others, such as certain mammalian cells, may have cell cycle times in excess of a year. Typical cells move through their cell cycles in a matter of hours or days.

Take a look below at the diagram of the typical cell cycle:

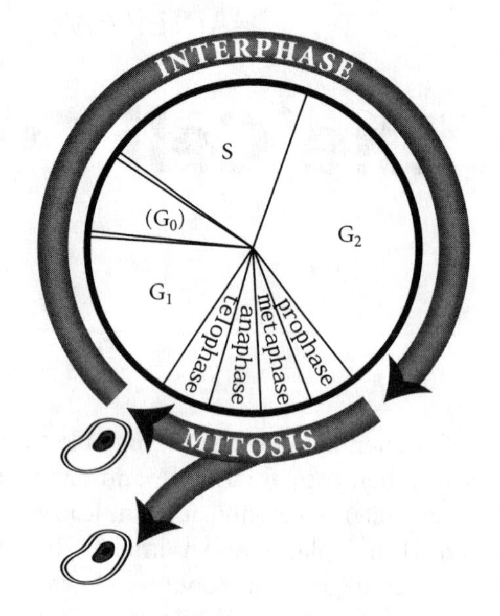

Interphase is the main phase of a cell's life, during which a cell's DNA is replicated. This occurs in S phase, so-called for the DNA **S**ynthesis that takes place. Before S phase, the cell is in G_1 phase, a **G**rowth phase in which the cell carries out its normal routine: getting food, using energy, and growing in size. At the end of G_1, the cell undergoes an actual chemical change that literally commits it to continuing on to S phase and, later, to actual division. There are some cells, notably mammal nerve and muscle cells, which usually do not cross this "point of no return" between G_1 and S once adulthood is reached. They remain active, working cells until they die, in a resting state known as G_0. Some cells may enter G_0 for an indefinite period and then resume their normal cell cycles.

HOW CAN YOU TELL WHICH PHASE OF THE CELL CYCLE A CELL IS IN?

You can measure the amount of DNA in the cell, since doubled DNA means the cell is either in S or G_2 phase. Cell cycle phases can actually be timed and measured using large groups of cells selected to start their life cycles all at one time. You cannot, it appears, measure protein content or RNA content of a cell to determine cell cycle position, since protein translation and DNA transcription are almost constant throughout interphase.

G_2 is a **G**rowth phase just before the actual mitosis, in which the cell can make sure that its volume and DNA have indeed doubled before division takes place. In some rapidly dividing cells, G_2 is quite short, and cells may not actually double in size at all before the next division takes place. We see this in rapidly proliferating embryonic cells, such as when cleavage is taking place. In this case, a quickly dividing ball of cells may remain nearly the same size even though the number of cells in the clump is doubling every division.

Mitosis

Among the key aspects of mitosis are the following:

i. Chromosomes shorten and thicken in the nucleus and the nuclear membrane dissolves.
ii. The mitotic spindle of microtubules is formed.
iii. The contractile ring of actin develops around the center of the cell.

In order for mitosis to work, a single pair of **centrioles** will copy themselves during S phase and the two pairs will move to opposite poles of the cell. These centrioles pairs form the foundation for **centrosomes,** microtubule organizing centers that will shoot linked tubulin proteins across the cell as mitosis begins. Keep in mind, however, that the centrioles themselves are not necessary for microtubules to form from the centrosome areas of the cell. In fact, *plant cells have centrosomes without centrioles.* The thing to remember for the GRE, though, is that the centrosome regions form the two poles (like north and south) on opposite ends of the cells, between which microtubule spindle fibers will form.

M phase is divided into six stages, which are described below next to the corresponding images. Despite the conventional division into distinct stages, mitosis is a continuous process that does not stop between each phase. Five of these stages comprise mitosis, and the sixth stage—**cytokinesis**—completes the M phase of the cell cycle as the cell pinches in two.

Prophase

- Chromatin shows up under the microscope as well-defined chromosomes.
- Chromosomes are seen as an "X" shape, two sister chromatids connects by a centromere, a specific DNA sequence.
- Mitotic spindle begins to form and elongate from the centrosome regions.

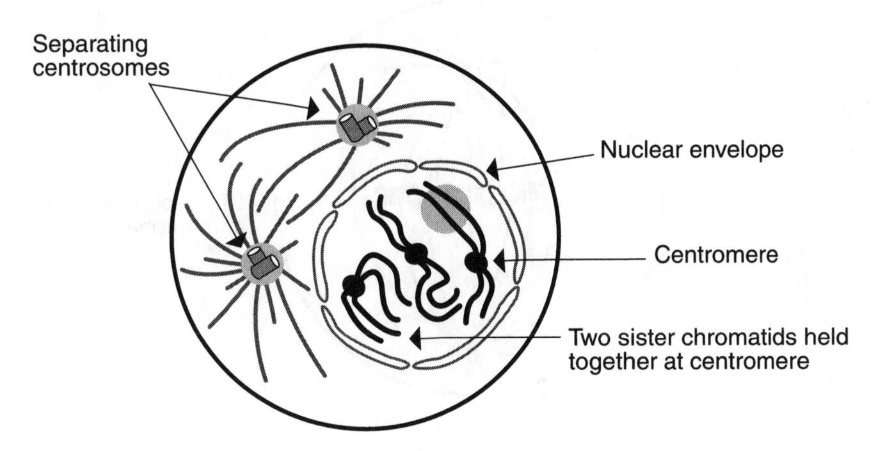

Prometaphase

- Nuclear membrane dissolves.
- Spindle microtubules enter nucleus and some attach to chromosomes within.
- Those that do attach are called kinetochore microtubules named after the kinetochore proteins that can be found at the centromere region of each copied chromosome.
- The other microtubules are called non-kinetochore or polar.

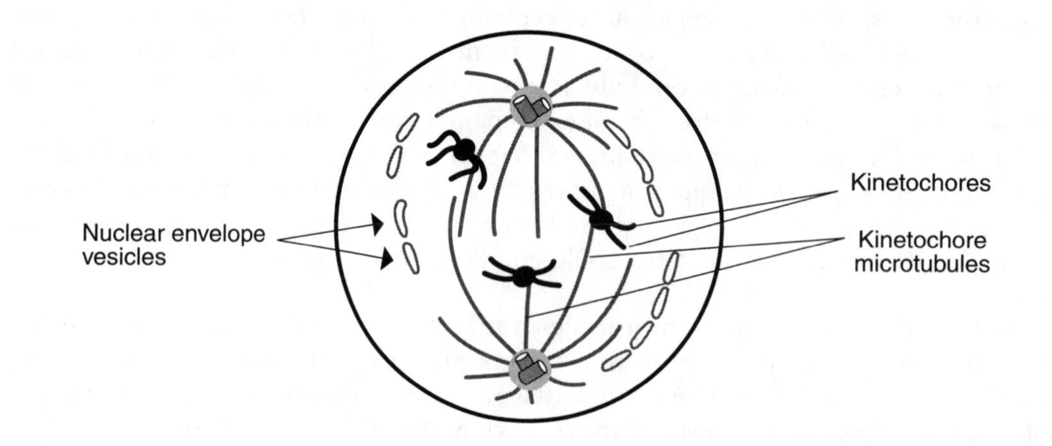

Metaphase

- Kinetochore microtubules push from opposite poles equally so that chromosomes are aligned in middle of the cell.
- This center area where the alignment occurs is called the metaphase plate.

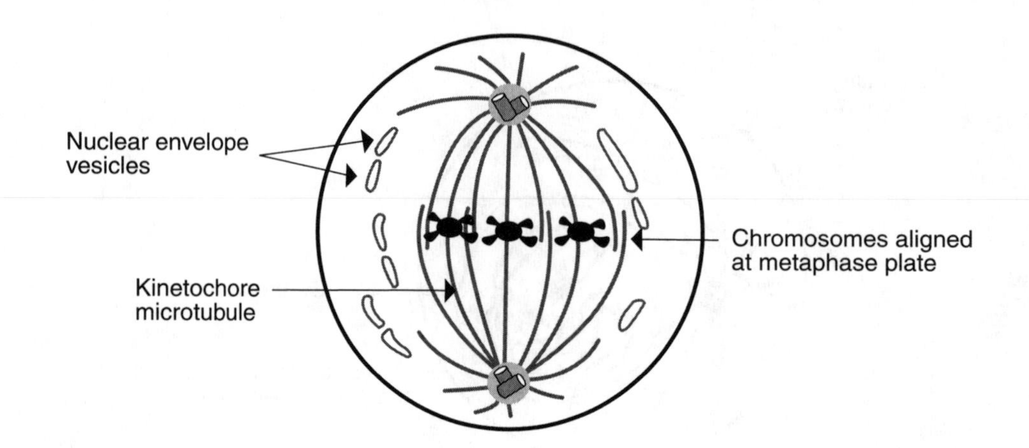

Anaphase

- Paired sister chromatids separate as kinetochore microtubules shorten rapidly.
- Polar microtubules lengthen as kinetochore microtubules shorten, pushing poles of cell farther apart.

Increasing separation of the spindles

Kinetochore microtubules shorten as the chromatid (chromosome) is pulled toward the pole

Telophase

- Separated sister chromatids group at opposite ends of cell, near the centrosome region, having been pulled there by the receding microtubules.
- New nuclear envelope reforms around each group of separated chromosomes.
- Mitosis has ended at this point.

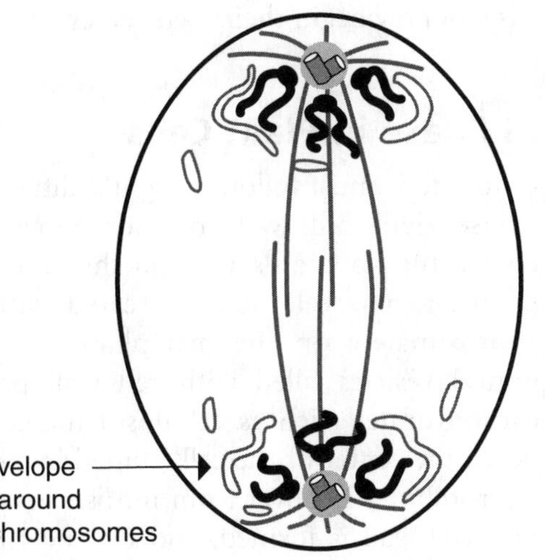

Cytokinesis

- Contractile ring of actin protein fibers shortens at center of cell.
- Creates a cleavage furrow, or indentation, where ring contracts.
- As one side of the cell contacts the other, the membrane pinches off and two cells now exist where one did before.

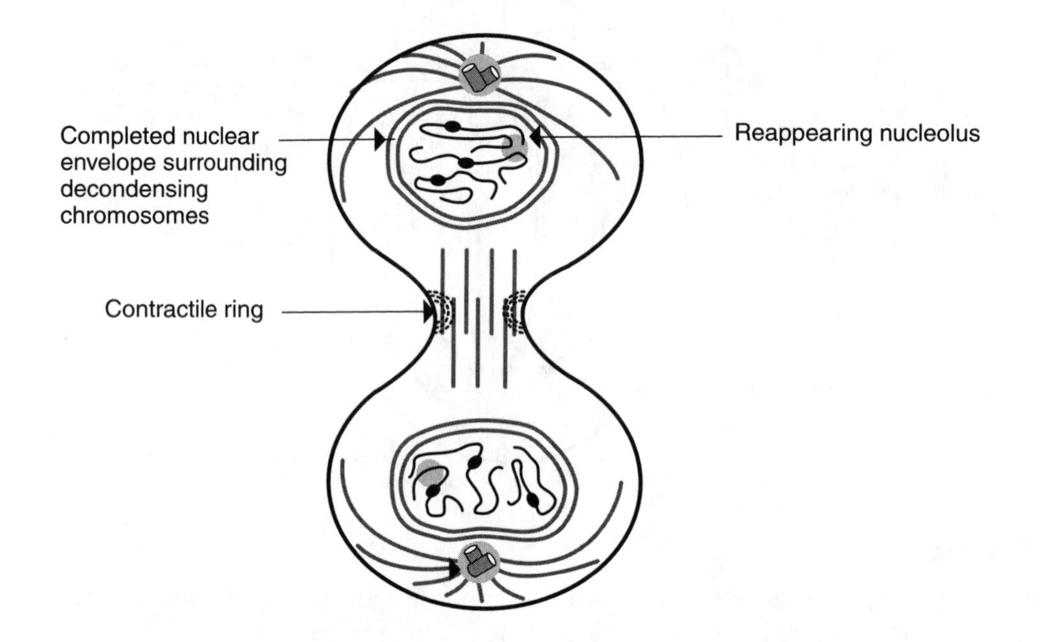

By the end of M phase, each new daughter cell is identical to the parent cell from which it came. The daughter cells have identical copies of DNA, a single pair of centrioles, and the same organelles. These new cells will now begin their own cell cycles in G_1 phase.

HOW MIGHT A CELL BECOME POLYPLOID DUE TO AN ERROR IN CELL DIVISION?

Consider a cell that undergoes S phase to copy its DNA and then does not enter or complete M phase. Thus, this cell remains with a double complement of DNA, and even if it survived and successfully divided after the next cell cycle, it would remain polyploid.

M Phase in Plant Cells

Cytokinesis must follow a slightly different path in plant cells, whose rigid cell walls do not allow for actin to form a contractile apparatus that pinches the plant cell in two. In plants, a new cell wall is created within the dividing cell, approximately on the metaphase plate. Small, membrane-bound vesicles filled with cell wall precursors (most likely carbohydrates such as cellulose) merge along a straight line down the center of the cell, guided by polar (nonkinetochore) microtubules left over from mitosis. As these vesicles merge, a new cell wall is formed, and two new plant cells are created from one. This merged vesicle structure is known as the **cell plate** when it is forming.

Cell Cycle Regulation

The cell cycle has decision points, or **checkpoints**, that determine whether the cell will proceed from one stage of the cell cycle to the next. We have already discussed the "point of no return" that occurs between G_1 phase and S phase. Yet another checkpoint determines when cells move from G_2 to M, and still others determine progression through prophase, metaphase, anaphase, and telophase.

Two kinds of proteins regulate passage through the cell cycle: **cyclin-dependent protein kinases (Cdk)** and **cyclins**, which are proteins that bind to Cdk's in order to activate them. Keep in mind that there are many levels of control throughout the cell cycle to ensure that cells divide in an orderly and regulated fashion. Different kinds of cyclin proteins are active at different points in the cycle; at every step of the way, a certain kind of cyclin binds to a certain Cdk, facilitating the cell's movement from one stage to the next.

Recall that protein kinases are enzymes that add phosphate groups to phosphorylate other proteins. In the case of cyclin-dependent protein kinases, these enzymes–once activated by cyclins–add phosphates to serine and threonine residues on certain proteins in order to activate those proteins.

The diagram below shows different stages of the mammalian cell cycle and the levels of each type of cyclin present in cells during those stages:

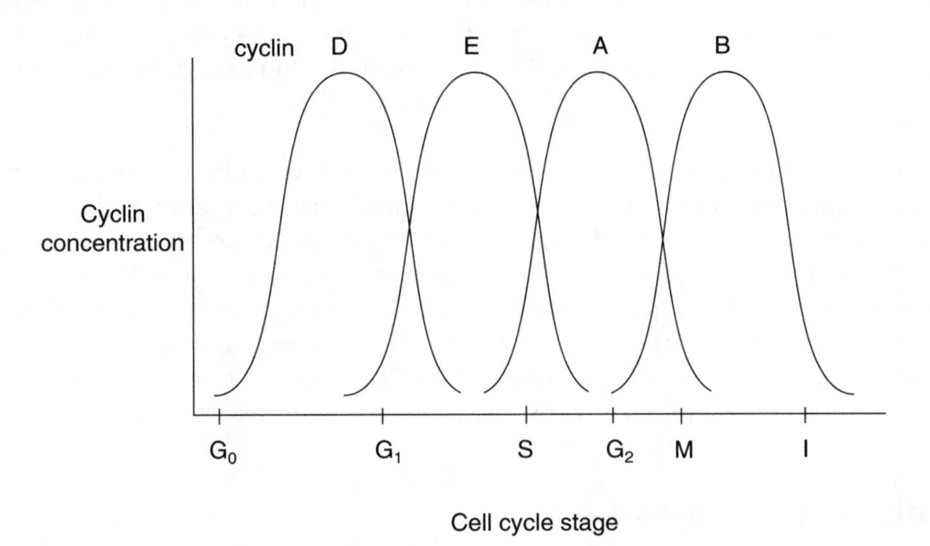

Once activated by cyclin binding, Cdk's activate other proteins to initiate the next step in the cell cycle. As an example, increased levels of cyclin B, the main mitotic cyclin, activate a specific Cdk called Cdc2, which in turn activates MPF (M-phase promoting factor), a protein crucial to a cell's passage into mitosis from G_2. MPF can be measured in a rapidly dividing group of embryonic cells to show that concentrations peak every half hour, just as the cells begin to divide again.

CANCER AND ONCOGENES

Cancer is characterized by a loss of control over cell proliferation. There are four types of genes that are associated with cell division: growth factors, growth-factor receptors, intracellular signal-transducing proteins, and nuclear transcription factors. Normal genes involved in the control of cell growth or division are known as **proto-oncogenes.** When mutations occur in these genes such that they no longer maintain control over a particular aspect of growth or division, these genes become known as **oncogenes.** *Remember that difference on test day!*

Growth Factors

These proteins are often released by cells and circulate through the blood or nearby tissue. They act on cell surface receptors that bind to them in a specific manner, causing those cells to initiate growth and cell division. Normal cells are able to suppress their growth when they are near other cells, a phenomenon known as **density-dependent inhibition** of cell division. When studied in a petri dish, normal cells multiply only until they form a single layer of cells on top of a layer of nutrient agar. If some are removed, the cells will divide enough to close the gap, but will then stop. Cancer cells, however, will continue to divide until a large clump of cells has thoroughly exhausted the available space and nutrients.

Different growth factors are released by different kinds of cells: epidermal growth factor (EGF) incites skin cells to divide; nerve growth factor (NGF) does the same for nerve cells; and, platelet-derived growth factor (PDGF) causes cell division of certain cells lining blood vessel walls.

If genes coding for the production of growth factors are affected by mutations, it's possible they could begin to produce more growth factor than normal, a condition which could cause uncontrolled cell division of nearby cells of the same type. In addition, mutated growth factor genes might produce growth factors that bind more strongly to growth factor receptors on other cells, so that they cause a growth signal to be propagated for a longer time. While research suggests that *increasing growth factor production alone is not sufficient for inducing cancerous transformations in cells,* extensive cell proliferation as a result of increased growth factor release can increase the chances of spontaneous or induced mutations in cells.

Growth Factor Receptors

These are membrane-spanning proteins that have an area within their amino acid sequence that can bind to growth factors like NGF and EGF. They transmit a signal through second messengers (see chapter 2) when growth factors bind to them. Problems with growth factor receptors arise when mutations allow for receptors to form with structural alterations that increase their affinity for growth factor proteins or when receptors are overexpressed (produced in too high a quantity). Both of these conditions may cause *persistent activation of second messengers,* which can lead, down the line, to continued activation of genes involves in cell growth and division.

Intracellular Signal Transducing Proteins

These are the proteins that carry signals from the receptors on the cell surface to the nucleus. These second messengers relay signals that will turn on or off certain genes. If produced in too high a quantity or are too active for a variety of reasons, these proteins can lead easily to an increase in the transcription of certain genes involving cell growth and division. Examples include the **Ras** protein and protein kinases such as **Src**. Mutations in the gene coding for Ras signaling protein are implicated in over 30 percent of all cancers. Overactive protein kinases (Src and others) increase the rate at which other intracellular proteins are phosphorylated and, therefore, amplify any growth signals coming from receptors.

Nuclear Transcription Factors (NTFs)

These proteins facilitate the activation of particular genes. With abnormally activated transcription factors, genes involved in cell growth and division can be constantly activated. Transcription of mRNA increases and production of cyclins and other mitosis-inducing proteins may be increased. Examples include proteins known as **fos** and **myc**.

Protective Mechanisms

Proto-oncogenes can be turned into oncogenes through DNA damage. Sunlight (UV radiation), certain viruses, X rays, and toxins can all lead to changes in DNA structure. Cells, however, have error-correcting machinery that can act to splice out error-filled DNA or can even cause precancerous cells to kill themselves, a process known as **apoptosis**.

One example of error-protection is the ***p53*** protein, coded for by a gene present in all mammalian cells. *p53* is known as the "guardian of the genome," accumulating in the cell in response to DNA damage. It is able to shut down the cell cycle in the middle of G_1 by activating another protein, *p21*, which is an inhibitor of a Cdk needed for the cell to move onward in the cell cycle. Because of this capability, *p53* is known as a "tumor suppressor gene," and there are many others like it. Mutations in *p53* have been extensively studied and implicated in a large number of cancers. Half of all human tumor cells lack a functional *p53* gene.

Cancer is not always the end result of DNA damage. Cells need many insults before becoming cancerous. Remember that the only DNA mutations that are passed onto offspring are those that involve DNA in sperm cells or egg cells, nowhere else!

Another tumor suppressor is the protein ***pRB***, a regulator of transcription, named after the disease retinoblastoma (RB), a rare eye cancer caused when two copies of the defective pRB gene are present. *pRB* works differently than *p53*: It appears to inhibit a transcription factor called *E2F*, crucial to many cells in their progression from G_1 to S. When *pRB* is bound to *E2F*, a cell cannot move into S, yet mutated *pRB* causes continual and rapid movement of cells into S phase with little control over the cell cycle. Defective retinoblastoma protein has been found in lung, bladder, and breast cancer cells.

Intracellular Signal Transducing Proteins

[illegible]

Nuclear Transcription Factors (NTFs)

[illegible]

Protective Mechanisms

[illegible]

CHAPTER SIX

The Basics of Molecular Genetics

Around 1865, based on his observations of seven characteristics of the garden pea, **Gregor Mendel** developed the basic principles of genetics. Mendel first described traits in pea plants as pairs, such as wrinkled seed and smooth seed. He then mated combinations of these plants and examined their offspring to determine the quantitative nature of the inheritance of these traits. As such, he arrived at an understanding of genetic inheritance that forms the foundation of modern-day genetics. Although Mendel formulated these genetic principles, he was unable to deduce a mechanism for *why* traits were passed on in the patterns he saw: That understanding would have to wait until the 1900s, when **Thomas Hunt Morgan** would rediscover Mendel's work and lend new evidence to his assertions.

MENDELIAN GENETICS AND PEDIGREES

Some basic rules of gene transmission you should know for the GRE:

- Diploid organisms have two copies of each chromosome—one from mom and one from dad—and, therefore, two copies of each gene (except on the X and Y chromosomes, for which the vast majority of genes exist as single copies on the X *only*).

- The two copies of each gene may not be identical. The different forms of a gene (e.g., the gene for eye color can be blue, brown, etc.) are called **alleles**.

- The type of genes an organism has makes up the organism's **genotype**.

- The physical features of an organism make up its **phenotype**.

- **Homozygous** organisms have two copies of the same allele (e.g., two blue eye genes).
- **Heterozygous** organisms have one copy of a dominant allele and one recessive allele.

Mendel's Laws of Inheritance:

- **Dominance**: Only one dominant allele is needed from the pair of alleles for the dominant trait to show up as the physical feature, or phenotype (WW or Ww). For recessive traits to show, individuals must have two recessive alleles (ww only). Recessive alleles are sometimes mutations to a dominant allele that result in loss of function in that allele; however, the phenotype often will not change if one allele remains dominant and active.
- The Law of **Segregation**: The two alleles for a given trait separate from each other during meiosis, and thus, end up in different sperm or egg cells. So if an organism is Ww, it can pass on *either* a "W" or a "w" but never both to an offspring.
- The Law of **Independent Assortment**: Genes for one trait separate independent of genes for other traits. Meaning, if you were Ww for one trait and Yy for another, just because your sperm or egg cells end up receiving a "w" doesn't mean you'll necessarily receive a "y" too—you could get a "Y."

Gene linkage (when genes are found close together on the same chromosome) can create results in matings not predicted by Mendel's Laws. That's because linked genes do *not* independently assort. The closer they are on a chromosome to each other, the more likely it is they will both be found in the same gamete after meiosis.

The law of segregation can be demonstrated by studying the cross between a *Pp* parent and a *pp* parent. Possible offspring according to Mendel will show a combination of one allele from one parent and one allele from the other:

	P	p	
p	Pp	pp	50% Pp offspring
p	Pp	pp	50% pp offspring

The law of independent assortment can be illustrated by examining a cross between two organisms heterozygous for two traits: height and seed shape. Notice in the following table that the tall and round parents will pass only one allele for height each onto their offspring along with one allele for seed shape each. However, the alleles for height and seed shape independently assort: that is, just because an offspring gets a particular allele for height from one parent does not mean it will also receive a particular allele for seed shape from the same parent.

Parent 2, Tall-Round seed Tt Rr (Gametes) \ Parent 1, Tall-Round seed Tt Rr (Gametes)	TR	Tr	tR	tr
tr	Tt Rr	Tt rr	tt Rr	tt rr
tR	Tt RR	tT Rr	tt RR	tt Rr
Tr	TT Rr	TT rr	Tt Rr	Tt rr
TR	TT RR	TT Rr	Tt RR	Tt Rr

Gene Linkage

Many genes in organisms are "linked," as mentioned in the previous sidebar. Thomas Hunt Morgan made a major contribution to the study of modern-day genetics with his investigation of gene linkage: When he crossed fruit flies heterozygous for gray, normal wings (GgWw) with flies homozygous for black, vestigial (useless and small) wings (ggww), he expected a 1:1:1:1 ratio of four different phenotypes in the offspring:

	Gw	GW	gW	gw
gw	Ggww	GgWw	ggWw	ggww

The results, however, showed many more offspring with gray, normal wings *or* black, vestigial wings than any combination of the two (i.e., there were very few gray, vestigial and black, normal offspring). In fact, his numbers looked something like this:

- 41% gray, normal "wild-type"
- 42% black, vestigial
- 8% black, normal
- 9% gray, vestigial

Notice how the majority of the offspring's phenotypes correspond *exactly* to the parental phenotypes, rather than to any combination of parental phenotypes. Morgan reasoned that genes for color and wing shape must be on the same chromosome. In order for offspring to be born with a recombinant phenotype, one that is a combination of the parental phenotypes, the genes for color and wing shape must be separated somehow during gamete formation. Yet the closer the linked genes, the less likely that crossing-over can split them up, and the less likely that offspring will express phenotypes that are not parental. The frequency of the recombinant phenotype can be used to map the distances between genes that are on the same chromosome.

Pedigrees

Pedigrees are family trees that enable us to study the inheritance of a particular trait across many related generations. Males are typically designated as "squares" on the pedigree, while females are designated as "circles." Those who phenotypically show a trait are shaded, while those who carry a trait but do not show it are either half-shaded or given a dot inside their circle or square.

Genetic traits can be classified as to whether they are transmitted on autosomal chromosomes (numbers 1–22) or on sex chromosomes (the X almost always, since the Y chromosome carries limited genes). In addition, traits can be dominant or recessive. For the GRE, you should be familiar with the characteristics of different inheritance patterns so you can easily spot patterns.

Autosomal Dominant

- Males and females are equally likely to have the trait.
- Traits do not skip generations.
- The trait is present if the corresponding gene is present.
- There is male-to-male and female-to-female transmission.

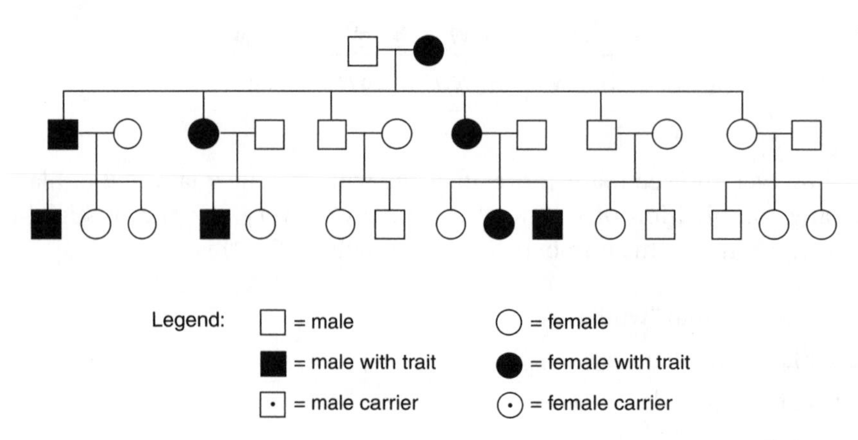

Autosomal Recessive

- Males and females are equally likely to have the trait.
- Traits often skip generations.
- Only homozygous individuals have the trait.
- Traits can appear in siblings without appearing in parents.
- If a parent has the trait, those offspring who do not have it are heterozygous carriers of the trait.

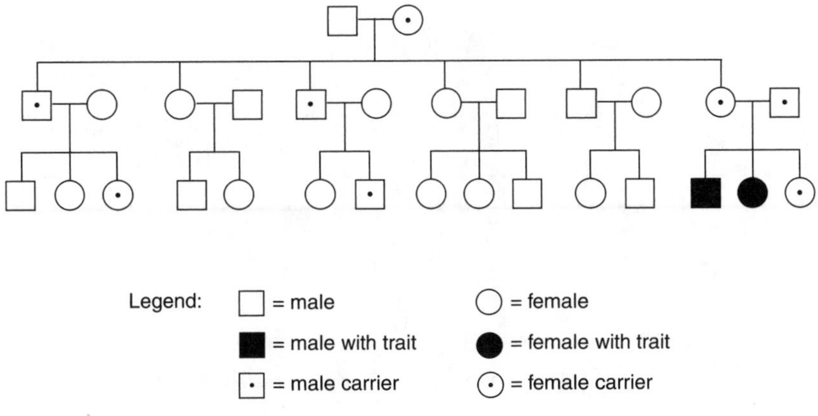

X-Linked Dominant

- All daughters of a male who has the trait will also have the trait.
- There is no *male-to-male transmission.*
- A female who has the trait may or may not pass on the affected X to her son or daughter (unless she has two affected X's).

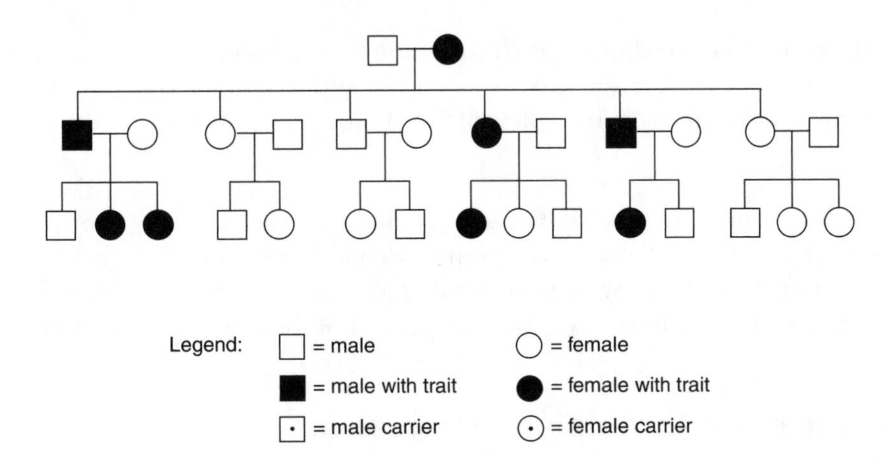

X-Linked Recessive

- Trait is far more common in males than in females.
- All daughters of a male who has the trait are heterozygous carriers (assuming their mother did not also pass them a recessive allele).
- There is *no male-to-male transmission.*
- Mothers of males who have the trait are either heterozygous carriers or homozygous and express the trait themselves.

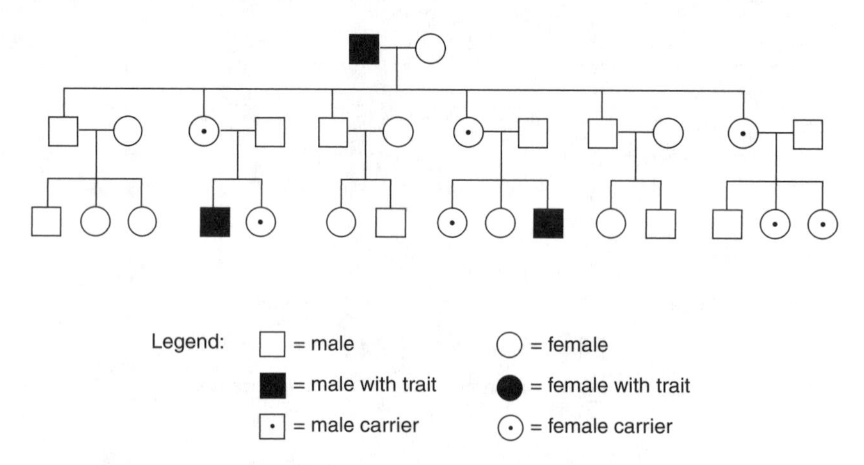

Remember the following points about inheritance:

Males cannot be carriers of sex-linked (X-linked) traits. Males have only one X chromosome in addition to their Y chromosome. If this X has the mutant gene on it, there is no normal gene to override. Females have a chance of having a normal gene on the other X that can produce enough normal protein that the defective gene on the first X is effectively overridden.

You cannot be a carrier of a dominant trait. Dominant means that a trait is expressed even if there is only one copy of the allele. Both sex-linked and autosomal traits will be expressed if coded for by a dominant gene, regardless of the other copy of the gene on the homologous chromosome.

Dominance has nothing to do with frequency. Dominant alleles are not more common in populations than recessive alleles. Two commonly cited human examples are the allele for six-fingers (a dominant allele), which exists only rarely in our population, and the recessive trait of O-Type blood, which occurs in over 60 percent of the human population.

Codominance and Incomplete Dominance

While we might like to think of a dominant allele as one that overpowers a recessive allele, "dominance" does not mean this at all. Recessive alleles are often those that, when transcribed, produce defective proteins that cannot perform their given function. Dominant and recessive alleles do not interact at all, however, which is why cells with both a dominant allele and a recessive allele for a given trait will produce both "normal" protein and defective protein.

In some cases, a dominant allele simply does not code for *enough* of a given protein; you see the full effect of the gene only when two copies of the dominant allele are present in the cell. In certain flowers (and you were likely taught this in high school biology), alleles exist for both red petal color and for white petal color. When two dominant alleles make up a flower's genotype, the flower is red; with two recessive alleles, it is white; and, with one dominant and one recessive it is pink. Yet what you may not know is that the pink color has nothing to do with the "white" allele and the "red" allele blending, even though that is what it looks like. It results because one dominant "red" allele is not sufficient to produce enough red-pigmented protein that the flower looks red—so, the flower looks light red (50% red), or pink. In addition, the recessive allele does not code for white color, but rather no color at all. The recessive allele codes for a defective protein that gives no color to the flowers, thus the white recessive phenotype that occurs. This phenomenon where a single dominant allele cannot by itself produce the full phenotype is known as **incomplete dominance.** We see it expressed as a blending, but on a molecular level, it has nothing to do with blending.

The extent to which a given gene expresses itself in an individual or population is called **penetrance.** An age-dependent penetrance, for instance, means that a mutation tends to show up early on in life or later in life. Some mutations can be carried and show up only rarely in individuals who carry the mutation: This would be considered low penetrance.

Codominance, on the other hand, occurs when the different alleles for a trait each code for enough protein so that, when they occur together, two different, normal proteins are made and show up in the organism's phenotype. A commonly cited example here is human blood type, for which three alleles exist: I^A, I^B, and i. These alleles are responsible for the building of certain "identity" proteins on the cell surfaces of red blood cells: A-type proteins and B-type proteins. Notice that there are three alleles for this trait, though each individual's genotype is made up of two of the three alleles. If you have two I^A alleles, you have A-type blood, two I^B alleles, B-type, and two ii alleles, O-type (where your red blood cells have neither A nor B proteins on their surfaces). The I^A and I^B alleles are codominant, meaning that if they are both present in someone's genotype (I^AI^B), that person will express both A-type proteins and B-type proteins on their red blood cell surface.

Some traits, such as skin color in humans, are continuous across a broad spectrum and are coded for by multiple genes. The degree to which one expresses the trait depends upon the ratio of dominant to recessive alleles in that group of genes. These traits are known as polygenetic traits; each one of the series of genes coding for a polygenetic trait is called a **polygene**.

Other Non-Mendelian Inheritance

In addition to codominance and incomplete dominance, other genetic phenomena occur regularly that cannot be explained by Mendel's Laws. Among these inheritance patterns are *epistasis, mitochondrial (maternal) inheritance, genomic imprinting,* and *triplet repeat extension*. Although it's unlikely that you will see these on the GRE, it is necessary to know a little about each.

Epistasis occurs when a second gene determines whether a first gene is expressed or not. Fur coloration in a certain mammal might be determined by a purely dominant-recessive gene combination: *B* for black fur and *b* for brown fur. Yet a second set of genes, call them *C* for "color expression" and *c* for "no color expression" may, in fact, determine if fur color will be shown at all *regardless of one's genotype* for fur. A *BBCc* individual would have both genes for black color as well as a dominant phenotype for color expression, and so would express black fur. An individual whose genotype was *BBcc* would have the same black fur alleles yet express no fur color (seen as albino) simply because of the second set of genes.

Mitochondrial (maternal) inheritance: Because one's cellular organelles derive entirely from the mother (sperm contains no organelles that are transferred to the egg upon fertilization), all genes present in one's organelles have a maternal, rather than paternal, origin. We know that mitochondria, once probably their own free-living prokaryotic cells, possess their own complement of DNA, distinct from the nuclear DNA of cells in which they reside. One's mitochondrial DNA, then, is entirely maternal in origin. Though rare, there are diseases that can result from defects in mitochondrial DNA, particularly those involving cellular respiration pathways and cell metabolism. These are passed onto offspring from mothers 100 percent of the time, and from fathers zero percent of the time.

Genomic Imprinting: This genetic inheritance pattern results from the fact that certain alleles seem to be encoded (perhaps by DNA methylation—methyl groups selectively attached to DNA bases) differently depending upon which parent the allele comes from. In some cases, a disease can result if the recessive allele is passed down from the mother to her offspring, but not if the same recessive allele is passed down from the father! Thus, seemingly identical genotypes can express very different phenotypes according to which parent they inherited the mutant version of a gene from. Well-studied diseases resulting from genetic imprinting (also known as **epigenesis**—not to be confused with epistasis!) include Prader-Willi Syndrome and Angelman Syndrome.

Triplet Repeat Extension: Some genes normally contain terminal regions with C-A-G nucleotide repeats (e.g., CAGCAGCAGCAGCAG...) or other repeats that can extend up to 50 repeats long in normal, healthy individuals. But enzymes that are responsible for copying DNA can make mistakes in copying repeat regions, so over several generations of replication, a region 50 repeats long can become much longer, with the number of repeats increasing dramatically. In some cases, this does not cause a problem, but certain diseases such as **Huntington's disease** and **Fragile-X** syndrome can result from an excess of repeats on certain chromosomes. These neurodegenerative diseases are thought to be caused by the accumulation of abnormally shaped proteins in neurons and in other cells of the body, and can occur in one generation without ever having appeared before. Because parents can possess genes with elongated triplet regions and not show any disease, they can pass onto their offspring triplet regions that can drastically increase in size from parent to offspring as DNA is replicated and passed on in sperm or egg cells. This building up of a trait over the course of several generations is also known as **genetic anticipation**.

BACTERIAL GENETICS AND OPERONS

Bacterial cells are the smallest and simplest on Earth, yet despite possessing only a single chromosome of DNA and reproducing by mitosis (binary fission), they are amazingly adaptable creatures that can vary their genotypes significantly from one generation to the next.

Bacteria have a single chromosome of double-stranded DNA found in the **nucleoid** region of the cell, an area of tightly packed DNA not contained within a nuclear membrane. Some bacteria can copy their DNA and divide every 20 minutes, producing large colonies within hours. The single chromosome unwinds and unzips at a single origin of replication, is copied, and then the cell divides by mitotic cell division. In addition, small loops of double stranded DNA exist within the bacterial cell, but not as part of the single, large chromosome. These **plasmids** can replicate independently of the larger chromosome and can be freely exchanged from bacterial cell to bacterial cell via extensions of cytoplasm from one cell to another (cytoplasmic bridges) during conjugation.

Genetic Recombination in Bacteria

Bacterial cells do not reproduce via meiosis and, therefore, cannot shuffle genes from chromosome to chromosome like many eukaryotic cells can. Yet, bacteria are able not only to exchange DNA with other bacterial cells through cytoplasmic bridges, as mentioned above, but also to pick up free-floating DNA from the extracellular environment or from viruses.

The process of **transformation** occurs when bacteria pick up "naked" DNA (free-floating DNA) that has spilled out of lysed bacterial cells nearby or has been placed into the surrounding environment. Many bacteria possess unique proteins to aid in the binding and transport of naked DNA. In fact, scientists have taken advantage of this ability of bacteria: They artificially introduce genes into bacterial cultures so that the bacteria will begin to express a gene of interest after taking up the foreign DNA.

Conjugation occurs between bacterial cells through cytoplasmic extensions that allow plasmids to move between cells. Since plasmids replicate at their own rate, copies of a given plasmid can be sent to other bacterial cells without the original parent cell losing its complement of those genes. Many plasmids contain antibiotic resistance genes (R-plasmids), which code for enzymes that break down or interfere with antibiotics. In fact, plasmid exchange via conjugation is how large numbers of bacteria can become antibiotic resistant from only a few initial resistant cells. Long-term exposure to antibiotics can make the bacteria that naturally live in one's body resistant to the chemicals, and this resistance can be passed on by conjugation to other, invading species of bacteria.

Transduction is the process by which viruses (phages—see chapter 9) infect certain bacterial cells, allowing new baby viruses to trap bacterial genes. These viruses then can go on to infect other bacterial cells, transferring the bacterial genes from their last infectious cycle to the newly infected cells. The transferred DNA can integrate into the host cell's chromosome, remaining there indefinitely and copying itself along with the other bacterial genes.

These three processes are powerful ways in which *bacteria recombine genes and shuffle their gene pools from generation to generation in the absence of sexual reproduction.*

Control of Bacterial Gene Expression

The most current model of how bacteria turn on and off genes is called the **operon** model. An operon is a functional unit of gene expression comprised of up to several related genes, an operator, and a promoter.

Below is pictured the lac operon, a combination of three genes, a promoter region, and an operator region that is responsible for the breakdown of lactose sugar in *E. coli* bacterial cells. Each of the three genes, *lacZ*, *lacY*, and *lacA*, produces a different enzyme used in a step of lactose metabolism:

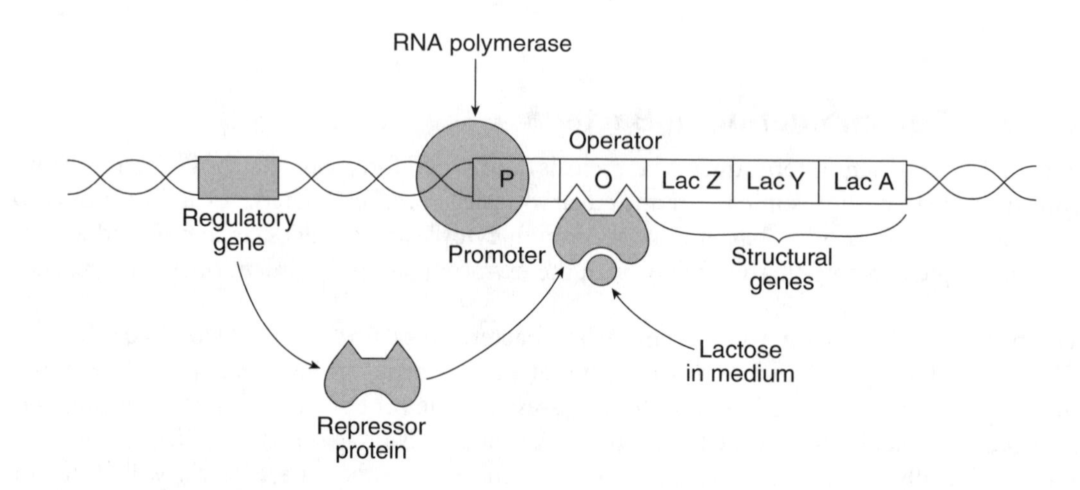

RNA polymerase must have access to genes on DNA strands in order to transcribe the DNA into mRNA. The enzyme attaches to the **promoter** sequence immediately prior to the gene that is to be transcribed and follows along the DNA strand, reading DNA into mRNA as it moves along. Yet, in this model of bacterial gene control, a protein known as the **repressor protein** can bind to the **operator** sequence in such a way that the RNA polymerase cannot move along the DNA to reach the structural genes it is trying to transcribe. This repressor protein is coded for by a regulatory gene often far upstream from the structural genes.

In the picture above, you can see that when the repressor protein is actively made, its binding sites allow it to attach to and sit on the operator sequence to block transcription of the *lac* genes. If these genes are not transcribed, the enzymes to digest lactose are not made. According to the model, however, this repressor protein also has binding sites for lactose, the binding of which causes the repressor protein to fall off of the operator sequence and allows RNA polymerase to transcribe the *lac* genes. This allows for the *lac* operon to be turned on when *lactose is present but remain off when lactose is not present.* The genes in the *lac* operon code for **inducible enzymes**, enzymes that can be induced—turned on—by the presence of a particular substance, in this case lactose.

There are other operons with genes for **repressible enzymes**, in which a repressor protein attaches to the operator sequence only when a certain compound is present in the environment. Thus, it is the presence of a certain molecule, rather than the absence of one, which causes structural genes to be turned off. This is the opposite scenario to the one pictured above. It is clear that there are advantages to both types of negative gene control through operons. Inducible enzymes are often found in cell pathways that deal with the breakdown of substances into smaller molecules, otherwise known as digestive pathways. Repressible enzymes are often found in pathways that synthesize substances so that synthesis can be shut down as soon as enough of the substance is produced. In this case, the molecule being produced will itself interact with the repressor protein that binds to the operator region to block transcription.

CHAPTER SEVEN

Eukaryotic Chromosome Structure and Genome Organization

Bacterial gene control is fairly simple compared to that of eukaryotes. The main reason for this is that eukaryotes have more DNA than prokaryotes do, and unlike the single bacterial chromosome that holds almost all of a bacterium's genes, this DNA is organized into many chromosomes. Because of the large amount of DNA that eukaryotes possess, they have evolved strategies to pack their DNA tightly into the nucleus. In addition, over time organisms have acquired large stretches of DNA that are simply evolutionary leftovers from ancestors, regions of DNA that may have once been actively transcribed but no longer are used. This chapter looks at the structure of eukaryotic chromosomes as well as methods of gene control found in the eukaryotic world.

NUCLEOSOMES, HISTONES, AND CHROMOSOMAL ALTERATIONS

Chromosome Packing

The nucleus of a human cell holds more than six feet of DNA, if the double helices that make up the 46 chromosomes were stretched end to end. For protein synthesis, this DNA must constantly unwind and rewind. The key to packing the DNA in such a way as to wind tightly yet unwind easily is the nucleosome. Nucleosomes are essentially spools of DNA wrapped around small histone proteins.

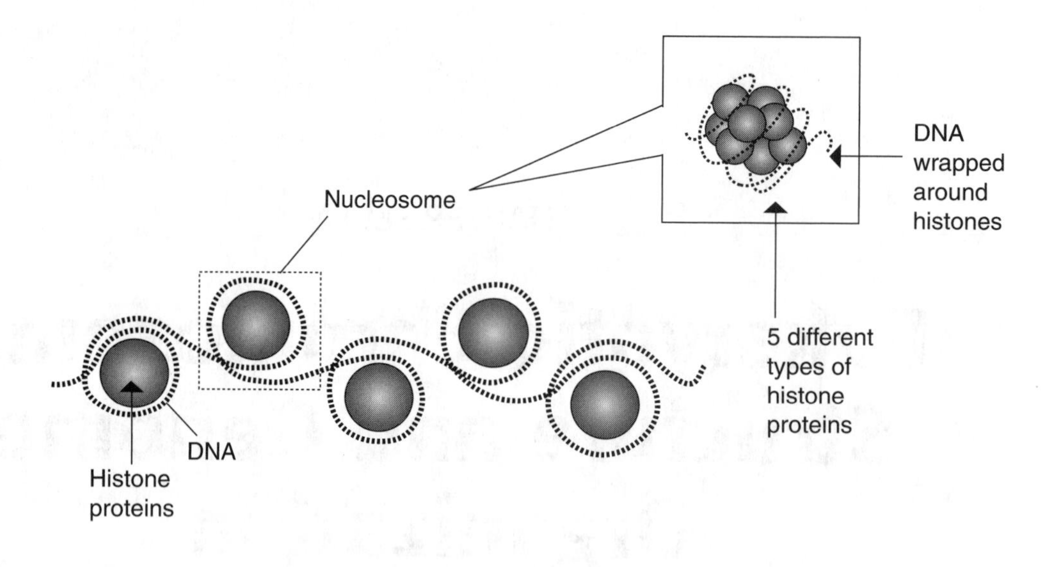

Each **nucleosome** is linked to the next nucleosome by a small protein strand and these spools of DNA can be packed together into larger loops with up to six nucleosomes per turn. The DNA is further packed into supercoiled loops and finally into chromosomes, which are generally large enough to be seen under simple light microscopes when condensed during mitosis.

You may see the words **"heterochromatin"** and **"euchromatin"** on the GRE. Recall that portions of DNA remain very tightly packed, while other sections are much less compact. The more tightly packed heterochromatin is not transcribed, while the looser euchromatin is.

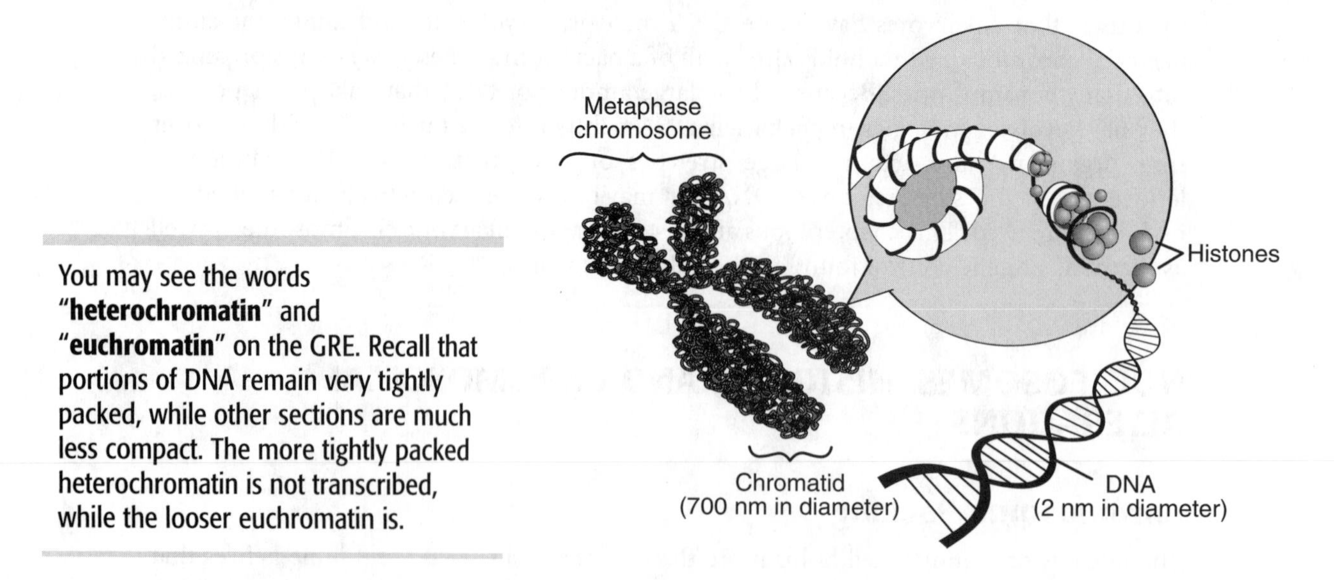

Chromosome Alterations

Several types of chromosomal breakage can occur in the course of DNA replication or at other points in the cell cycle. Some of these alterations can cause drastic changes in the ability of a cell to produce certain proteins: Many birth defects can be traced to defects in chromosome structure passed down through sperm or egg cells in which one of the following alterations has taken place:

Deletion: A chromosomal fragment, either from the end of a chromosome or from somewhere in the middle, is lost as the chromosome replicates.

Duplication: If the fragment that detaches from a chromosome during a deletion event reattaches itself to the homologous chromosome, that chromosome will then have two sets of identical genes in a particular region (a duplication), while the original chromosome where the deletion occurred will be shorter than normal.

Inversion: This deleted fragment could also attach back into the chromosome from which it came, but in a reverse direction.

Translocation: A common alteration in which the piece of DNA that breaks off a chromosome attaches to the end of another chromosome, most commonly not the homologous chromosome. In some cases, there is a reciprocal translocation, in which the chromosome giving a piece of DNA also receives one back that is comparable in size from the receiving chromosome.

In many cases, these alterations in structure render large groups of genes useless, especially because most genes need neighboring regulatory genes to work properly. After certain alterations, genes may be far enough away from their regulatory elements (see later on in this chapter) that they cannot be transcribed.

INTRONS, EXONS, AND TRANSPOSONS

Transcription is the process by which information coded in the base sequence of DNA is transcribed into a strand of mRNA. The mRNA is synthesized from a DNA template in a process similar to DNA replication. The DNA double helix unwinds at the point of transcription, and synthesis occurs in the 5' → 3' direction, using only one DNA strand as a template for mRNA construction. The base-pairing rules are the same for mRNA as they are for DNA, with the exception that RNA uses the nitrogenous base uracil (U) in place of thymine (T), so that A in DNA binds to U in RNA. **RNA polymerase** I and RNA polymerase II are the main enzymes responsible for making this transition between DNA codes and RNA codons. In general, RNA polymerase must bind to a promoter sequence before it can gain access to the structural genes to be transcribed (remember the bacterial operon mentioned earlier), and transcription occurs until the RNA polymerase reaches a termination sequence at the end of the structural genes. The DNA helix reforms and mRNA leaves the nucleus after several key processing steps.

Introns, Exons, and Other Post-Transcription Modifications

Most eukaryotic DNA does not code for proteins (only about 5 percent does). Noncoding, or "junk," sequences are found among the coding sequences. A typical gene consists of several coding sequences called ***exons*** and several noncoding sequences called ***introns***. The mRNA that is initially transcribed off a DNA template is a "precursor" molecule known as hnRNA (heterogeneous nuclear RNA), which contains both introns and exons. During hnRNA processing, introns are removed and exons are "spliced together" as the cell prepares an mRNA strand made up of only coding sequences.

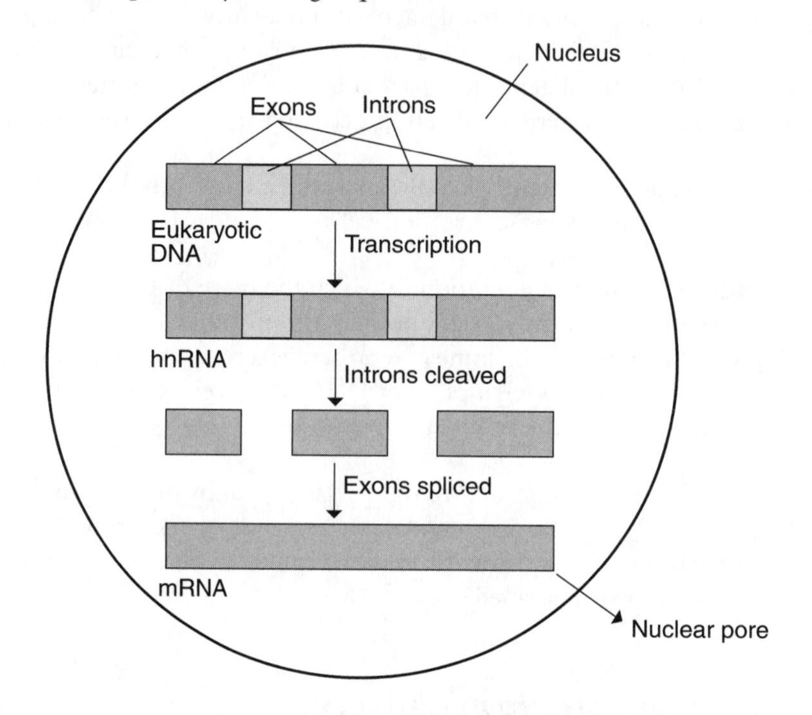

In addition to the splicing together of exons and the removal of introns, the mRNA strand undergoes the following modifications within the nucleus:

5' capping: methylated guanines (G) are added to the 5' end of the mRNA. This cap seems to serve two purposes: first, to protect the 5' end against degradation and damage as the mRNA passes through the nuclear membrane and travels around the cell; and, second, in the later initiation of protein synthesis at the ribosome.

3' poly-A tail: after being cleaved at a specific site, the mRNA precursor receives a poly-A tail made of 100–300 adenine nucleotides (A). Like the 5' cap, the poly-A tail seems to aid in stabilization and protection of the mRNA transcript as well as to help in later translation at the ribosome.

To remember the difference between exons and introns, recall that introns remain IN the nucleus while exons EXIT the nucleus to go to the ribosomes for translation and protein synthesis.

RNA in the nucleus is also stabilized by proteins known as snRNPs ("snurps" for short). These **s**hort **n**uclear **r**ibo**n**ucleo**p**roteins are similar to histone proteins that help wrap DNA into tightly packed chromosomes, and RNA molecules in the nucleus can often be found wrapped around these snRNPs. The large ribonucleoprotein that forms during the excision of introns and the splicing together of exons is known as a **spliceosome.** Because RNA itself can act as an enzyme, it is possible that the RNA participates in its own splicing.

Because many genes have evolved by gene duplication over hundreds of generations, different exons within a single gene can often code for different proteins. The protein that is eventually built at the ribosomes is determined by which exons are left in the mRNA transcript and the order in which the exons are spliced. For example, the family of immunoglobulin genes, responsible for building antibodies, are encoded for by genes that have undergone multiple duplications and can be spliced together to code for an amazing diversity of antibody structure (see chapter 10).

Transposons

First discovered in corn, *transposable genetic elements* are common features in many genomes. Transposons, as they're often called, are pieces of DNA that can move from place to place within an organism's genome. Bacterial transposons move from the main chromosome out to the plasmids or vice versa, and even from spot to spot on the plasmid or main chromosome.

The mechanism by which these DNA segments move involves the **transposase** enzyme and the use of **inverted repeats** to correctly guide the transposon into place in its new location. Transposons can end up in a variety of locations, and the swapping of a transposon from one place to another is not accompanied by an equal and opposite transfer of DNA to the place from which the transposon moved. Some transposons, however, will be copied before they move locations, so that it is the copy that inserts in the new location while the original gene is left in place.

Transposons are always found in between two inverted repeats, nucleotide sequences that are upside-down mirror images of each other. The simplest transposon is one that has two inverted repeats surrounding the gene for the transposase enzyme. In other words, this transposable genetic element consists of simply the gene needed to move the DNA segment elsewhere, and no other genes.

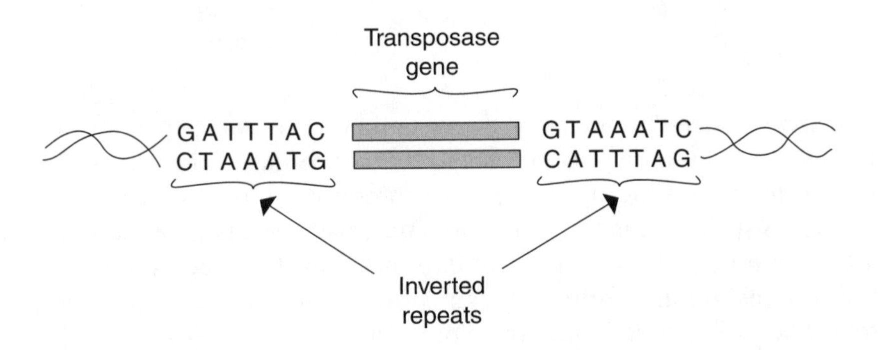

When the transposon above is cut out by the enzyme that it itself produces, it will move to another location in the genome and set up shop there. The movement of this small gene can disrupt other gene sequences that it lands in the middle of, or may affect regulatory elements of other genes. You can see below how the enzyme product of the transposase gene is responsible for the pinching off and separation of the moveable gene sequence, which includes its inverted repeats as well.

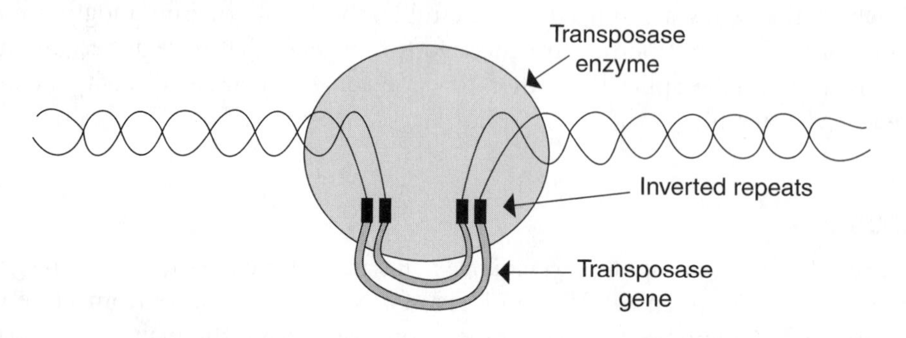

This simple transposon is called an **insertion sequence.** It is comprised of two inverted repeats surrounding a transposase gene. More **complex transposons** have not only the inverted repeats and the transposase gene, but also one or more genes that move as well. In the diagram below an *antibiotic resistance* gene is found in between two different sets of inverted repeats, each of which surrounds a transposase gene. In other words, the antibiotic resistance gene is found between two insertion sequences, which are responsible for guiding it into another section of DNA when it detaches from its current location.

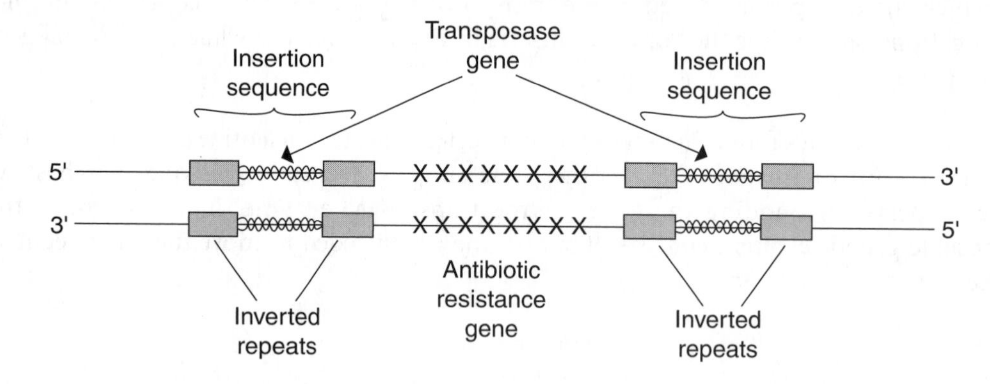

Complex transposons like the one above allow antibiotic resistance genes to move from a bacterial chromosome onto one of the many plasmids found in the cell, and can then be copied and transferred to other bacteria. Once transferred into another bacterial cell, the insertion sequences surrounding the antibiotic gene usually result in the transposition of that gene from the plasmid back into the main chromosome. Thus, because of the ability for complex transposons to jump from one segment of DNA to another, bacteria can rapidly transmit traits like antibiotic resistance from one cell to another.

A commonly cited example of transposons in eukaryotic cells is "Indian corn," where kernels come in a variety of colors and splotchy patterns. These color patterns are due to transposable elements within the cells that make up each kernel. The transposons jump from certain regions on the corn chromosomes into regions near genes encoding kernel color. These color genes are either activated or inactivated by the insertion of the transposons, and the kernels each reflect different colors because of this.

PROMOTERS, ENHANCERS, AND TRANSCRIPTION FACTORS

The major difference between prokaryotic genes and eukaryotic genes is the presence of introns. Another difference is the presence of **enhancers**, noncoding regions of DNA that influence the activation of genes. These enhancers are often located a considerable distance upstream from the genes they exert control over.

When eukaryotic DNA is transcribed, RNA polymerase must bind to a promoter sequence ahead of the structural genes to be "read," just like in prokaryotes. The enhancers are regions of DNA that bring certain **transcription factors** into contact with the promoter regions of genes and act to "enhance" transcription. Experimental removal of enhancers can cause drastic decreases in gene transcription. It is thought that these enhancers, even though they may be thousands of nucleotides away from the structural genes they affect, are brought into contact with those genes or with their promoter sequence by a loop in the DNA structure.

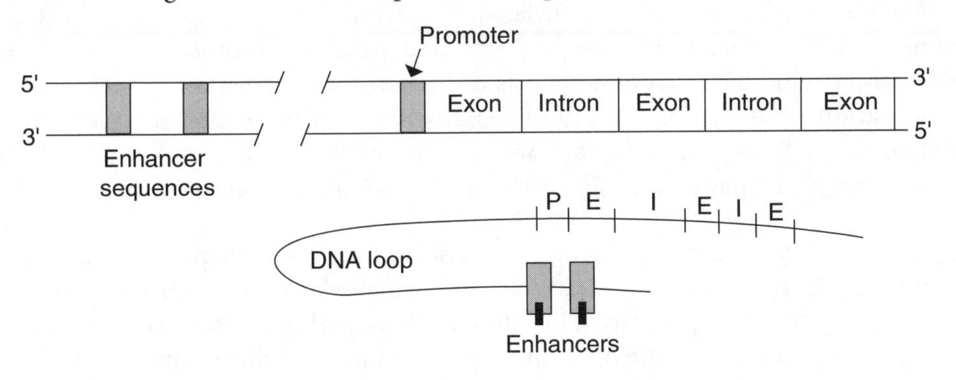

Promoters

One of the most common elements in eukaryotic promoters is the "**TATA box.**" Although not present in "housekeeping genes" and other developmental genes, such as the homeotic genes (covered in later chapters), TATA boxes are A-T rich regions of DNA that are involved in positioning the start of transcription. The reason for this is that regions of DNA rich in adenine and thymine tend to separate more easily than those rich in C's and G's. Adenine and thymine form only two hydrogen bonds across the double helix, while C and G form three. Separation of DNA at the TATA boxes in the promoter regions of genes allows DNA to unzip at those regions for RNA access to the DNA template. Other types of promoters known as *internal promoters* also exist that can be found within the introns of genes. These promoters occur especially in genes that encode rRNA and tRNA molecules.

Transcription Factors

There are hundreds or thousands of proteins that exert transcriptional control over the genome, and they are known as transcription factors. Some bind to enhancer sequences, others to promoter sequences. All transcription factors help RNA polymerase find and bind to a given promoter region. Each of these proteins has a highly conserved DNA-binding domain, which allows it to attach to DNA nucleotides. Most are specific for certain enhancer or promoter sequences.

Other Mechanisms of Eukaryotic Control of Gene Expression

In addition to the above mechanisms to control gene expression within the nucleus, other means of control exist as well. We have discussed DNA packing into nucleosomes, which themselves pack into chromosomes, and this DNA packing and unpacking affords a degree of control that is not found in bacterial cells because of their relatively small amount of DNA. In addition, eukaryotic DNA is subject to **methylation**, the attachment of methyl (CH_3—) groups to nitrogenous bases. When methylated, bases cannot be transcribed by RNA, and thus methylation serves to inactivate certain genes at certain times.

When scientists search for genes in an unknown genome, one thing they can look for to point them toward regions containing coding segments of DNA are "CG islands," promoter regions rich in CG repeats that are present upstream of genes. These CG regions remain unmethylated in order to protect the ability of repair enzymes to correct mismatches when the DNA replicates. If they were methylated, these errors would not be corrected. The mutations would then remain in these very sensitive regions—promoter regions—in which any DNA errors could have profound effects on the synthesis of certain proteins down the line. Methylation of these CG promoter regions would also reduce transcription since methylation blocks RNA polymerase action. Thus, unmethylated CG repeats are often clues to scientists that genes may be lurking nearby in the genome they are studying.

It also seems that **genomic imprinting**, discussed earlier, is regulated through selective methylation of DNA, so that maternal alleles are methylated with certain patterns and paternal alleles with other patterns. This allows differential expression of identical alleles depending upon which parent the offspring inherited a mutant allele from.

Another method of control over gene expression that you may see on the GRE is **gene amplification**, a phenomenon in which certain genes are rapidly copied by the organism, so that multiple copies of these genes exist within the genome. Imagine a gene for which only one or two copies exist on a particular chromosome. Gene amplification could result in the creation of up to 100 or more copies of this gene. The process occurs most often during embryologic development and has been extensively studied in plants and amphibians. It is believed that gene amplification exists to provide an increased amount of a particular protein product at certain times in the life cycle of an organism. With large numbers of a particular gene, a cell is able to construct a large amount of mRNA encoding the same protein, and much more of this protein can be made by the ribosomes than if the gene only existed as a single copy. Amplification is developmentally or environmentally regulated and usually results in a temporary increase in the number of copies of a certain gene within a cell.

Once mRNA reaches the cytoplasm, each strand can be read by hundreds of ribosomes before the mRNA gets degraded, which means that *a single mRNA transcript can encode the production of hundreds of molecules of a given protein*. The speed at which mRNA gets degraded is yet another aspect of gene expression control. More controls exist when the protein molecules themselves are made at the ribosomes. These proteins are modified in several steps that will be detailed in the next chapter.

CHAPTER EIGHT

DNA Replication and Protein Synthesis

All organisms are able to copy enormous quantities of DNA with almost no mistakes in very short amounts of time. Many enzymes and other proteins play a role in this replication process. DNA replicates both for cell division (meiosis and mitosis) purposes, as well as for protein synthesis. As a review, the basic unit of DNA is the nucleotide, composed of a deoxyribose sugar attached to a phosphate group attached to a nitrogenous base.

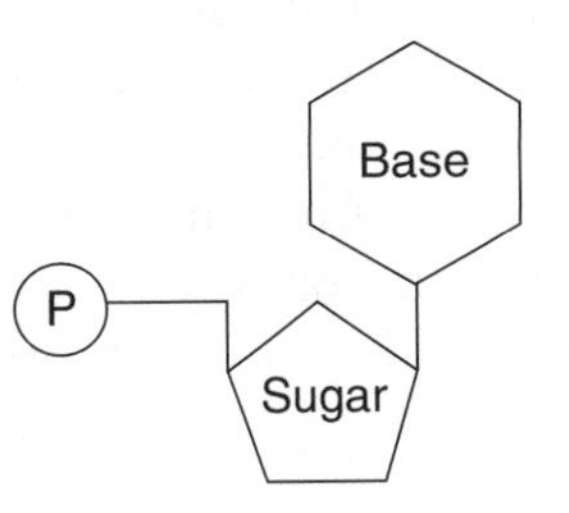

The bases of DNA come in four varieties: **adenine** (A), **cytosine** (C), **guanine** (G), and **thymine** (T). Two of the bases, G and A, are known as purine bases and can be recognized by their two fused rings made of nitrogen and carbon. C and T, the **pyrimidine** bases can be recognized by their single rings of nitrogen and carbon. Because the DNA double helix has a uniform width, **purine** bases must always pair across the double helix to pyrimidine bases. Two single-ring or two double-ring bases paired together would narrow or widen the width of the double helix.

To remember that the purine bases, G and A, are bigger (double rings) than the pyrimidine bases, C and T (single rings), recall that the state of Georgia (GA) is bigger than the state of Connecticut (CT).

Nucleotides

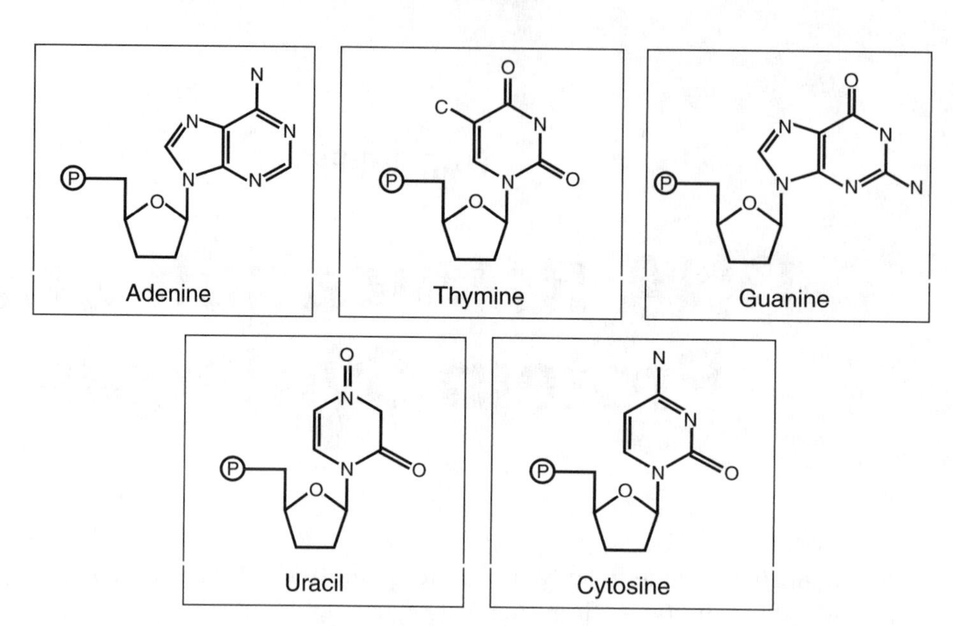

Nucleotides bond together to form polynucleotides. A DNA molecule is a double-stranded helix with a sugar-phosphate **"backbone"** and **nitrogenous bases** as "rungs of a ladder" within the backbone. Thymines (T) are always across the helix from adenines (A), and guanines (G) are found across from cytosines (C). As can be seen in the diagram of the double helix below, A's and T's form two different hydrogen bonds with each other across the helix, and C's and G's form three. That is the reason that TATA boxes (described in the last chapter) are found in many promoter regions located upstream from structural genes: the A-T rich regions have fewer hydrogen bonds, and it is easier to separate the complementary strands of DNA in those areas than in G-C rich areas.

DNA molecule

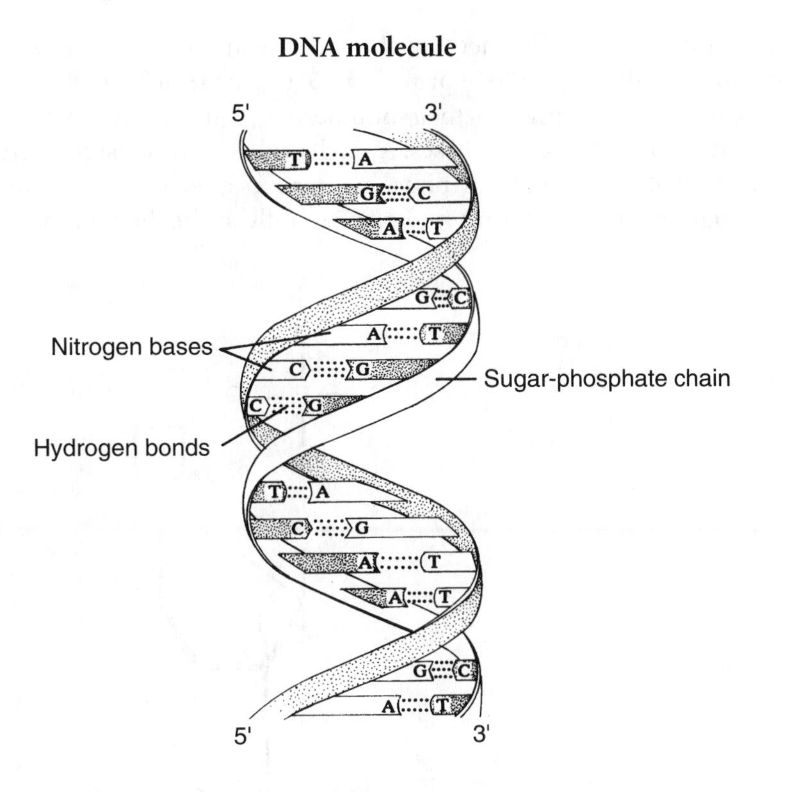

DNA REPLICATION AND DAMAGE CONTROL

If you look at the structure of a DNA molecule carefully, you will note that the two strands are not only complementary (for every A, a T exists across the helix and for every G, a C exists across the helix) but also *oriented in an "anti-parallel" arrangement.*

Structure of DNA

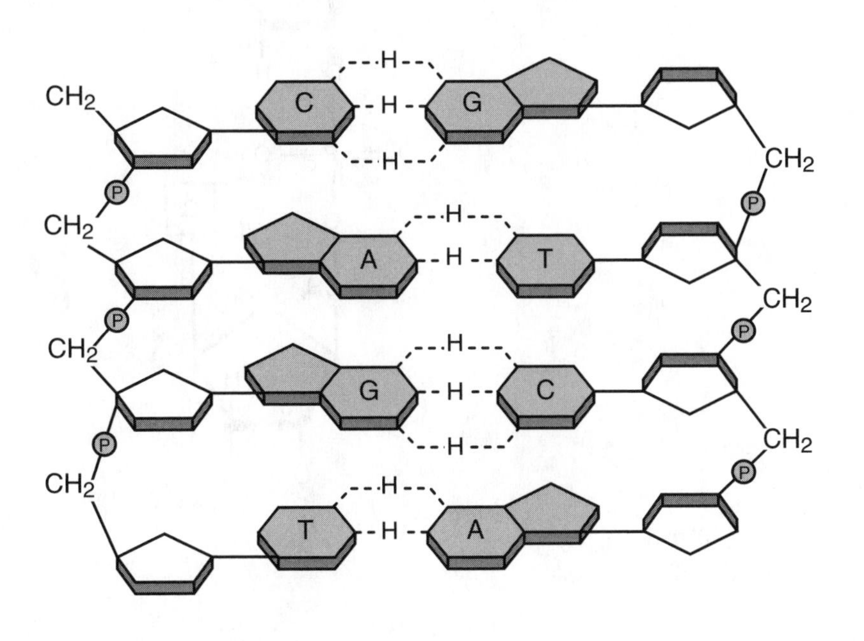

This terminology comes from the fact that the carbons in the rings that make up the sugar backbone are numbered from 1' ("one prime") to 5' ("five prime"). The "prime" sign is used for the sugar carbons to distinguish their numbering from the numbering of the carbon atoms that make up the nitrogenous bases. A nucleotide's own phosphate group is attached to the 5' carbon of its own sugar. A phosphate group attaches to the sugar of the next nucleotide through that sugar's 3' carbon. Look carefully at the diagram below:

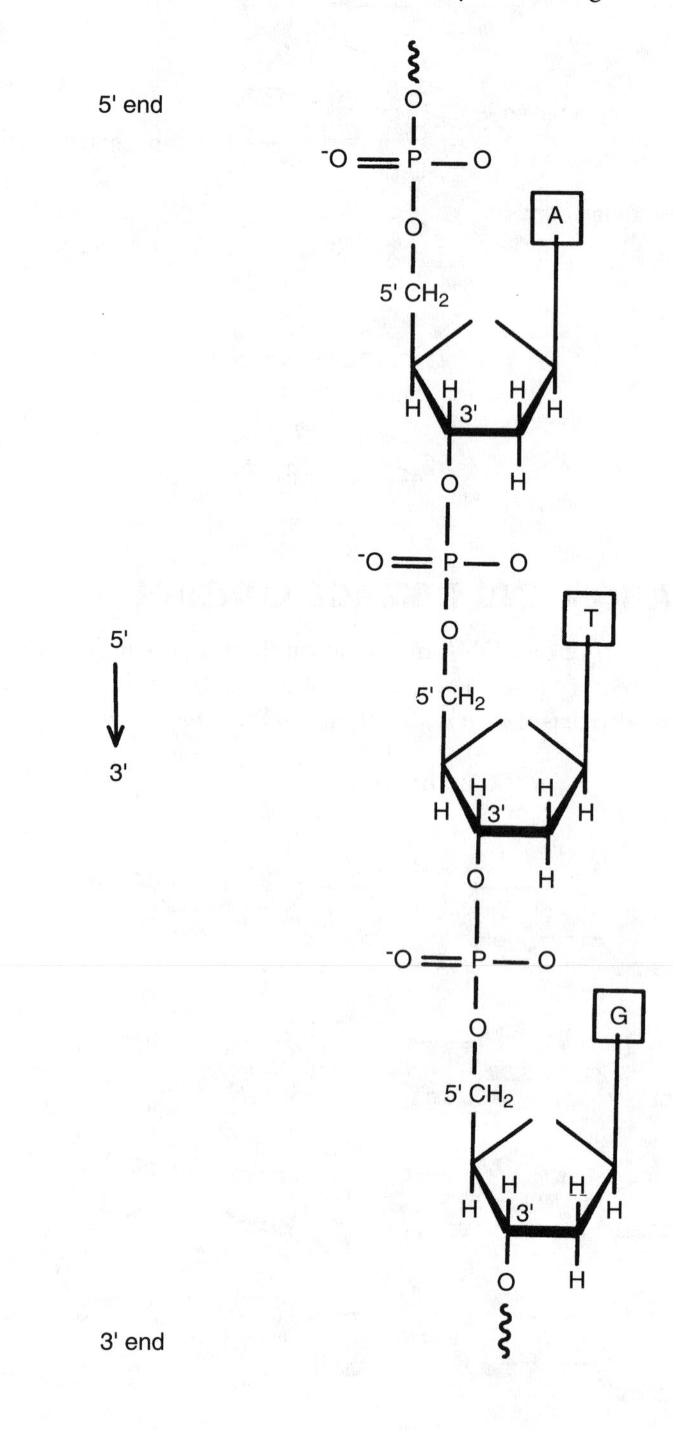

One side of the helix is oriented upside-down to the other side so that the helix can fit together properly in order to conserve space within the nucleus. Because of this, if one were to look from the top of a double helix downward, one side of the double helix would run from the 5' end of the sugar molecules down toward the 3' end (known as the 5' → 3' direction), and the opposing strand would run from the 3' sugar ends to the 5' ends (3' → 5' direction).

DNA Replication

In bacteria, there is a site within the single bacterial chromosome or plasmid where replication begins and proceeds in both directions at once until the whole chromosome has been copied. This "**origin of replication**" is generally found at an AT-rich region, where the separating DNA forms V-shaped "**replication forks.**" Replication forks move bidirectionally from the replication origin. In eukaryotes, many origins of replication open up at the same time, speeding the process of copying any given chromosome.

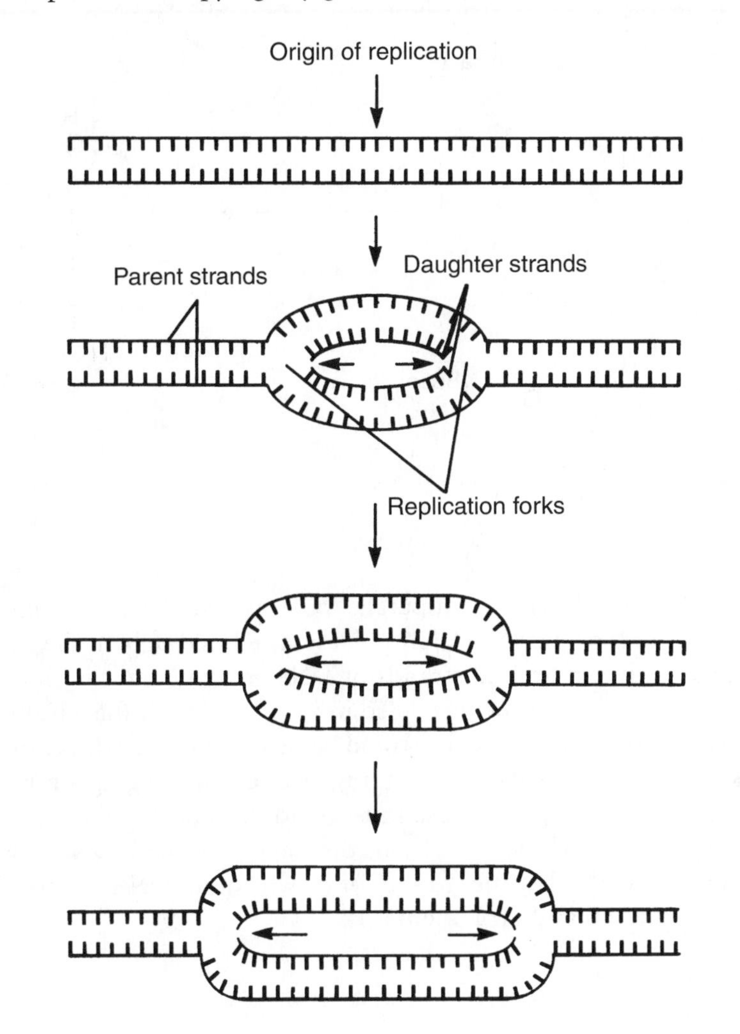

The enzyme responsible for attaching together new nucleotides as they arrive at the original (parental) strands is **DNA polymerase**. Each molecule of the polymerase enzyme is able to add approximately 500 nucleotides per second to a growing strand of new (daughter) DNA in bacteria, and 50 nucleotides per second to a growing DNA strand in eukaryotes.

Every nucleotide arrives at the growing strand as a triphosphate nucleotide (**dNTP**, or deoxynucleotide triphosphate—recall ATP from a few chapters ago). When it joins the strand, only one phosphate is used in binding to the sugar of the neighboring nucleotide. Therefore, each dNTP loses two terminal phosphates as it is bound into the new strand, and these phosphates are known as "leaving groups." The reaction is exergonic and provides the energy needed for the nucleotides to attach to the growing strand of DNA.

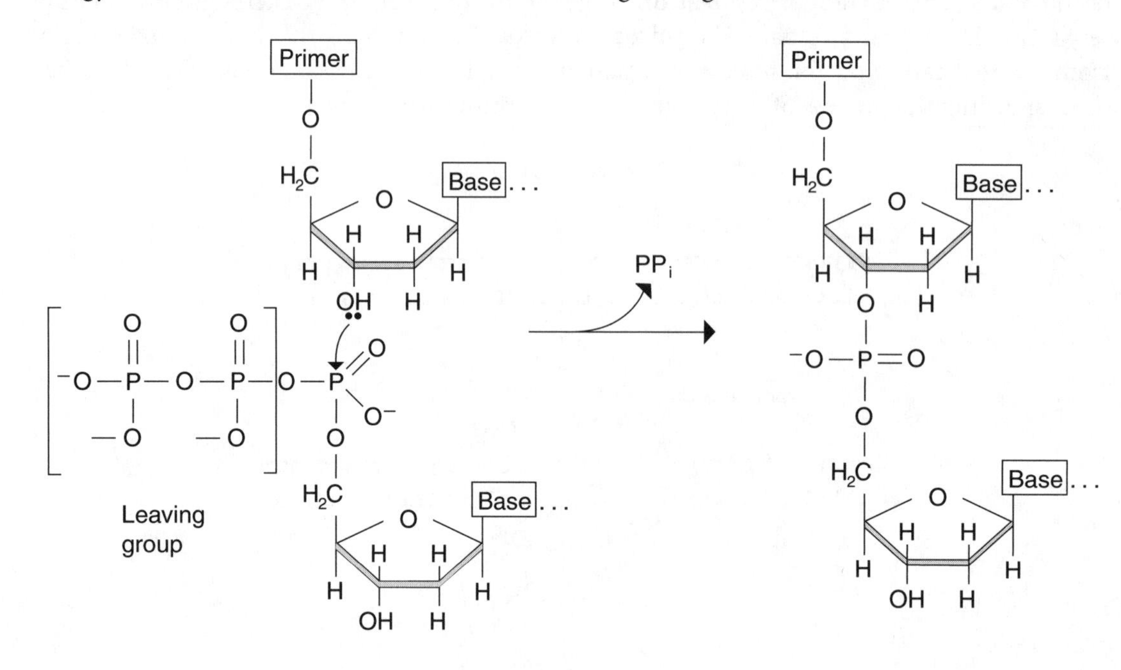

Recall that the DNA double helix has antiparallel polarity, with bases running 5' → 3' in one direction and 3' → 5' in the other. Yet, DNA polymerases build only in the 5' → 3' direction. Therefore, the two newly synthesized strands of DNA must be built in opposite directions simultaneously. The strand that is being copied in the direction of the advancing replication fork is called the **leading strand**, while the strand being copied in the direction away from the replication fork's movement is called the **lagging strand**. The lagging strand is synthesized discontinuously because DNA polymerase must build in the 5' → 3' direction as the DNA unzips and, then, jump back to build on the newly opened DNA. Because of this discontinuous synthesis, the lagging strand of newly built DNA is built out of small fragments of DNA called **Okasaki fragments**.

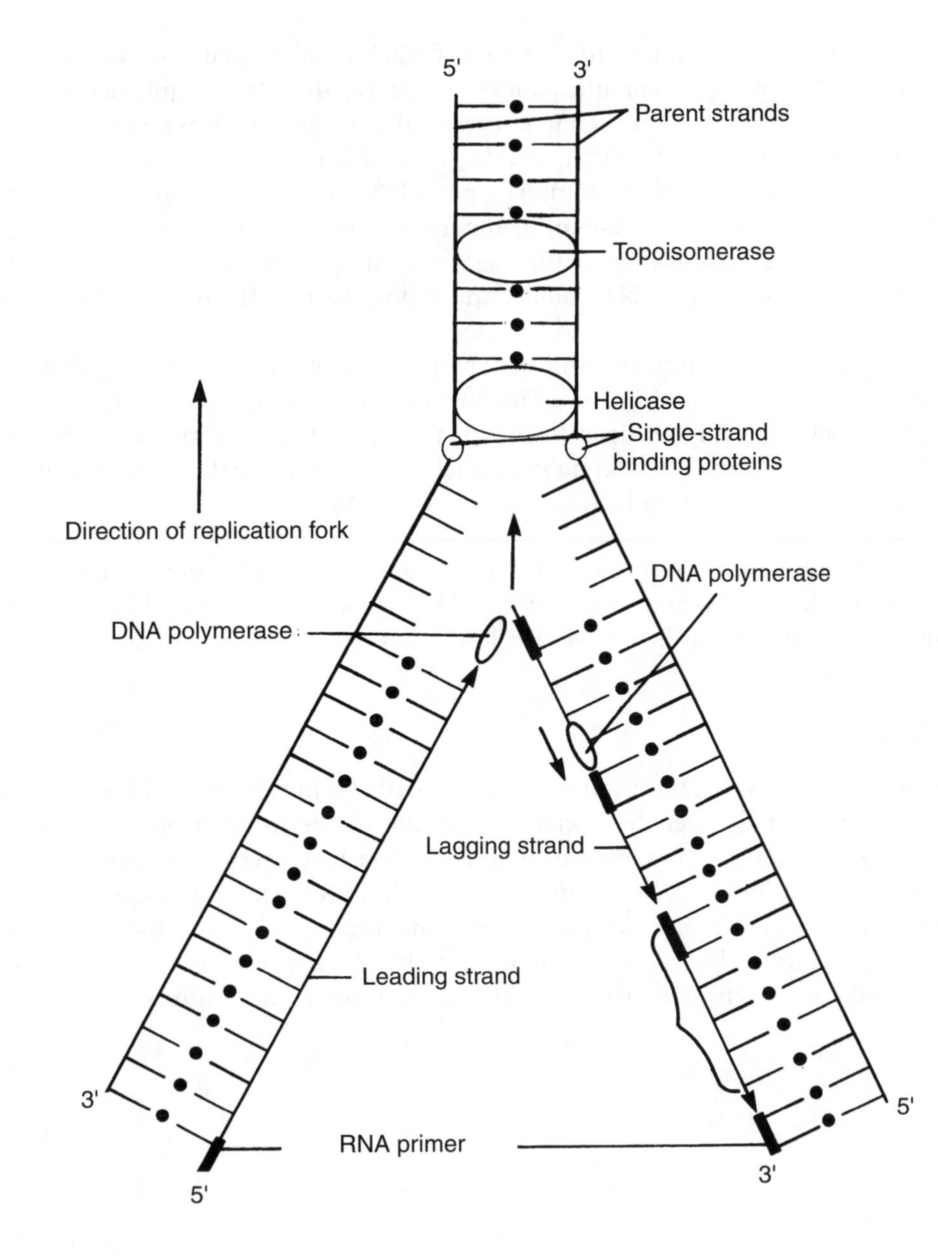

Yet one more restriction applies to DNA polymerases: they cannot initiate the building of a complementary DNA strand. In other words, DNA polymerases can only add nucleotides to new nucleotides already in place. Thus, these enzymes require a **primer** to put the first nucleotides into place next to the parental DNA template. The primer is actually made of RNA and replaced later by DNA, but it provides a free –OH group on the 3' end of the growing chain, which allows free dNTPs to join.

The enzyme responsible for building this primer of RNA is called **primase**, and it builds short strands of RNA, perhaps ten nucleotides in length, that are complementary and antiparallel to parental DNA that has been first unwound by enzyme **helicase.** Helicase works at the replication fork to unzip the DNA double helix, which is then stabilized by single-strand DNA binding proteins until the complementary DNA strand falls into place. Another class of enzyme to be familiar with is the **topoisomerases**, which regulate the supercoiling of DNA into chromosomes. Topoisomerase I causes single strand breaks and ligations in the DNA, which affect the nature of DNA coiling, and topoisomerase II causes double strand breaks. They aid in DNA unwinding for both transcription and replication, and they may also have a role in crossing over and recombination. The final enzyme you should know for the GRE is **DNA ligase**, which is responsible for ligating, or connecting, the Okasaki fragments left over from discontinuous synthesis on the lagging daughter strand of DNA.

RNA primers are needed constantly at the lagging strand because each time an Okasaki fragment is made, an RNA primer must be laid down first.

At the end of replication, the RNA primers must be cut out of the newly formed strands. This is accomplished nucleotide-by-nucleotide with DNA polymerase, which replaces the RNA primer with DNA nucleotides as it removes the RNA ones.

Replication Errors

There is an error rate during replication of about 1 in 10,000 nucleotides, which would be disastrous if not corrected, given how many nucleotides have to be copied in order to replicate entire genomes. The errors made during replication are corrected by a process called DNA mismatch repair. When a DNA polymerase molecule adds the wrong base to a growing strand, it will usually cut this nucleotide out of the strand and replace it with the correct one. Although DNA polymerase has 5' $\rightarrow$ 3' building activity, it corrects mistakes in a 3' $\rightarrow$ 5' direction. When it corrects mismatches, it is said to be working as an **exonuclease.**

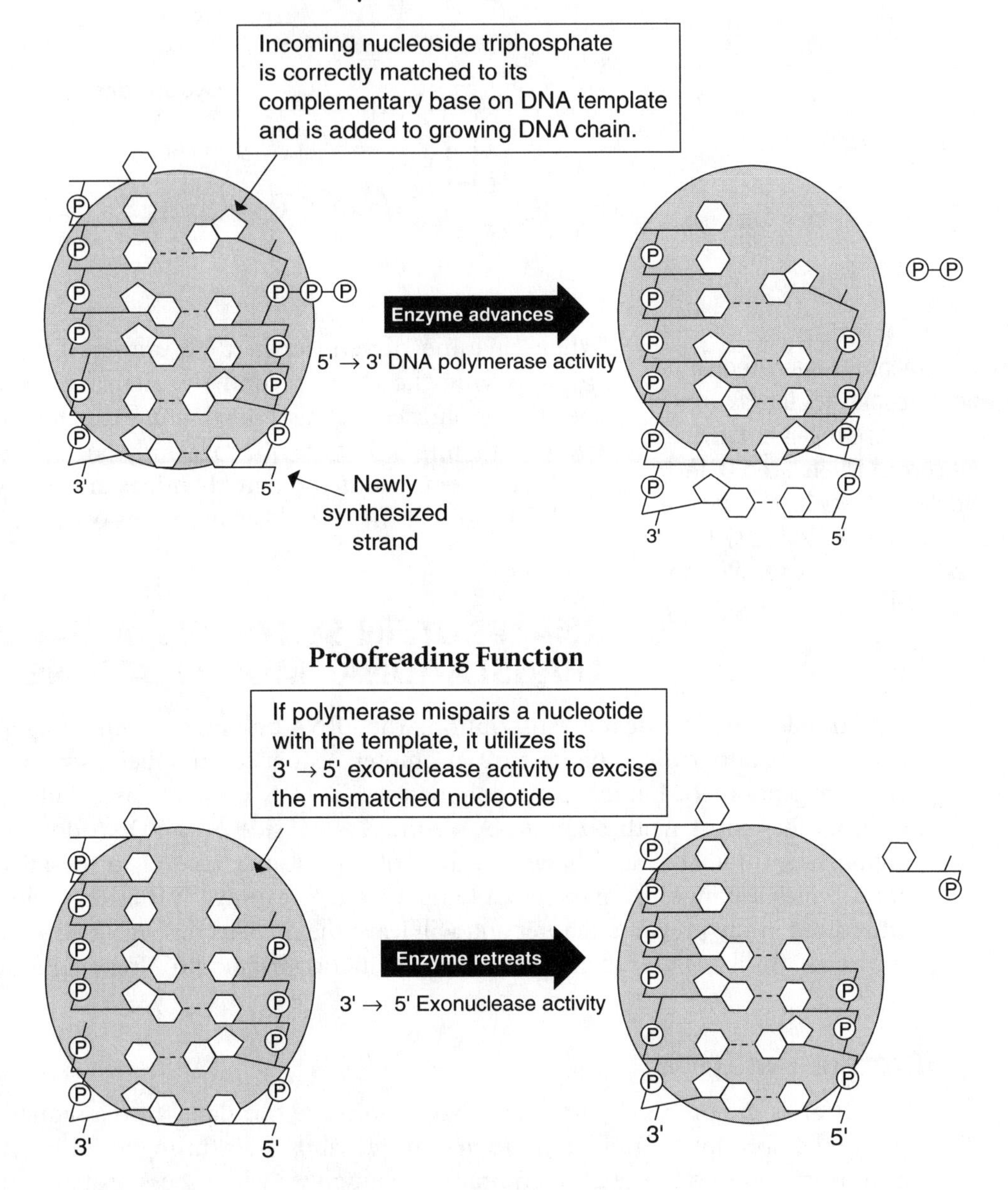

Other exonucleases exist as well, but they can correct mismatches only from the end of a growing strand. Yet, errors can occur within a strand as well, and these must be corrected by **endonucleases**, which can cleave out and replace damaged DNA from the middle of strands. The repair enzymes that recognize these DNA errors are very specific for the kind of damage they repair. One example is the enzyme that repairs ultraviolet (UV) light damage caused by excessive exposure to sunlight. UV light causes **thymine dimers** to form, which are adjacent thymine nucleotides that become bonded covalently due to UV energy.

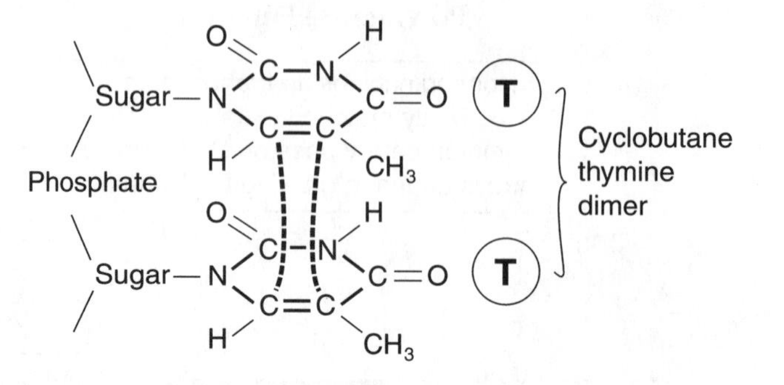

Xeroderma pigmentosa is a hereditary disease where people lack the UV-specific endonuclease repair enzyme. They cannot correct UV-induced DNA damage and have a very high incidence of skin cancer as well as severe skin lesions. They are extremely sensitive to sunlight as a result.

When thymine dimers form, they prevent DNA polymerase from copying the DNA beyond the site of the dimer. A UV-specific endonuclease cleaves the thymine dimers out and inserts two new thymines that are not covalently bonded to each other. DNA polymerase adds the new thymines and DNA ligase seals the "nick" so that the strand is continuous once again.

RNA, PROTEIN SYNTHESIS, AND POST-TRANSLATIONAL MODIFICATIONS

DNA strands also separate for transcription, when **RNA polymerase** pushes apart the double helix in a certain region, binds to a promoter sequence, and beings to transcribe the structural genes there. Because RNA polymerases, like DNA polymerases build in the 5' → 3' direction, they must synthesize mRNA off the 3' → 5' side of the DNA double helix. RNA polymerases are able to add bases at a rate of 50–100 per second to the growing mRNA strand, which separates from the DNA template after forming. This mRNA will be modified as discussed in the previous chapter and will leave the nucleus through a pore in the nuclear membrane. Unlike DNA polymerase, RNA polymerase has no proofreading function.

Protein Synthesis

Translation is the process by which mRNA codons are translated into a sequence of amino acids at the ribosomes. Translation occurs on free ribosomes throughout the cytoplasm and on ribosomes bound to the ER membrane. Although we will discuss protein synthesis at the ribosome in a general way, it is important to note the differences between protein synthesis at free ribosomes and at those on the ER membrane.

- ER-bound ribosomes secrete polypeptides (proteins) into the ER lumen (pocket within the ER membrane), while free ribosomes secrete polypeptides directly into the cytoplasm.

- Proteins synthesized on the ER membrane are destined to be modified within the ER and Golgi and shipped in vesicles to reside within the cell membrane, within the membrane of other organelles (including the ER and Golgi themselves), or at points outside the cell. These proteins include membrane-spanning proteins (e.g., ion channels) found in the cell membrane and secreted enzymes.

- Proteins synthesized on free ribosomes will be used within the cytoplasm or back in the nucleus. Those that end up back in the nucleus have a certain sequence of amino acids—the **nuclear localization signal** (NLS)—that allows them to pass through the nuclear membrane from the cytoplasm and remain in the nucleus.

Proteins that end up within the ER lumen initially have amino acid sequences, known as **signal sequences**, that tell the ribosomes synthesizing the proteins to remain bound to the ER membrane until synthesis is over. These signal sequences are cut off as modifications within the ER membrane shape the newly made protein. Yet, proteins that have signal sequences pointing them toward the nucleus (NLSs) retain these sequences even after they are modified. In all likelihood, these sequences must remain on the polypeptide because these nuclear proteins can be found back in the nucleus even after the nuclear membrane dissolves away during prophase of cell division. These nuclear proteins include the DNA and RNA polymerases already discussed, as well as histone proteins and others.

Translation of proteins at the ribosomes involves three key steps: Initiation, Elongation, and Termination. All three stages require both the use of ATP as well as a number of enzymes. Translation involves "reading" the mRNA nucleotides and attaching together a sequence of amino acids specified by these codes. Whereas mRNA was used in the nucleus to make a complementary copy of DNA bases, **tRNA** (transfer RNA) is used at the ribosomes to carry the correct amino acid into place for protein building. Transfer RNA is not a long strand like mRNA, but rather looks a bit T-shaped as the single RNA strand doubles back on itself. The tRNAs that float freely in the cytoplasm are "loaded" with amino acids by an enzyme called **aminoacyl-tRNA synthetase**. Each tRNA is able to be loaded with a single amino acid, and once loaded is called an **aminoacyl-tRNA.** The amino acids that get loaded onto tRNAs come from digestion of food or from the environment, usually from more complex polypeptides that are broken down by enzymatic processes outside of the cell or in the lysosomes.

There are *20 naturally occurring amino acids* found in the proteins of living organisms, and the sequence of mRNA bases (called "**codons**") needed to code for each one of these 20 amino acids was painstakingly discovered back in the 1960s through the experiments of Marshall Niremberg. Although you do not need to know the history of the genetic code's discovery for the GRE, it is worth reading about in any introductory biology text. The genetic code is summarized in the chart below:

Although there are exceptions in some bacteria and protists, the genetic code is essentially the same for all organisms on Earth.

The Genetic Code

Second Base

First Base (5')	U	C	A	G	Third Base (3')
U	UUU, UUC } Phe UUA, UUG } Leu	UCU, UCC, UCA, UCG } Ser	UAU, UAC } Tyr UAA, UAG } *Stop*	UGU, UGC } Cys UGA } *Stop* UGG } Trp	U C A G
C	CUU, CUC, CUA, CUG } Leu	CCU, CCC, CCA, CCG } Pro	CAU, CAC } His CAA, CAG } Gln	CGU, CGC, CGA, CGG } Arg	U C A G
A	AUU, AUC, AUA } Ile AUG } *Start* or Met	ACU, ACC, ACA, ACG } Thr	AAU, AAC } Asn AAA, AAG } Lys	AGU, AGC Ser AGA, AGG } Arg	U C A G
G	GUU, GUC, GUA, GUG } Val	GCU, GCC, GCA, GCG } Ala	GAU, GAC } Asp GAA, GAG } Glu	GGU, GGC, GGA, GGG } Gly	U C A G

WHAT IS THE SOURCE OF THE GENETIC CODE'S REDUNDANCY?

This idea that the third base is not needed as much as the first two bases is called the "**wobble** hypothesis," so-called because the tRNA dropping off the proper amino acid does not bind as tightly to the third base of the mRNA codon as it does to the other two. For example, the base U in a tRNA anticodon can bind to either an A or a G in the mRNA. This wobble permits 45 types of tRNAs to carry all the amino acids necessary for translation rather than the expected 64 (4^3).

You should note the following characteristic: despite only 20 amino acids that are coded for, there are *64 possible codons* that can be made from combinations of three bases at a time (4 DNA or RNA bases—A, C, G, T (U)—to the third power represents four bases taken three at a time. $4^3 = 64$ total codons). A quick look at the above chart shows that the genetic code is redundant, and it is called a "degenerate code" for this reason. Many different codons can code for the very same amino acid. The reason for this seems to be that the "reading" of each codon at the ribosome is done mainly with the first two bases, and the third base is less important for accurate translation. Notice that the codons coding for the same amino acid differ only in their third base, but not in their first two bases.

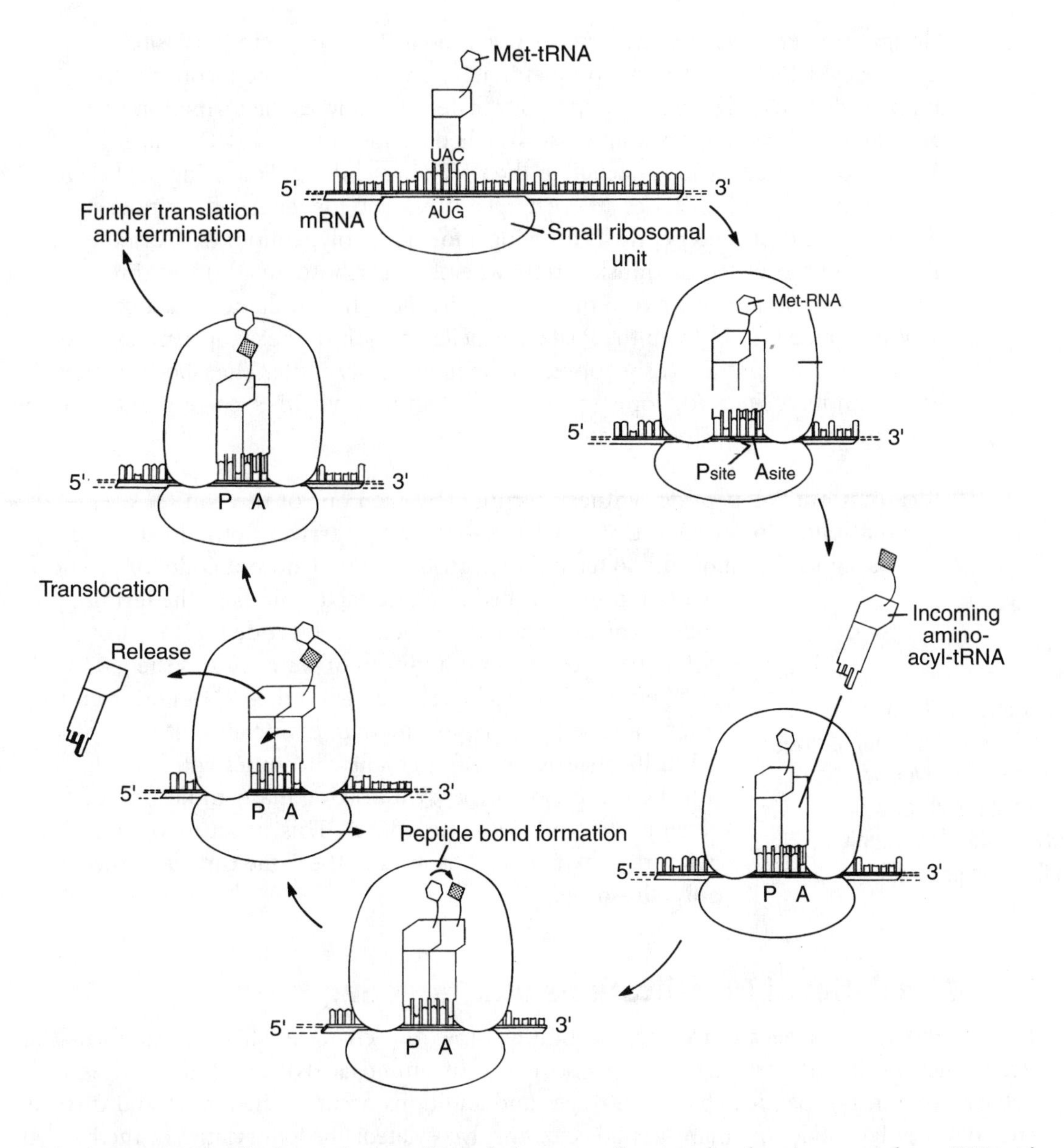

Initiation: mRNA binds to the smaller of the two ribosomal subunits near the mRNA 5' terminus and proteins called initiation factors help stabilize the mRNA to the ribosome. An initiator tRNA, carrying the amino acid methionine, binds to the initiator (start) codon, always AUG. The *anticodon* of the initiator tRNA (the place that actually binds temporarily to the mRNA codon AUG as it is held in place by the ribosome) is UAC, a sequence complementary to the mRNA sequence. The large ribosomal subunit then binds to the small one, creating a complete ribosome with the methionine-tRNA complex sitting in the P-site (peptidyl-tRNA site). The mRNA strand is now effectively locked into place between the ribosomal subunits and will be read much like swiping a credit card through a magnetic card reader.

Elongation: Hydrogen bonds form between the mRNA codon in the A site (aminoacyl-tRNA site) and its complementary anticodon on the incoming aminoacyl-tRNA. The enzyme peptidyl transferase catalyzes the formation of a peptide bond between the amino acid attached to the aminoacyl-tRNA sitting in the A site and the methionine attached to the tRNA in the P site. Following peptide bond formation, a ribosome carries an *un*charged tRNA (no longer called aminoacyl-tRNA) in the P site and peptidyl (carrying a forming polypeptide) tRNA in the A site. The cycle is completed by translocation, whereby the ribosome advances three nucleotides along the mRNA in the 5' → 3' direction. In simultaneous action, the now-uncharged tRNA from the P site is expelled back into the cytoplasm and the peptidyl-tRNA from the A site moves to the now-empty P site. The ribosome then has an empty A site ready for entry of the next aminoacyl-tRNA corresponding to the next codon.

Termination: Polypeptide synthesis terminates when one of three mRNA termination codons (UAA, UAG, or UGA—see chart) arrives in the A site. These codons signal the ribosome to terminate translation; they do not code for amino acids. A protein called a release factor binds to the termination codon, causing a water molecule to be added to the polypeptide chain. This addition of water causes the polypeptide chain to be released from the tRNA and from the ribosome itself. The ribosome then dissociates into two subunits once again. Often, many ribosomes will read an mRNA at the same time, synthesizing many molecules of a given protein off the same mRNA. This structure of an mRNA being read by many ribosomes at the same time is known as a **polyribosome**.

If the details above seem confusing, go through them again slowly while looking back and forth between the words and the diagram of the process found above. For the GRE, it is unlikely you will be tested on the small details of protein synthesis, but you should certainly be familiar with the "big picture."

Post-Translational Modifications and Targeting

During and after its release, the polypeptide assumes the characteristic 3-D conformation determined by its primary structure (its sequence of amino acids). Disulfide bridges form between certain polypeptide chains, cleavages and additions occur at the amino and carboxyl ends of the polypeptide, and many amino acids are glycosylated, phosphorylated, or methylated.

Yet in order for the final protein product to be made and for the protein to find its way to its final destination, the protein must move from place to place in the cell, often being packaged in vesicles in order to travel through various organelle membranes. The signals that direct the proteins to various places are contained within their amino acid sequences, usually toward the ends of the amino acid chain. The sorting signals at the end of the protein chain are called **signal peptides**, while those that are made from various amino acids that bunch together in the middle of the proteins 3-D conformation are called **signal patches**. Different signal peptides or patches are used to send the proteins to various destinations.

As mentioned before, almost all the proteins that will be secreted by the cell, as well as those that will remain in the ER or Golgi, are produced by ER-bound ribosomes and elongate directly into the ER lumen. It is here that many protein products are placed into vesicles, small membranelike sacs that hold the protein and can merge with other membranes to deliver the proteins within. This can be especially important, for example, in building lysosomes, because the cell needs to transport digestive enzymes directly from the ER lumen into the lysosome without having those enzymes running around loose in the cytoplasm.

Proteins that will remain in the ER have a special **ER retention signal** added to their structure. This group of four amino acids added at the carboxyl terminal restricts the protein to the ER lumen and is used for enzymes and other proteins that have functions in the ER itself. Proteins known as **chaperones** aid in the folding of new ER proteins.

One of the major functions of the ER in protein synthesis is to add sugars to proteins, a process known as **glycosylation.** Glycosylation seems to signal that the protein is destined for export out of the cell or for one of the many membranes within and around the cell, because cytosolic proteins are rarely, if ever, glycosylated. The most common sugar group added to a protein in the ER to make a glycoprotein is called an **N-linked oligosaccharide**. The sugar is added to the —NH_2 group of an asparagine amino acid somewhere within the protein's structure, which gives the sugar the name "N-linked." The oligosaccharide is large, containing a mixture of fourteen different sugars covalently bonded to each other: glucoses, mannoses, and N-acetylglucosamines.

Enzymes in the ER lumen can also switch some amino acids on the carboxyl terminus of a protein and replace them with a special glycosylphosphatidylinositol **(GPI) anchor**, which allows the proteins eventually to be anchored into the exterior of the cell membrane:

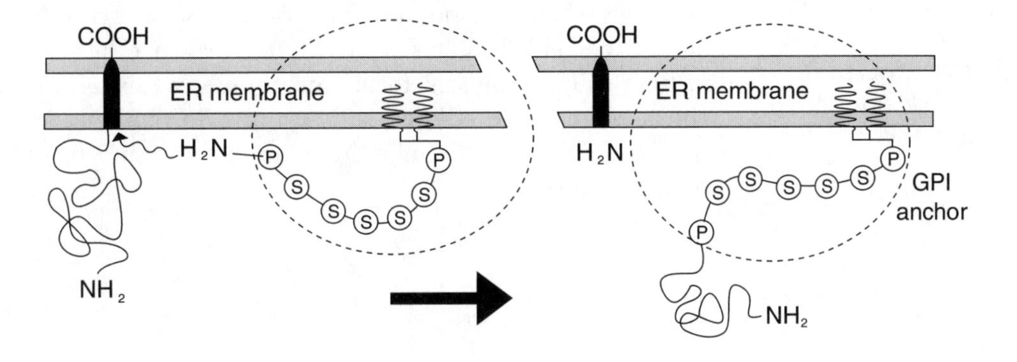

All proteins, *even ones that receive an ER retention signal*, are automatically sent to the Golgi after glycosylation. They are folded into vesicles that "bud" off of the ER surface, literally taking with them a piece of the ER membrane. When these vesicles reach the cis face of the Golgi, the side facing the ER, they merge with the Golgi membrane and dump the protein into the maze of sacs that make up this organelle. Proteins that have an ER retention signal of four particular amino acids are recognized by the Golgi and repackaged in vesicles to return to the ER where they will remain. For the remaining proteins, the Golgi is responsible for further sugar modifications, including tinkering with the N-linked oligosaccharide initially added in the ER. These modifications include the removal of mannose sugars and the addition of more N-acetylglucosamine as well as sialic acid and galactose. Some new sugars are added to the —OH groups of certain serine and threonine side chains, a process called **O-linked glycosylation.**

All of this glycosylation of proteins makes the proteins more resistant to being digested by proteases and peptidases as well as helps to limit the access of many substances to the cell membrane. The sugars form somewhat of a selectively permeable layer above the cell membrane. Once the proteins reach the far end of the Golgi, the *trans* end, they are packaged into vesicles again, this time made of a piece of Golgi membrane. Proteins are secreted out of the cell in these vesicles or sent to lysosomes. Lysosomal proteins are packaged in Golgi vesicles coated with a protein called **clathrin**, which then merge with lysosomal membranes and dump their enzyme contents inside. Other proteins are either secreted automatically or wait to be secreted in secretory vesicles that will merge with the cell membrane and dump out their contents only when they receive some other signal (e.g., secretory vesicles store neurotransmitters at the ends of axons and only release the chemicals when an action potential makes its way down to that part of the cell). Proteins with GPI anchors or other types of anchors will also be sent to the membrane in this fashion but will remain anchored into either the cytosolic side or the extracellular side of the membrane when the vesicle fuses into the membrane.

Postranslational modifications are essential to the proper functioning of proteins. I-cell disease is a rare, hereditary disease that results from lysosomal enzymes being secreted outside the cell instead of being sent to the lysosomes. The defect lies in the gene that codes for an enzyme responsible for proper glycosylation in the Golgi. Because certain enzymes are not present in the lysosomes, certain undigested substances build up and eventually kill the cells.

Mutations

Changes in the sequence of DNA bases can either be inherited or caused by **mutagens**, external cancer-causing agents such as UV light, asbestos, and radon. Remember that only mutations that affect germ-line DNA (DNA that is in sperm and egg cells) can be passed down to offspring. Common mutations include:

Point Mutations occur when a single nucleotide base is substituted by another. If the substitution occurs in a noncoding region, or if the substitution is transcribed into a codon that codes for the same amino acid as the previous codon, there will be no change in the resulting amino acid sequence of the protein. This type of point mutation is a "silent" mutation. However, if the mutation changes the amino acid sequence of the protein, the result can range from an insignificant change to a lethal change depending on where the alteration in amino acid sequence takes place.

Frameshift Mutations involve a change in the "reading frame" of an mRNA. Because ribosomes and tRNAs "read" the mRNA in sections of three bases (codons), if a base is inserted or deleted due to faulty transcription or a mutation in the actual DNA, the reading of the resulting mRNA will shift, and this called a frameshift mutation. Base insertions and deletions, particularly toward the start of the protein's amino acid sequence can render the remaining structure nonfunctional as almost every amino acid along the sequence gets changed.

Nonsense Mutations produce a premature termination of the polypeptide chain by changing one of the codons to a stop codon. Beta-thalassemia is a hereditary disease in which red blood cells are produced with little or no functional hemoglobin for oxygen carrying. The different forms of this disease can be produced by a variety of mutations, including point mutations, frameshift mutations, and nonsense mutations.

CHAPTER NINE

Phages and Other Viruses

Viruses defy much of the logic with which we approach our study of life on Earth. Their genomes often strikingly differ from those of other living organisms, their life cycles depend upon their ability to enter and replicate within a living host cell, and they have no cell membranes or organelles as part of their structure.

Virus particles are surrounded by a **capsid**, or protein shell, which can come in a variety of shapes (cones, rods, polyhedrons). Capsids are built out of proteins, many of which have various sugars attached to them, poking upwards from the viral surface. These glycoproteins are used to gain entry into a living cell by binding with surface proteins on the living cell's membrane. Most viral capsids are made up of only one or two different types of protein. In addition, some viruses are able to surround themselves with an envelope of cell membrane as they burst out of a cell they have just infected. This **viral envelope** can help them avoid detection by the host's immune system, since the viral particles resemble (at least on the outside) the host's own cells.

VIRAL GENOMES AND VIRAL REPLICATION

Viral genomes can range from quite small (5–10 genes in all) to fairly large (several hundred genes). The genetic material found within a virus may be held on DNA or RNA, both of which can be found in either a double-stranded or single-stranded state. The nucleic acid can be linear or circular, and although the DNA is always found together in a single "chromosome," the RNA can be in several pieces. Viruses are considered to be **obligate intracellular parasites**, meaning that they must "live" and reproduce within another cell, where they act as a parasite, using the host cell's machinery to copy themselves or to make proteins encoded for by their DNA or RNA. Independent of host cells, viruses conduct no metabolic activity of their own.

Lytic Cycle

Some viruses contain the enzymes they need for replication of their genome, stuffing these proteins into each baby virus as it is produced by the host cell. Others do not travel with their proteins but rather make them only when they are inside a host. The typical viral growth cycle includes the following ordered events:

Attachment and penetration by parent virion (viral particle): The specificity of the proteins coating the capsid of the virus determines the host range of the particular virus, or how many kinds of cells the virus can infect. Those viruses with a wide range of surface proteins (unusual) or with proteins that can bind to many kinds of cell surface receptors (more common) are said to have a wide host range.

Uncoating of the viral genome: As the viral particle penetrates, the cell traps it in a vesicle at which point the virus will break open its capsid. When the vesicle breaks due to viral uncoating, the inner core of the virus with its genetic material can dump itself into the cytoplasm.

Viral mRNA and protein synthesis: DNA viruses replicate their DNA in the nucleus of the cell and use the host cell's RNA polymerases to make their proteins. There are a few exceptions to this (the pox family of viruses that causes smallpox, chicken pox, etc.) that cannot enter the nucleus and actually carry the necessary RNA polymerase with them. RNA viruses replicate in the cytoplasm, sometimes using their RNA as mRNA directly, and sometimes using their RNA as a template for mRNA synthesis. One group of RNA viruses, the retroviruses (e.g., HIV), converts its RNA into DNA first, using the enzyme *reverse transcriptase*. The DNA copy is then transcribed into mRNA by the host's RNA polymerases. Many of the viral proteins that are first synthesized on the host's ribosomes are the ones needed for replication of the genome.

Replication of the genome: All virus particles will first replicate the DNA or RNA they came into the host cell with; the complementary copy that is created is then used as a template for all new copies of the genome. This is analogous to making a negative from a photograph, and using the negative for all subsequent copies rather than using the photograph itself. Replication proceeds using host cell DNA and RNA polymerases as cells normally would.

Assembly and Release: Viral nucleic acid is packaged within a protein capsid, and "baby" virus particles are released from the cell either en masse or slowly. *En masse* ruptures and kills the host cell; *slowly* means the particles "bud" out of the host cell surface and envelop themselves in host cell membrane as they push their way out. Budding in a slow and deliberate fashion will usually allow the host cell to live.

Lysogenic Cycle

The cycle described above is typical of most viruses: It is called a **lytic cycle** (because the host cells usually lyse, or burst, in the end). Some viruses, however, use a **lysogenic** cycle, whereby the viral DNA gets integrated into the host cell's DNA and can remain there indefinitely, allowing viral DNA to be replicated for generations alongside the host cell DNA. The infection continues so that all "offspring" of the host cell carry the viral DNA. This DNA can simply remain a part of the host cell's genome forever. For example, certain bacteria (e.g., *Corynebacterium diphtheriae*, which causes diphtheria, and *Clostridium botulinum*, which causes botulism) secrete toxins that are coded for by viral genes that were acquired from a **bacteriophage** (a virus that infects bacteria). In some cases, however, certain environmental events may cause this integrated viral DNA to begin a full replicative cycle (lytic cycle), accompanied by the production of "baby" viruses and the eventual death of the cell.

BACTERIOPHAGES

The viruses that infect bacteria are known as bacteriophages, or **phages** for short. They are a diverse group of organisms that are best characterized by their "head" and "tail" structure, which is *unique to phages*:

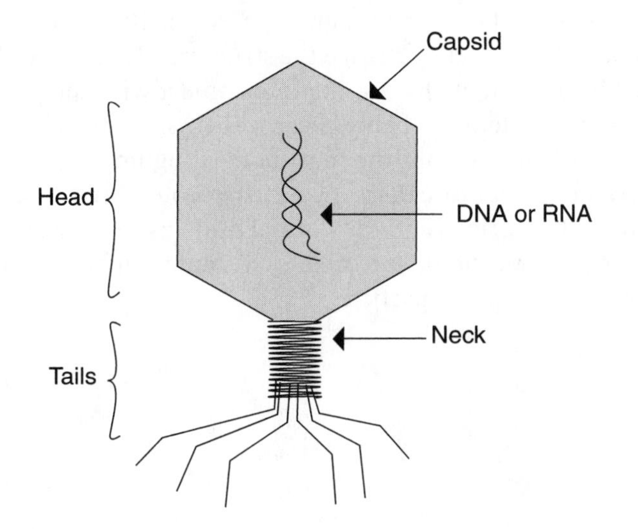

The tails are used to latch onto the host cell surface after which enzymes digest a portion of the cell membrane to insert viral genetic material (either DNA or RNA) into the host cell. The capsid and the tails are left outside of the host cell membrane, differentiating phage infection from that of other viruses. Often, the viral DNA or RNA is rapidly destroyed by powerful bacterial enzymes called **restriction enzymes** (which will be described in detail in chapter 11). These enzymes are a primitive type of immune system in bacteria and chew up foreign genetic material.

For the viral genomes that do survive, they will either cause the host cell to produce new viral particles as described above or can integrate into the bacterial chromosome. Phages that infect bacteria in a lytic way and cause active viral replication are called **lytic phages**, while those whose DNA gets integrated into the host cell's DNA (in a lysogenic fashion) are called **temperate phages**. While integrated, the bacteriophage is known as a **prophage**.

Atypical Viruslike Forms

There are a few exceptions to the typical virus as described above, and these particles (which certainly cannot be considered living organisms according to the standard definition of "life") are infectious to many living cells:

Viroids are viruslike particles that are composed of a single molecule of circular RNA without any surrounding capsid or envelope. The RNA can replicate using a host's machinery, but it does not seem to code for any specific proteins. Despite this, viroids have been *implicated in some plant diseases*, though not in diseases in any other organisms.

Prions are simply pieces of protein that are infectious. They have gained fame through the recent spread of "Mad Cow Disease," and they are connected with diseases that are typically slow to form, taking years before symptoms develop. Despite once being called "slow viruses," prions are not viruses at all and do not contain any DNA or RNA. Recent evidence suggests that prions are infectious because they change the structure of one's own "normal" proteins, a switch not encoded in genes at all, by coming into contact with the proteins. They are not recognized by the immune system, probably because their structure is similar enough to the structure of "normal" proteins. Perhaps the most fascinating implication of studying prions is that their existence shows that more than one tertiary protein structure may form from the same primary structure. Both prions and the normal proteins from which prions derive have identical amino acid sequences, and no traceable DNA mutation has yet been discovered that could lead to the formation of prion particles.

CHAPTER TEN

Immunobiology

The task of fighting infections is quite demanding: The human immune system must be able to respond to disease-causing organisms that are as small as a few nanometers in diameter (e.g., the poliovirus), and as large as 10 meters (e.g., a tapeworm). Since the major cells of the immune system are all approximately 10–30 micrometers in diameter, the cells must be able to fight off a range of organisms from 1,000 times smaller than they are, to a million times larger! Because of this, the immune system is made up of many different divisions that can perform a variety of simultaneous tasks.

A VIEW OF THE IMMUNE SYSTEM: OVERALL PHYSIOLOGY

The immune system can be divided into two major divisions: nonspecific and specific. The nonspecific immune system is comprised of defenses that are used to fight off infection in general, and are not targeted to specific pathogens. The specific immune system is able to attack very specific disease-causing organisms by protein-to-protein interaction, and is responsible for our ability to become immune to future infections from pathogens we have already fought off.

Nonspecific Defenses

The skin and mucous membranes form one part of the nonspecific defenses that our body uses against foreign cells or viruses. Intact skin cannot normally be penetrated by bacteria or viruses, and oil and sweat secretions give the skin a pH that ranges from 3–5, acidic enough to discourage most microbes from being there. In addition, saliva, tears, and mucous all contain the enzyme **lysozyme**, which can destroy bacterial cell walls (causing bacteria to rupture due to osmotic pressure) and some viral capsids. Mucous is able to trap foreign particles and microbes and transport them to the stomach (through swallowing) or to the outside (by coughing or blowing the nose).

Certain white blood cells are also part of the nonspecific defenses. **Macrophages** are large white blood cells that circulate looking for foreign material or cells to engulf, which they do through **phagocytosis**. Macrophages are known as **monocytes** when they are circulating through the blood, and are able to transport themselves through capillary walls and into tissues that have been infected or wounded. Once in the tissues, monocytes are referred to as macrophages, and they use their pseudopodia (like amoebas) to pull in foreign particles and destroy them within lysosomes. Macrophages are called **antigen-presenting cells (APCs)** because of their ability to "display" on their own cell surface the proteins that were on the surface of the cell or viral particle they have just digested. Because macrophages and other APCs do not distinguish between "self"-proteins destined for their cell membrane, and "non-self" proteins that used to be on another organism's membrane, both types of proteins get shipped to the macrophage's cell surface. The advantage of this is that macrophages are able to display to other, more specific, immune system cells the **antigens** (foreign proteins) they have just encountered. That in turn often spurs on a more intensive immune response from these more specific cells (see B and T cells later in this chapter).

Neutrophils are white blood cells that are actively phagocytic like macrophages, but are *not* APCs. Our bodies normally produce approximately one million neutrophils per second and they can be found anywhere in the body. They usually destroy themselves as they fight off pathogens. People who have decreased numbers of neutrophils circulating through their blood are extremely susceptible to bacterial and fungal infections. Other white blood cells that secrete toxic substances without fine-tuned specificity include the **eosinophils**, **basophils**, and **mast cells**.

Encapsulated bacteria and ones that can form **spores**, such as *Mycobacterium tuberculosis* or *Bacillus anthracis*, can resist getting phagocytized by macrophages and can live within macrophages, using them to "decloak." As the bacteria fill the macrophages, the white blood cells burst and release nonencapsulated bacteria into the bloodstream.

The Inflammatory Response

Both basophils and mast cells release large quantities of a molecule called **histamine**, responsible for dilating the walls of capillaries nearby and making those capillaries "leaky." For this reason, histamine is considered a *potent vasodilator*, which can lower blood pressure across the whole body if enough is released at once (blood pressure is maintained by the integrity of the capillary walls). Notice below the amino functional group on histamine, which helps give it its name:

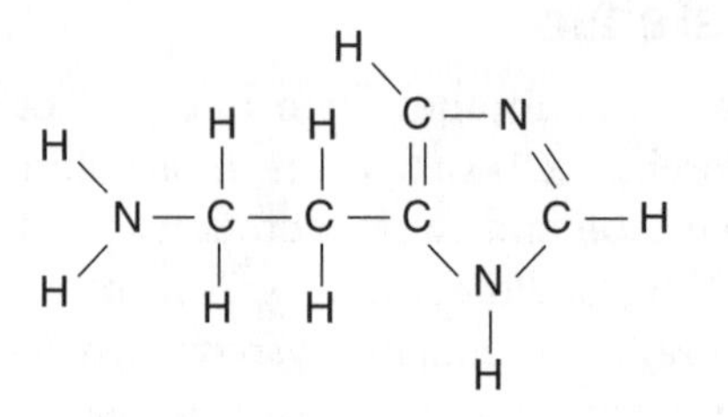

Leaky capillary walls allow macrophages and neutrophils to more easily reach the site of an injury. This nonspecific defense increases overall blood flow to areas of tissue injury, and is responsible for the characteristic redness and heat felt in areas of injury. Basophils and mast cells have large secretory vesicles filled with histamine molecules, but they also release **cytokines**, chemicals that excite specific immune defenses to activate. Other responses to injury that are more systemic (body-wide), rather than local (at the injury site), include fever and increased production of all types of white blood cells.

Antihistamines are compounds that can be taken to limit immunologic reactions. They bind to histamine receptors and shut down their ability to cause leakage at the sight of injury. For certain reactions, particularly allergic ones where the immune system is overreacting to a seemingly harmless substance, antihistamines can be lifesaving since swelling caused by leaky capillaries in the lungs and face area can make breathing difficult or impossible.

Specific Defenses

The major specific defense of the immune system is the use of specialized white blood cells known as **lymphocytes.** These lymphocytes come in two varieties, B cells and T cells. Both are produced by stem cells in the bone marrow after embryologic development has finished. Although T cells mature in the **thymus**, B cells do not. The thymus is essential for "educating" T cells, and T cells that recognize "self" antigens (proteins found on one's own cell surfaces) are killed off, so that auto-immune reactions are less likely to occur. This so-called negative selection results in the development of T cell tolerance, a necessity of the specific immune system. Yet, a positive selection process also exists whereby T cells that do not react to a specific set of glycoproteins called **MHC (major histocompatibility complex)** proteins are killed off as well. The resulting T cells are highly capable of bonding to both self-MHC molecules and a variety of foreign antigens simultaneously, which is essential to the proper working of T cells.

Remember the antigen presenting macrophages that dot their surfaces with foreign proteins they have digested? A certain class of T cells, known as helper-T cells, is able to bind simultaneously to self-MHC proteins on the macrophage surface and the displayed foreign proteins. This combination of signals is needed to activate the helper-T cells.

There are three types of T cells: helper (T_H); cytotoxic (T_C); and suppressor (T_S). While T_H cells are mediators between macrophages and B cells (see sidebar above), T_C cells are able to kill virally infected cells directly. Because virally infected cells display some viral proteins on their cell surfaces, T_C cells can bind to self-MHC proteins and viral proteins on the cell surfaces and secrete enzymes that perforate the cell membrane and kill the cell. Cytotoxic T cells are an essential part of the body's defenses against viruses. T_S cells are involved in controlling the immune response so that it does not run out of control. They accomplish this by suppressing the production of antibodies by B cells. It seems likely that these T_S cells are not a separate class of T cells altogether but rather certain T_H cells that secrete inhibitory chemical messengers (cytokines).

The HIV virus that causes AIDS infects a particular type of helper T cell called a CD4 T cell, so named because of the CD4-type proteins on its cell surface. HIV uses these receptors and others to gain entry into the cell and, once inside, replicates and kills the cell. CD4 T cells form an essential part of the immune response, primarily used to activate B cells to produce antibodies; when they are killed off in large numbers, our immune systems cannot fight off infections adequately.

Keep in mind that T cells cannot detect free antigens; they can only respond to displayed antigens and MHC on the surfaces of cells. When they do recognize a displayed antigen, it is always in combination with a self-MHC protein displayed along with the antigen on the host cell surface. Interactions between T cells and APCs are enhanced by certain proteins that hold the T cell to the APC as it recognizes the antigen-MHC combination. One of these "holding" proteins is the CD4 protein mentioned earlier.

B cells make up about 30 percent of one's lymphocytes, and the average B cell lives only for days or weeks. About 1 billion are made each day in the bone marrow. Every B cell has on its surface receptors that are, in fact, identical in structure to a certain class of **antibody** protein that the B cell is able to produce and secrete into the bloodstream. In other words, the receptors on B cell surfaces are essentially bound Y-shaped antibody proteins that can recognize a specific set of foreign antigens (proteins found on the surfaces of foreign cells and viruses). B cells can be "activated" in one of two ways: either they can come into contact with a foreign antigen that can bind to the B cell surface receptors; or, they can engulf a pathogen, displaying its antigens on the B cell surface much as a macrophage would, and then getting activated to divide by chemicals released by a helper T cell (T_H) that recognizes the foreign proteins sitting on the B cell surface.

For reasons that will be discussed later, both B and T cells each have unique cell surface receptors. That means that *almost every one of the several billion B and T cells in the body is capable of responding to a slightly different foreign antigen.* When a particular B or T cell gets activated, it begins to divide rapidly to produce identical **clones**. In the case of B cells, these clones will all produce antibodies of the same structure, capable of responding to the same invading antigens. B cell clones are known as **plasma B cells**, and can produce thousands of antibody molecules per second as long as they live.

ANTIBODY STRUCTURE AND PRODUCTION

Antibodies are also known as **immunoglobulins** and are *produced solely by B cells.* As mentioned before, each immature B cell has specific antibodies stuck to its surface so that the B cell can be activated if it comes into contact with an antigen that its antibodies are specific for. A single B cell can produce billions of antibodies in its short lifetime.

Antibody Structure and Class

There are five types or "classes" of antibody: IgM (Immunoglobulin M), IgG, IgD, IgE, and IgA. Each class serves a different purpose in the body, and the same B cell can produce each type at different times during an infection. B cells always produce IgM class antibodies first, then *class-switch* to another type, usually IgG, depending upon where the B cell is in the body and the type of antigen it is responding to.

The structure of all five classes is based upon the same four polypeptide chains. Two of the chains are called **heavy chains** and two are **light chains**. Both light chains are identical to each other in amino acid sequence as are the two heavy chains:

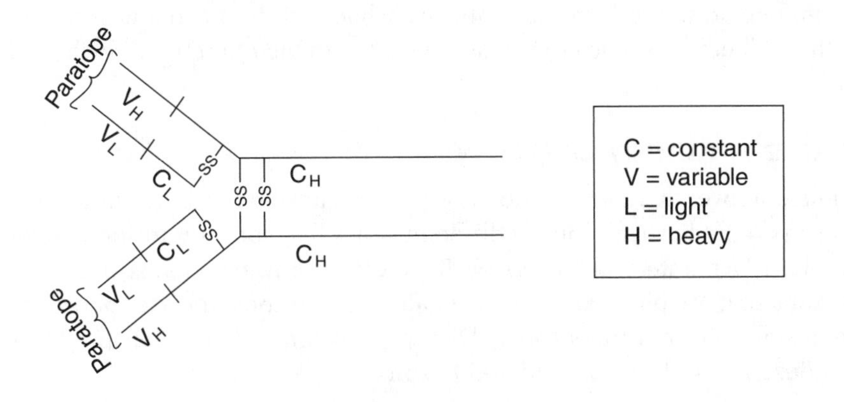

Notice in the diagram above that both the heavy chains and the light chains have **constant regions** and **variable regions.** It is the variable regions that are responsible for the specificity of a particular antibody for a particular antigen. The antigen-binding site, where antibodies can bind to foreign proteins, is also known as the **paratope.** The paratopes are each formed from a combination of variable amino acids in one heavy chain and one light chain. It should be clear from the diagram that each Y-shaped antibody has *two antigen-binding sites.*

Keep in mind that, while the B cells may have individual antibodies bound to their surface for use as receptors, the vast majority of antibodies that are produced are sent out to float freely through the bloodstream. IgG is the simplest type of antibody with a structure approximated above. IgA, however, is a dimer made of two Y-shaped antibodies placed back to back, and IgM is a pentamer of five Y-shaped antibodies placed outward-facing in a circle:

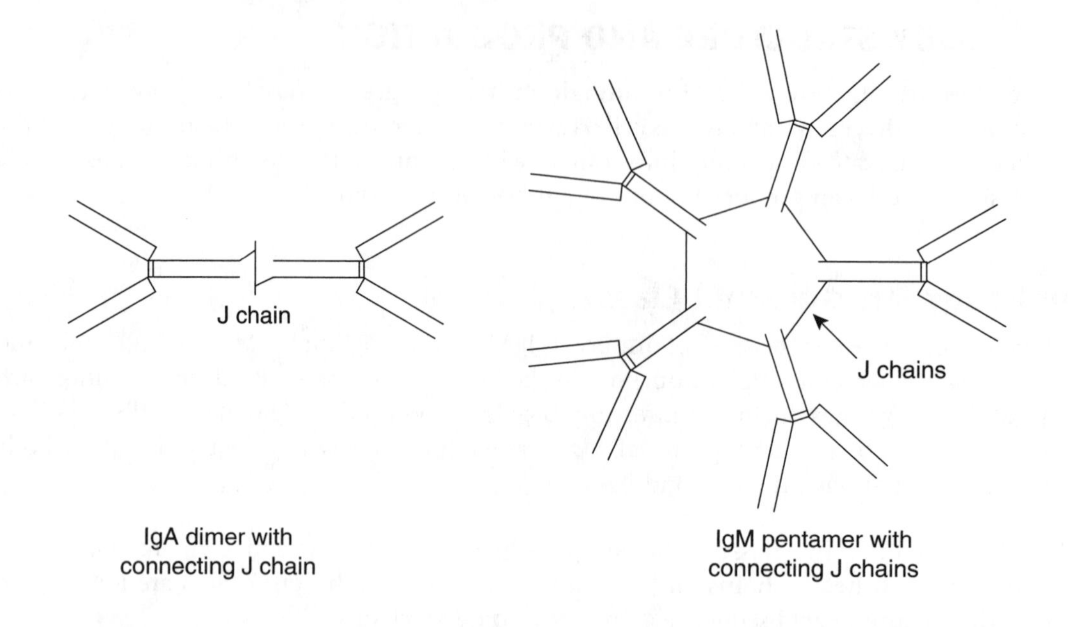

IgA dimer with connecting J chain

IgM pentamer with connecting J chains

Each IgA antibody can attach to four antigens, while each IgM can attach to 10 antigens. The ability of the antibodies to attach to antigens is central to their function, which we'll discuss later.

Generating Antibody Diversity

As mentioned above, there are millions of possible antibodies that one's B cells can make. Certainly you do not have millions of different genes that code for antibody structure, so how is it possible to generate such diversity? To create the heavy chains of an antibody, three separate exons that are physically dispersed along a chromosome are spliced together. These three regions are referred to as the V, D, and J elements. There are hundreds of different V exons, a dozen or so D exons, and 4–5 J exons.

$$5' - V_{H_1} -- V_{H_2} -- V_{H_3} -- V_{H_N} -- // D_{H_1} -- D_{H_N} -- // J_{H_1} -- J_{H_2} -- J_{H_3} -- // \text{Constant genes} - 3'$$

Which exons a particular B cell uses to produce antibodies depend on a random selection process that takes place as the B cell differentiates and matures after being released from the bone marrow. mRNA from random V, D, and J exons is spliced together to create a unique antibody amino acid sequence. Constant genes are added downstream from the VDJ regions, and the final mRNA transcript to leave the nucleus for translation into antibodies is: 5'-VDJC-3'. The light chain is generated by much the same process, except that the chromosome regions which encode light chain regions do not have D exons, so the light chain mRNA transcript is 5'-VJC-3'.

So, how much diversity is generated by this process of rearranging multiple gene segments in random order?

Heavy chain mRNA: (300 possible V genes) × (20 D genes) × (4 J genes) = 24,000 VDJ combinations. If you add an approximate factor of 10 for random loss of nucleotides during gene splicing or from mutations and another factor of 10 for random gain of nucleotides, you get 2.4 million VDJ combinations in heavy chains of antibodies that can be generated by a given B cell.

Light chain mRNA: (100 V genes) × (4 J genes) (factor of 10 for random gain or loss during splicing) = 4,000 VJ combinations.

The total number of possible antibody permutations: 2.4 million × 4,000 = approximately 10 billion unique antibody molecules! That's more than enough to deal with any potential invader. Yet, we still fall sick and die at the hands of microbes because, although we have the potential to create all these antibodies, many events must fall into place in order for each type of antibody to actually be created. Therefore, many possible combinations are never made during a lifetime.

How Antibodies Work

Agglutination/Neutralization: antibodies cross-link adjacent antigen molecules (on bacteria and other organisms) so that these invaders literally get stuck together by the antibodies circulating in the bloodstream. Because each Y-shaped antibody can stick to two different organisms, antibodies can cause the clumping together of many pathogens in a short time. These agglutinated bunches of bacteria or virus particles form large, insoluble masses that are no longer able to invade cells and can be easily engulfed by circulating macrophages.

A word you might see on the GRE is "**opsonization,**" meaning the coating of a foreign cell with antibodies. Opsonization stimulates macrophages to engulf and digest these invaders.

Precipitation: similar to agglutination but used for soluble antigen molecules such as small bacterial toxins which dissolve in the bloodstream. Antibody binding allows rapid phagocytosis and destruction of small proteins by macrophages.

Complement activation: antibodies bound to the surfaces of foreign cells activate a system of twenty different "complement" proteins that circulate in the bloodstream. These proteins are turned on in a cascadelike fashion with each one activating the next, allowing for a great deal of control over the process.

- *Classical pathway*: Requires antibodies bound to antigens. Complement proteins bridge the gap between two adjacent antibody molecules and use a protein complex called the membrane-attack complex to lyse the cell membrane of the invader. Complement proteins also activate mast cells to release histamine, which brings more blood cells to the area.

- *Aclassical pathway*: occurs independently of antigen-antibody binding. Cell surface molecules of many bacteria, yeasts, viruses, and protozoan parasites can cause membrane-attack complexes to form without the help of antibodies.

Hybridomas and Monoclonal Antibodies

Antibodies that arise in the natural course of fighting many pathogens are considered **polyclonal**—that is, they are produced by several different clones of plasma B cells and cover a wide range of specificity. Antibodies arising from a single clone, a single B cell that has rapidly divided into identical B cells, are called **monoclonal**, and they have important scientific uses. The specific nature of antibody binding makes them attractive research targets for disease cures. Imagine for example being able to target specific cancer cells with an injection of antibodies that seek out and destroy only those cancer cells. Such antibodies can be made in the lab by fusing a myeloma cell (a cancerous, always-dividing B cell) with an antibody-producing cell from a mouse. The resulting cell, called a **hybridoma** because it is a hybrid cell from two different species, can produce almost unlimited quantities of a particular monoclonal antibody, which can be used in research.

HUMORAL VERSUS CELL-MEDIATED DEFENSES

The specific arm of the immune system, which uses B and T cells to target individual microbes, can work either by secreting antibodies into the bloodstream (considered one of the body's "humors") or by directly killing cells. While antibody secretion (humoral immunity) is the job of the B cells, direct killing of cells and overall activation of the immune system (cell-mediated immunity) is the job of T cells.

The Humoral Response

The first time someone is exposed to an antigenic stimulus, many events must take place before a protective antibody response is made. The period after exposure to a pathogen, but before helpful levels of antibodies have been made by B cells, is called the "**lag period**." During this period, APCs such as macrophages and neutrophils process and display antigens to T_H cells, which rapidly grow and divide. These T cells activate B cells, some of which have also contacted antigens, and these B cells grow and divide as well.

After the first exposure to a microbe, the lag period lasts 7-10 days before enough antibodies are present in the blood to noticeably slow the infection. Within a week or so, the infection will have likely subsided. If the person is exposed to the same antigen a second time, there is a very quick **secondary response** (with a lag period of 1–4 days) and levels of antibody in the bloodstream typically reach much higher levels than they did in the **primary response**:

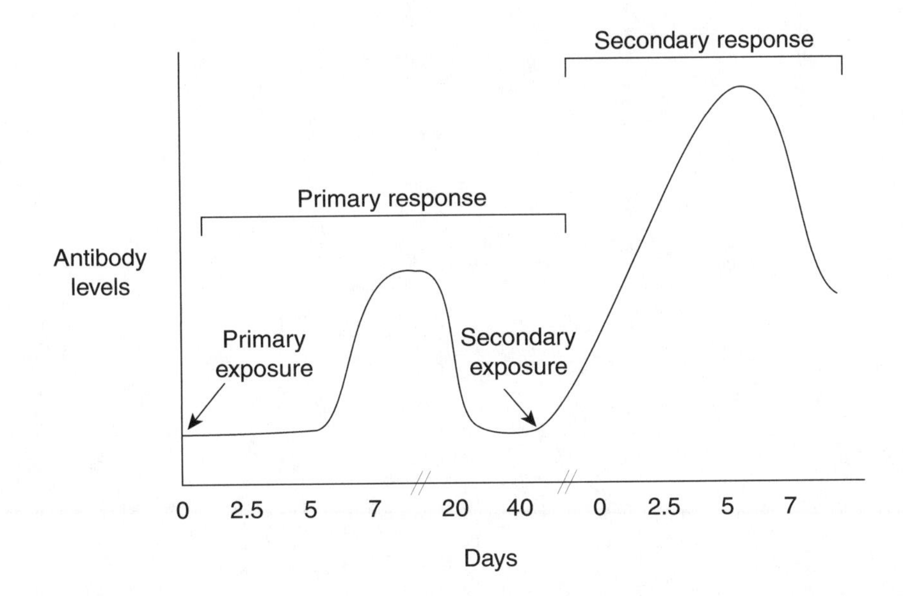

The reason for this is that certain antigen-specific "memory cells" remain after a primary infection. Both B and T cells form **memory** cells. While a typical B or T lymphocyte may live only for days or weeks, memory B and T cells can live for decades. The reasons for this are unclear, but the formation of memory cells is the basis of **immunity** and forms the concept behind **vaccination**. Small doses of antigen given as an injection or swallowed allow the body to recognize and form a primary response against the antigen so that upon actual exposure to the pathogen carrying that antigen the body will mount a quick and sufficient response.

The Cell-Mediated Response

The timing of the T cell response follows almost the same pattern as shown above in the graph of the humoral response. After primary exposure to an antigen, specific T cells rapidly divide and form clones of identical cells. Upon secondary exposure, memory T cells left over from these clones will divide much more rapidly than they did at the first exposure. The cell-mediated response severely limits the ability of viruses to proliferate within cells, since cytotoxic (killer) T cells will destroy any cells harboring viruses. Helper T cells participate in antigen recognition and B cell activation, and natural killer cells (related to cytotoxic T cells) destroy infected and cancerous cells directly.

Altogether, nonspecific defenses such as skin, fever, and macrophages along with specific defenses such as B and T lymphocytes and the complement system make up a highly efficient and adaptable machine that serves to protect the body from a wide assortment of invaders.

Because the "lag period" can be unacceptably long for some pathogens, doctors often provide preformed antibodies (usually made in horses or chickens) to people who have been exposed to a microbe or who face probable exposure. This **passive immunity**, to be distinguished from the **active immunity** that forms when your B cells make their own antibodies, is temporary. Individuals who travel abroad are often given a shot of "gammaglobulin" before they leave. This mixture of preformed antibodies to several tropical diseases is like a soup of temporary protection, though within a week or two the antibodies will disintegrate.

CHAPTER ELEVEN

Biotechnology

In the 1950s, scientists realized that foreign DNA was often broken up by something in bacteria. This observation remained just that, a bizarre observation, for nearly two decades until scientists at Johns Hopkins University in the 1970s discovered the first **sequence-specific restriction endonucleases**. Remember from the DNA replication chapter that *endo*nucleases cut DNA from *within* a strand rather than at the end of a strand, and these enzymes found in bacteria can do just that. Restriction enzymes were the first major step toward manipulating DNA. The science of biotechnology, whose techniques are discussed here, is all about manipulation of DNA, cloning (copying) specific pieces of DNA or entire genomes, DNA analysis, and creating transgenic organisms. The possibilities are endless for drug design and disease therapies.

RESTRICTION ENZYMES AND DNA SEQUENCING

Restriction enzymes are a special class of protein that recognize *specific* DNA sequences and cleave them. As mentioned before, they are essential to bacteria in breaking down foreign, usually phage, DNA. Each enzyme gets its name from the bacterial strain in which it was first found. For example, the enzyme *EcoRI* (pronounced "Eco-R-one") was the first enzyme found in the bacterium *Escherichia coli*. With few exceptions, restriction enzymes cut at *palindromic* DNA sequences, complementary sides of the double helix that have the same sequence of bases when read 5' → 3' on their respective sides. They are very specific for the sequences where they cut. Notice in the diagram below that EcoRI cuts at a palindrome of the bases 5'-GAATTC-3'.

Restriction Enzymes

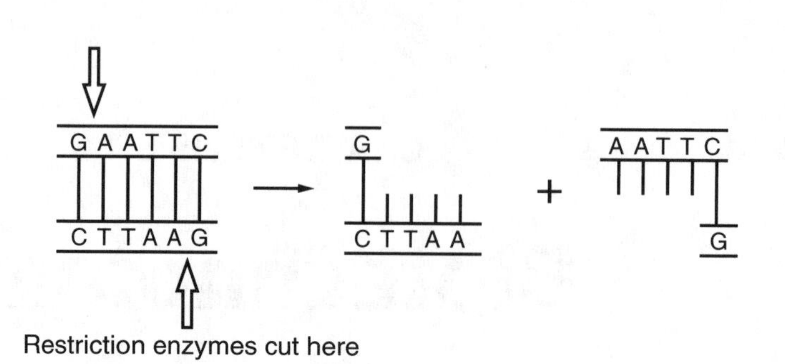

Some enzymes cut the sequences asymmetrically so that overhanging ends exist, as happens when EcoRI cuts GAATTC sequences. These overhangs are referred to as "**sticky ends.**" The hydrogen bonds between complementary bases are broken, so bases are left that want to hydrogen bond with other bases but cannot do so until they are brought near complementary bases. Other enzymes cut and leave "**blunt ends,**" without these overhangs. Sticky ends are hugely important to DNA technology because they allow DNA from different sources to be cut with the same restriction enzymes and stuck together using DNA ligase. This permits the manufacture of **recombinant DNA**, the combination of genes from different organisms, which can then be placed into a "**transgenic**" organism.

Plasmids and the Uses of Restriction Enzymes

Recall that plasmids are small, circular pieces of bacterial DNA independent of the main bacterial chromosome. These autonomously replicating mini-chromosomes are several thousand base pairs each in length, and there are a variable number of plasmids per bacterial cell. To get foreign DNA into a bacterial plasmid, you can *cleave the foreign DNA of interest by the same restriction endonuclease that is used to cleave the bacterial plasmids* you use and then place them both together in a test tube. Some of the cleaved plasmids will take up pieces of the foreign DNA, and the plasmids can then be introduced into bacterial cells growing in a petri dish or other medium. This form of bacterial **transformation** allows you to introduce small pieces of DNA into bacteria so that the bacteria can start to transcribe the foreign gene(s) of interest.

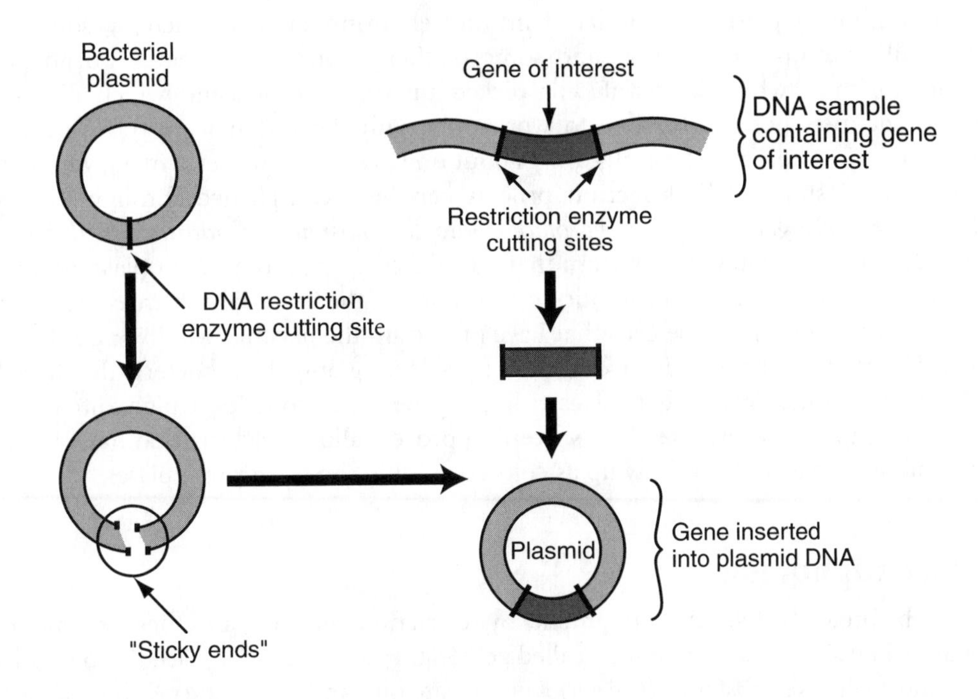

These plasmids, now called "**recombinant plasmids**" because they contain some bacterial genes and some foreign genes, can be copied by bacteria and exchanged from cell to cell. The plasmids are considered "DNA **vectors**," because they are the means by which foreign DNA can be introduced into the bacterial cells. There are other vectors, such as phages, that can transfer foreign DNA into bacterial cells, yet the transfer of DNA using plasmids can be carefully controlled in the lab. These techniques are extremely useful because they allow you to do the following things:

a. *Figure out the function of an unknown gene*: when placed into a bacterial cell, an unknown foreign gene will be transcribed and proteins made from it. These proteins can be analyzed and studied in a controlled lab environment. It would be much more difficult, for example, to study the function of a human gene in a living human with all the other life processes going on at the same time.

b. *Produce clinically useful substances*: bacteria have been used for years to produce human insulin by transforming bacteria with recombinant plasmids carrying the human insulin gene. These bacteria make enough insulin to supply diabetics with the drug they need to live. Bacteria have also been used to make tPA, or tissue plasminogen activator, a drug crucial for controlling brain damage in stroke victims.

c. *Give bacteria new functions*: bacteria have been "engineered" to digest oil by transforming them with a gene that produces enzymes to break down petroleum products. These bacteria can be made by the billions and poured out in the ocean over oil spills to help remove the toxic oil.

When recombinant plasmids are placed into an environment with bacteria, some of the bacteria will take up the plasmids and some of them won't. For research purposes, it is necessary to know which bacterial cells picked up the plasmids and which did not. In addition, scientists need to be able to work only with those that were transformed and remove those that were not; typically, only about one percent of bacteria in a given medium will pick up plasmids. This **selection** process can be accomplished through the use of *antibiotic-resistance genes that can be placed onto the plasmids in addition to the genes of interest.* That means that any bacterium that has picked up a recombinant plasmid not only gets a foreign gene but also an antibiotic-resistant one. If the bacterial culture is then exposed to a particular antibiotic, those cells that have picked up the plasmid will live and the others will die. This is an extremely effective method for selecting only those bacteria that have been transformed. Another method involves using a gene that produces a pigmented protein, rather than antibiotic resistance. This **screening** process allows bacteria that have picked up the recombinant plasmids to show up as colored colonies on the growth plates.

Gel Electrophoresis

DNA can be broken into smaller fragments by restriction enzymes, and these fragments can be separated by size using a technique called gel electrophoresis. DNA placed into small wells carved into a jellolike substance (the gel) will migrate into and across the gel when an electric field is applied at the end opposite the wells:

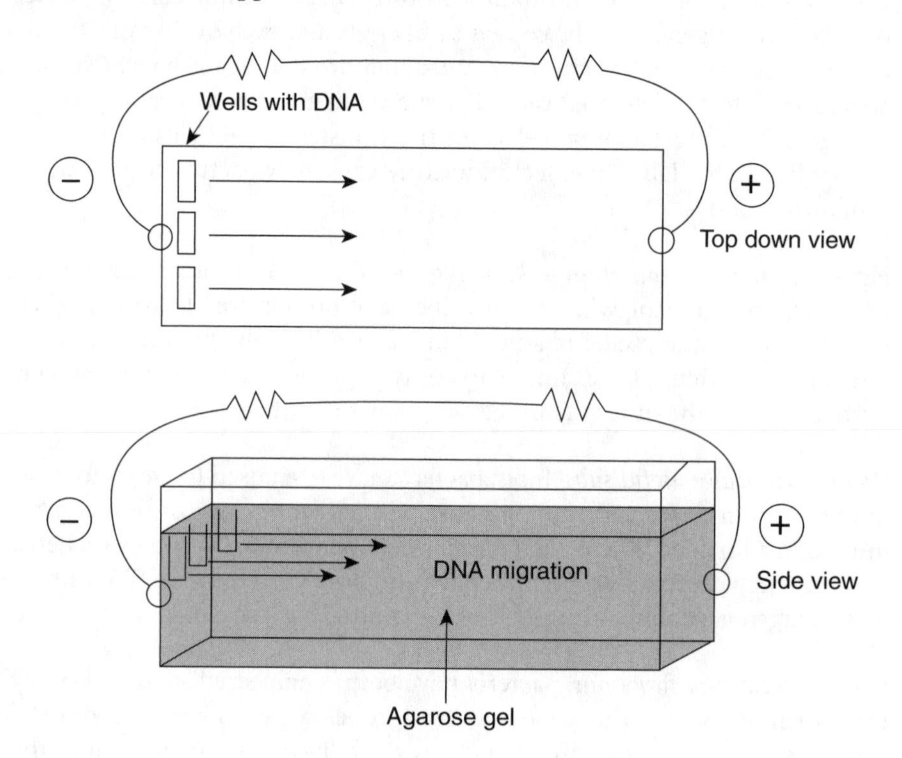

Because DNA is a negatively-charged molecule overall, it will be pulled toward the positive terminal of a circuit placed across the gel. The gel is made of **agarose**, a galactose polymer that has long, fine fibers of polysaccharide running through it. These fibers trap the fragments of DNA as they are dragged across the gel by the current. *Longer fragments get caught closer to the wells and smaller fragments can migrate further toward the positive end of the gel.* When the current is shut off, the DNA fragments can be stained to determine their positions relative to each other, and particular fragments can be removed from the gel for further study. Gel electrophoresis is used for many molecular genetics techniques including DNA sequencing, whereby the actual sequence of nitrogen bases in a gene or segment of DNA is determined.

DNA Sequencing

The most common method for sequencing DNA is known as the Sanger method. An unknown piece of DNA is heated to make it single-stranded. It can now be used as a template to match up complementary **oligonucleotides** (short sequences of DNA). The unknown piece of DNA is first copied multiple times using the **polymerase chain reaction** technique (PCR) discussed later. With primers that are radioactively or fluorescently labeled so that they can later show up easily on a gel, multiple copies of the unknown DNA are placed into *four* test tubes along with DNA polymerase (necessary for copying) and A, C, G, and T nucleotides in every tube. You can imagine now that the primers will allow the free nucleotides in each tube to line themselves up as strands of DNA complementary to the unknown strand. Yet, in each of these tubes, there are also *dideoxy*nucleotides, which look just like the normal A, C, G, and T bases but have no oxygen at all on the 3' carbon of their sugars, so that no further nucleotides can bind to them after they settle into a growing strand. That means that a complementary strand of DNA will be built along the unknown templates in each tube *only until* a dideoxy (dd) nucleotide comes along and stops the building. In tube 1, there are ddATPs (dideoxyadenosines), so that all oligonucleotides that are made in tube 1 get stopped with an A at their 3' end. In tube 2, there are ddCTPs, so that all oligonucleotides made in that tube terminate at a C. And so on for tubes 3 and 4. Because the synthesized strands get stopped at random places, you end up with lots of small DNA stands that have at their 3' end a nucleotide that is complementary to a nucleotide on the unknown strand.

For example, let's say that the unknown strand of DNA had a sequence with the base thymine 4, 6, and 9 bases away from the start (3' end) of the DNA segment (i.e., its sequence was 5'-xxxxxTxxTxTxxx-3'). In tube one, with the ddATPs, you'd end up with segments that were 4 base pairs long, 6 base pairs long, and 9 base pairs long, but not others—the synthesis is stopped by ddATPs at every instance that a ddATP binds in a complementary fashion to a T in the unknown strand. By using a gel to separate all the segments that come out of the tubes with dideoxynucleotides, you can literally read the sequence of the unknown piece of DNA directly off the gel. Take a look at the figure below and, if you need to, look at the figure again while rereading the above information.

The GRE will almost surely ask about Sanger DNA sequencing. Study the diagram in this section carefully to learn how to read a gel and determine a DNA segment's sequence.

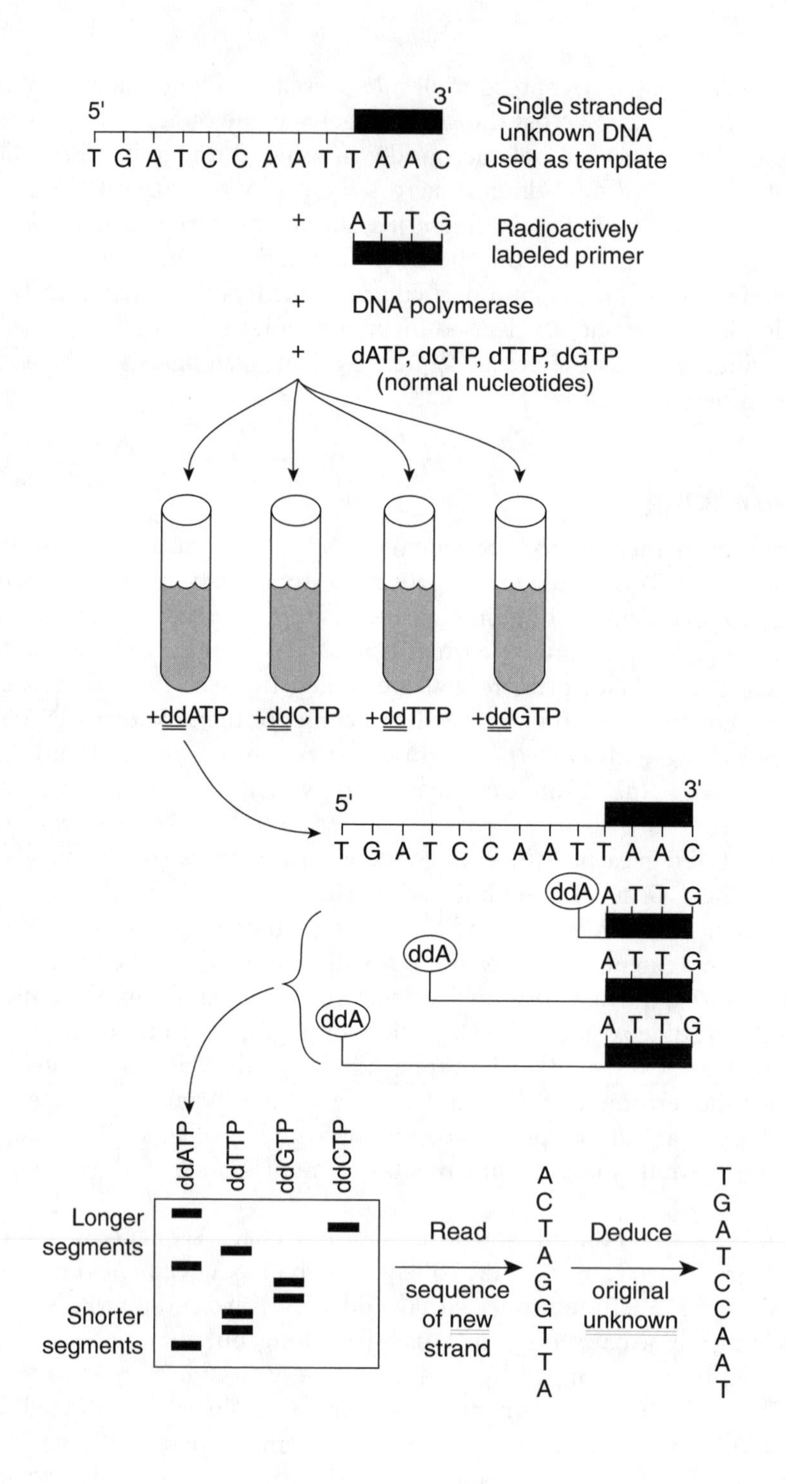
5'
3'
T G A T C C A A T T A A C
Single stranded unknown DNA used as template
+
A T T G
Radioactively labeled primer
+
DNA polymerase
+
dATP, dCTP, dTTP, dGTP (normal nucleotides)
+ddATP
+ddCTP
+ddTTP
+ddGTP
5'
3'
T G A T C C A A T T A A C
ddA
A T T G
ddA
A T T G
ddA
A T T G
ddATP
ddTTP
ddGTP
ddCTP
Longer segments
Shorter segments
Read
sequence of new strand
A C T A G G T T A
Deduce
original unknown
T G A T C C A A T

The Southern Blot and Probing

One of the most important and widely used molecular genetics technique is called the Southern blot, named after its developer, Ed Southern. It is used to "probe" DNA for certain sequences through the use of **nucleic acid hybridization**. Let's say you are a scientist looking for a particular gene or part of a gene whose nucleotide base sequence you know. In front of you is an organism's genome or part of a genome, and you'd like to know whether this sequence of interest is in the DNA you have in front of you. Your first step is to design a **radioactive probe**, or piece of DNA with radioactive nucleotides (usually labeled with phosphorus, ^{32}P), that is *complementary* in sequence to the DNA sequence you are searching for.

As an example, suppose you have a probe with the base sequence 5'-AGCCG-3'. Which of the following DNA fragments will this probe bind to (hybridize to) in a complementary fashion?

A. 3'-AAACAAGGAATCT-5'

B. 3'-CCGGAAATC-5'

C. 3'-AACTCGGCATAATACGGCAA-5'

The only fragment that has the sequence *complementary* to your probe, 3'-TCGGC-5', is choice (C).

Once you have cut up with a restriction enzyme of your choice the DNA you are examining, gel electrophoresis can be used to separate the DNA fragments. The DNA is stained and the gel is photographed to record the migration of larger and smaller fragments. The gel, which has the DNA caught in it, is now bathed in an alkaline solution to denature the DNA fragments and the DNA is transferred from the gel onto an absorbent nylon membrane. The DNA fragments can now be washed with a solution containing your radioactive probe, which will now bind in *complementary* fashion (as illustrated above) to any DNA fragments from the gel that contain the sequence of interest. The DNA can be examined so that the "glowing" radioactive probes illuminate fragments they have hybridized to. Southern blotting, then, is an *excellent means of finding small amounts of DNA to work with from a much larger sample.*

Probing can be used in a variety of research. In medicine, probes can be used to screen individuals for the presence of certain genes. Suppose you identify the sequences of both functional genes and disease genes (which the Human Genome Project and related ventures have aimed to do). Molecular probes could be designed from base sequences of both "normal" and "abnormal" genes and used to screen individuals' DNA for the presence of these genes.

RFLPs AND PCR

Because restriction enzymes cut DNA at very specific sequences, they can be used to compare the DNA of different organisms or to screen for the presence of various alleles. Once fragments of DNA are isolated, they can be rapidly amplified for use in experiments by a technique known as PCR, or polymerase chain reaction.

RFLP

If a restriction enzyme were to cut DNA, say, from two family members, the cuts would be made in similar places because the DNA of these individuals is extremely similar. Yet, slight differences exist between the individuals' DNA and this is the reason that the cuts would not be made in exactly the same place for both people. One individual, for example, might have a mutation at a certain point on a chromosome where the other individual had an EcoRI cutting site (GAATTC). If EcoRI were used to cut up DNA from these two people, the size of the fragments created would differ because cutting sites differed between the two stretches of DNA. These differences in the length of the fragments made by restriction enzyme digestion of two samples of DNA are called **restriction fragment length polymorphisms**, or RFLP for short.

RFLP analysis is used not only to compare DNA across species, but also to make a DNA "fingerprint" of an individual's DNA. This is widely used in criminal cases where DNA left at the scene can be cut with restriction enzymes and compared to restriction enzyme-digested DNA taken from an accused individual. If the fragments match when run on a gel, it is practically certain that the individual was at the scene of the crime. In addition, RFLP analysis can be used for paternity cases to clarify who is the father of a child. Using restriction fragments of the mother, child, and possible fathers, DNA fragments can be run on a gel and compared for similarity.

RFLPs can also help diagnose the presence of a disease gene on a chromosome, often because disease-causing alleles are shorter or longer than the normal allele. As seen in the diagram below, with the letter "R" representing a site where a particular restriction enzyme would cut, a diseased individual would show different RFLPs than a normal one, since the number and length of restriction fragments differs between them.

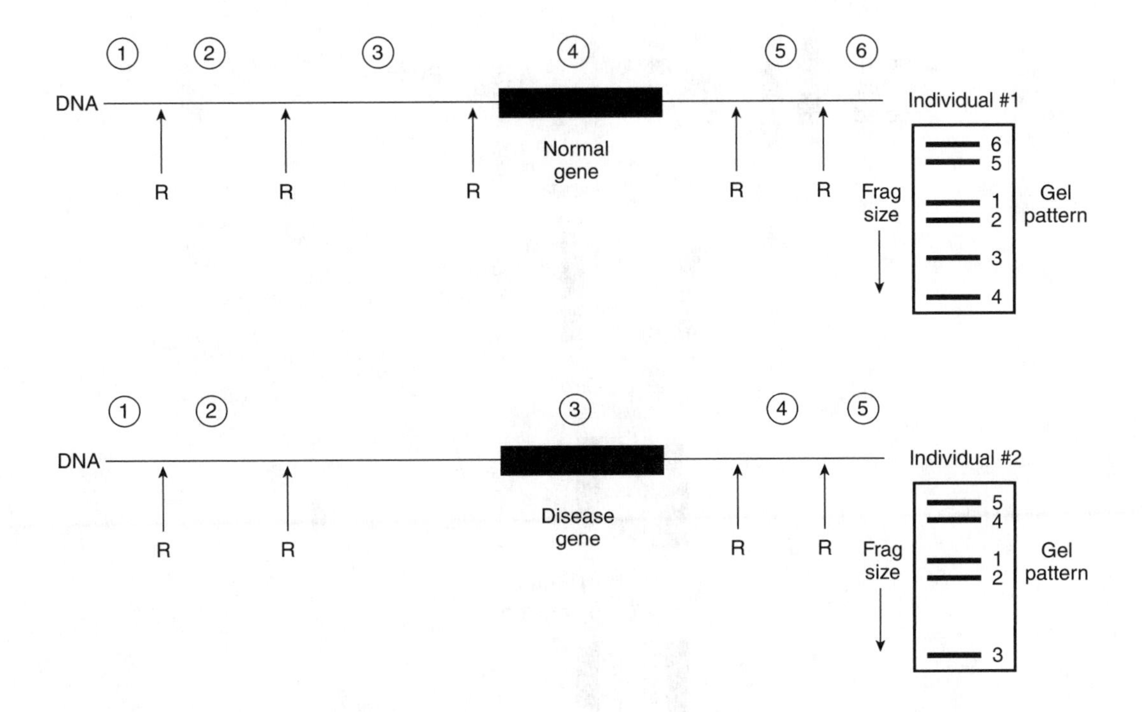

Notice from the figure above how the banding pattern of the restriction fragments on a gel would be different toward the bottom of the gel where individual #1 has two distinct bands of DNA and individual #2 has only one. This is an indication, in this case, that individual #2 has the disease allele if #1 is known to be normal.

Polymerase Chain Reaction

PCR is a way to rapidly amplify a piece of DNA so that there are many identical copies of this DNA to work with. There are slower way to copy given stretches of DNA, such as putting the DNA into bacteria using plasmids and letting the bacteria copy the DNA, yet PCR does not need a living organism to copy DNA and it is much faster and more reliable than any other methods. PCR starts with the *double-stranded* DNA sequence of interest that is to be copied. A supply of free nucleotides (A, C, G, and T) as well as DNA polymerase and primers are added to a test tube containing the DNA. The DNA is heated to separate its strands and then cooled to let the primers hydrogen bond to the ends of the strands. DNA polymerase builds on the primers and complementary strands are formed. Now, the amount of DNA has been doubled. This cycle of heating, cooling, and building complementary strands is repeated over and over (it takes only several minutes per cycle) as the amount of identical DNA grows exponentially.

The primers made for PCR are chemically synthesized with complementary DNA—not RNA—before PCR starts. They are made based on the ends of the sequence to be copied, so the only limitation to PCR is that you have to know a small piece of the gene(s) to be copied in order to synthesize the primers before you start.

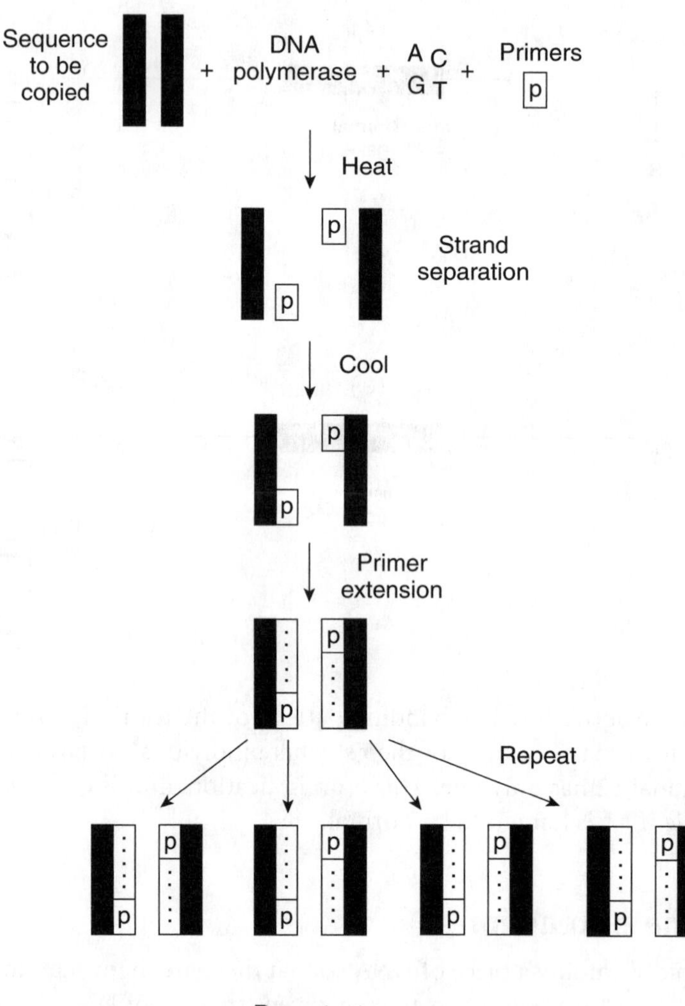

CLONING AND OTHER APPLIED MOLECULAR TECHNIQUES

Transgenic "Pharm" Animals

As we have discussed, bacteria can be used to produce proteins of interest when their genes are inserted into plasmids used to transform the bacteria. Yet transgenic animals, such as cattle, can be used as "bioreactors" for the production of recombinant proteins with widespread applications in the pharmaceutical industry. Transgenic goats, pigs, and cattle can be produced using microneedles to *inject DNA directly into fertilized egg cells*, some of which will take up this DNA and integrate it into their chromosomes. Although few products have been produced at this time, the idea is that transgenic farm animals could make the proteins necessary for some human therapies. For example, researchers have created transgenic sheep that produce human Factor IX, a blood-clotting factor used as a drug for hemophiliacs. The gene for Factor IX protein was placed onto a circular piece of DNA along with a promoter sequence that is specific for mammary gland expression (certain tissues have what are known as **tissue-specific promoters** guaranteeing expression of a particular gene in only one type of tissue). This allowed the transgenic sheep to shed the Factor IX protein directly into the milk they produced in their mammary glands. In theory, humans needing the protein could drink this milk.

Cloning

Several years ago, Scottish researchers cloned **Dolly**, the sheep, who was genetically identical to a 6-year old ewe (female sheep) from whose mammary glands DNA had been taken in order to be implanted into an egg. The technique of cloning, forming an identical copy of an organism, is fairly straightforward: Using an electric pulse, a cell from the mammary gland of the donor ewe was fused to an unfertilized sheep egg cell that had had its nucleus removed with a tiny needle. The nucleus from the diploid donor cell became the nucleus for the egg cell and because the egg cell was now diploid, it was implanted into a surrogate mother without any fertilization, who carried the egg cell to term. Dolly was the only sheep that survived out of 30 embryos that were created in this fashion. For the GRE, you should be familiar in a general sense with the "nuclear transfer method" of cloning as pictured below:

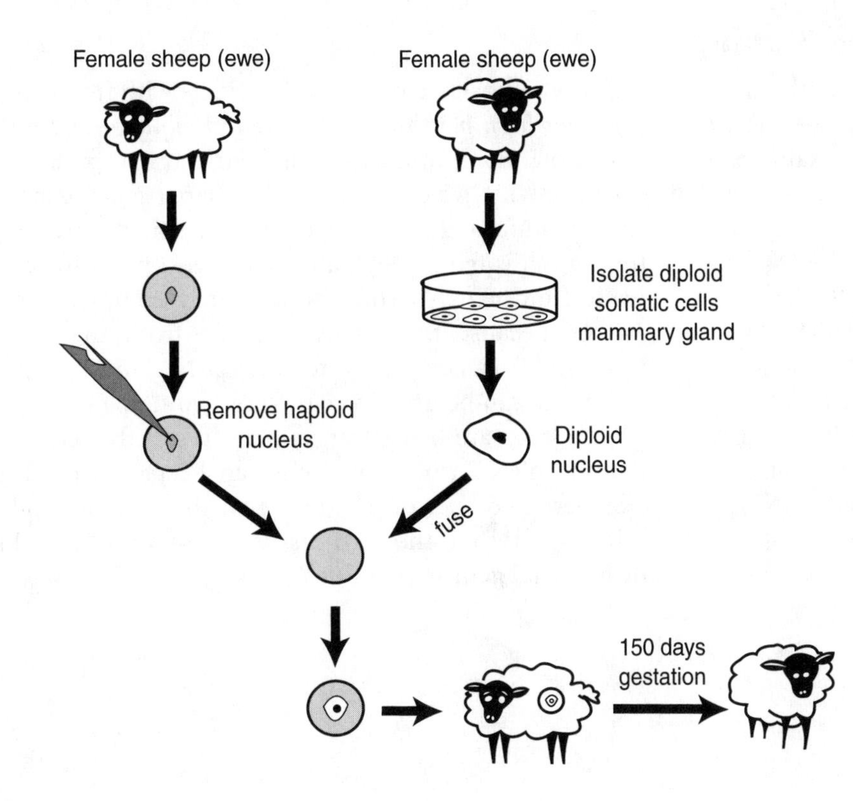

Genomic Libraries and cDNA

A genomic library is a large collection of bacterial colonies or virus clones, each containing copies of a particular foreign gene or DNA segment. Some genomic libraries are complete, meaning that they contain bacterial colonies, each on separate plates, which taken together represent the entire genome of a particular organism. Libraries are important for researchers who would like to work with one particular gene from an organism. They are constructed by creating pieces of cDNA (complementary DNA) and cloning the cDNA into bacterial cells that copy rapidly to make many copies of the cloned DNA. cDNA is, in fact, DNA constructed using the mRNA that codes for a given protein. Scientists work backward from this mRNA to create the DNA that matches it—hence, the term cDNA or complementary DNA. The reason for using cDNA is that a typical gene contains non-coding introns that are spliced out when the mRNA is made from that gene. *cDNA is the gene without the introns* because it has been created backward from the mRNA that has already been processed by the cell. In other words, cDNA in a genomic library gives researchers the pure gene of interest, not one that has to be tinkered with. cDNA is usually used by researchers instead of regular DNA because there are no introns to work with, and radioactive probes (as discussed above) are far more accurate and reliable if there are no introns getting in the way.

cDNA can be created from mRNA by using the enzyme reverse transcriptase to make double stranded cDNA from any mRNA transcript.

Northern Blotting

Similar to Southern blots, which use radioactive probes to hybridize with fragments of DNA separated by gel electrophoresis, *Northern blotting is the same technique to spot mRNA*. This is used, for example, to find out how much mRNA a particular cell is transcribing from a given gene, i.e., the level of genetic activity in a cell with regard to a particular gene (the more RNA produced, the more that gene is getting expressed). mRNA from a cell is fed through an agarose gel and radioactive probes made from cDNA (which is absolutely complementary to the mRNA because the cDNA is made from the same mRNA transcripts) can hybridize to this mRNA and light it up. This technique can be used in cells from two tissue types to discover which genes are being transcribed differently between the two tissues. mRNA from both cells can be placed on a gel and a Northern blot performed to assess differential production of mRNA between the two tissue cells (otherwise known as **differential gene expression**).

Because one typically needs a lot of RNA to do a Northern blot, the RNA can be amplified by PCR if it is first turned into cDNA using the reverse transcriptase enzyme. This technique is called RT-PCR (reverse transcriptase-PCR) and can produce ample cDNA from just a small bit of mRNA.

Reporter Genes

It is sometimes difficult to monitor levels of gene expression in a cell simply by measuring mRNA production or even production of a particular protein. Some genes just don't produce that much mRNA or protein even though they are active. So when scientists place a gene into a cell and want to make sure that the gene is actually producing mRNAs, they often fuse a "**reporter gene**" to the regulatory region of the inserted gene. Because of their placement, these reporter genes are controlled by the same promoter region as the gene of interest, but reporter genes produce copious amounts of easily measured protein. Some reporter genes produce a green, highly fluorescent protein that shines under special light; others produce a blue color when exposed to lactose sugar in their environment.

Reporter genes also allow scientists to add various mutations into promoter regions of genes so that they can see the effect of these mutations on gene expression. Mutations affecting the expression of a gene will also affect expression of a reporter gene fused nearby, which can be easily measured.

PRACTICE QUESTIONS FOR PART II

1. Plasmids (circular genetic elements present in bacteria) can

   (A) act like transposons.
   (B) be found in all eukaryotic cells.
   (C) replicate only when other plasmids replicate.
   (D) never move from cell to cell.
   (E) only be made from RNA.

2. It has been observed that neighboring sections of a polypeptide chain can take shape in such a way as to block the access of water to certain areas of the polypeptide surface. This indicates that

   (A) water bonds to protein surfaces using hydrogen bonding.
   (B) different conformations of the same protein may alter the protein's specificity or reactivity.
   (C) enzyme active sites do not work properly when exposed to water.
   (D) enzyme-substrate interactions depend upon exposure to water only on select areas of the tertiary structure.
   (E) the interior of the protein is entirely hydrophobic.

3. Reactivity of a skeletal muscle cell to the neurotransmitter acetylcholine may be increased by

   (A) increasing the concentration of acetylcholinesterase at the neuromuscular junction.
   (B) increasing the amount of myosin-II present within the skeletal muscle fibers.
   (C) downregulation of acetylcholine receptors on the surface of the T-tubule system.
   (D) increasing the number of calcium ions stored within the sarcoplasmic reticulum.
   (E) strengthening the attachment of the muscle to nearby bones.

4. It is important that certain free ribosomes bind to the outer surface of the endoplasmic reticulum (ER) in order to complete their protein synthesis because

   (A) the ER membrane will break down without the presence of numerous ribosomes.
   (B) it allows for the synthesis of certain proteins to be completed in the cytosol.
   (C) it prevents the possibility that the synthesis of certain proteins, such as lysosomal hydrolases, would go to completion in the cytoplasm.
   (D) mitochondrial ribosomes must transcribe proteins encoded for by mitochondrial DNA in this manner.
   (E) posttranscriptional modifications to the mRNA, such as the addition of a poly-A tail, could not take place outside of the ER lumen.

5. The hybridization kinetics of DNA double helices are closely related to the sequence of nitrogenous bases on those helices. At temperatures below the melting point (T_m) of complementary DNA strands, most DNA is

   (A) single-stranded due to the denaturation caused by cooler temperatures.
   (B) double-stranded due to the strength of hydrogen bonds.
   (C) hybridizing to form longer strands for added stability.
   (D) made of highly complex base sequences that simplify as the temperature rises.
   (E) able to absorb more ultraviolet light than at temperatures above T_m.

6. Heat-stable DNA polymerases that have 3' → 5' exonuclease activity would be most useful during polymerase chain reactions (PCR) for which of the following reasons?

   (A) The enzymes would stabilize the hydrogen bonds between the bases of each strand.
   (B) The enzymes do not break down during the gel electrophoresis process that typically takes place after PCR in order to isolate DNA segments of interest.
   (C) These polymerases replicate DNA much faster than normal polymerases that do not have exonuclease activity.
   (D) The amplification of a PCR product depends upon the attachment of PCR primers, which anneal to complementary sequences on the DNA template.
   (E) PCR uses heat to denature DNA and the rapid rate of DNA replication often results in copying errors.

7. Although all B-lymphocytes start out with antibody molecules bound to the outer surface of their cell membranes, an immune response often results in the secretion of free antibodies by B-cells rather than the attachment of antibodies into the plasma membrane. The mechanism causing this change most likely involves

   (A) a change in the types of lipids used to build new plasma membrane after B-cell activation.
   (B) alteration of the heavy and light polypeptide chains used to build the antibodies.
   (C) changes in RNA processing resulting in the addition or removal of certain signal sequences on the RNA coding for antibody proteins.
   (D) mutations in the V, D, and J regions of the chromosomes that code for antibody protein structure.
   (E) binding of B-cells to helper T-cells within lymph nodes.

8. The membrane-spanning regions of transmembrane proteins are frequently α-helical. This is best explained by the fact that

   (A) the α-helix is the native conformation with the greatest stability out of all possible 3-D configurations of transmembrane proteins.
   (B) β-pleated sheets are alternative conformations only when the membrane possesses many cholesterol molecules for added reinforcement.
   (C) the polarity of the space between the lipid bilayers requires maximal exposure to charged amino acid side chains within the protein's structure.
   (D) the α-helix shields polar groups on amino acid side chains within the core of the transmembrane region.
   (E) α-helices resemble phospholipids with their alternating polar and non-polar regions.

9. Many neighboring animal cells have connections between them that serve as direct passageways between their cytoplasms, allowing the movement of ions and small molecules back and forth. These connections also couple electrical responses in one cell with electrical responses of adjacent cells. These cell-to-cell linkages are known as

   (A) plasmodesmata.
   (B) gap junctions.
   (C) hemidesmosomes.
   (D) lamella.
   (E) occluding junctions.

10. The Cdk inhibitor *p16* binds to Cdk4/cyclin D complexes, which are normally responsible for allowing cells to pass through the restriction point from G_1 into S phase. Underexpression of *p16* protein could lead to

   (A) uncontrolled cell division
   (B) cessation of mitosis
   (C) increased inhibition of Cdk4/cyclin D complexes
   (D) overexpression of *p53* protein
   (E) a delay in the onset of mitosis

<u>Questions 11–14</u>

   (A) glycerol
   (B) glucagon
   (C) gluconeogensis
   (D) glycogen
   (E) glycolysis

11. Ubiquitous metabolic pathway in the cytoplasm in which sugars are degraded to produce ATP

12. Organic building block of phospholipids

13. Storage form of excess glucose in animals

14. Liver and muscle cells contain large granules of this molecule, a highly branched chain of glucose monomers

ANSWERS

1. (A)

Plasmids are able to reintegrate themselves back into the main bacterial chromosome, much as transposable elements in eukaryotic genomes can move from chromosome to chromosome using enzymes and insertion sequences. Plasmids are not found in eukaryotic cells, choice (B), nor are they made only out of RNA, choice (E). Plasmids can replicate themselves independently of the main bacterial chromosome and they can easily be transferred from cell to cell, so choices (C) and (D) are also wrong.

2. (B)

The fact that certain substances can gain access to certain areas of an protein based upon that protein's 3-D conformation suggests that different conformations can alter protein specificity. While water bonds to many surfaces by hydrogen bonding, choice (A), this is not an adequate explanation for the significance of the water-blocking conformational change described in the question stem; nor is there any evidence offered to support choices (C) or (D). In choice (E), it is possible for the interior of the protein to be entirely hydrophobic, but again there is no evidence for this in the question stem. Beware of extreme-sounding answers such as choice (E).

3. (D)

The neurotransmitter acetylcholine (ACh) causes the rapid release of Ca^{+2} ions from the sarcoplasmic reticulum in muscle fibers. This calcium allows myosin and actin filaments to slide past each other and causes the muscle to contract. Of the answer choices given, only choice (D) would allow for a greater response to neurotransmitter. Increasing acetylcholinesterase or downregulating ACh receptors would facilitate faster breakdown of ACh and cause less muscle stimulation, so (A) and (C) are wrong; increasing the amount of myosin would not help muscle contraction unless entire sarcomere units were also produced in greater quantities (choice B); and, (E) is incorrect because strengthening muscle attachment to nearby bones should not increase responsiveness to the nerve signal itself.

4. (C)

Many proteins must be deposited into the ER lumen (membranous sacs) as they are made. Some of these are to be secreted out of the cell and must start their journey in the ER; others are simply too dangerous to synthesize in the cell's cytoplasm (cytosol), such as lysosomal hydrolases that would digest away parts of the cell if allowed to freely float around the cell after synthesis. Thus, choice (C) is the only correct answer here.

5. (B)

The most important thing to read in the question stem here is the word *below*. If DNA is below its melting temperature, then it should remain in its native double-stranded conformation because of the hydrogen bonds holding its complementary helices together. As it gets warmer, or beyond the melting point, hydrogen bonds break allowing DNA to denature and become single-stranded. Choice (E) is incorrect because single-stranded DNA (ssDNA) absorbs more UV light in a test tube than double-stranded DNA (dsDNA) does, since ssDNA is more dispersed than its tightly packed counterpart. As long as DNA remains cool and double-stranded, it will absorb less UV than at temperatures above T_m.

6. (E)

Although PCR needs heat to break apart DNA for replication, the heat can often inactivate the enzymes needed for the replication. Therefore, heat-stable DNA polymerases are extremely beneficial, especially when they possess an exonuclease proofreading ability that cuts down tremendously on errors made during copying.

7. (C)

The most sensible answer here is that changes in RNA processing are what cause the release of free antibody proteins rather than their attachment into the cell membrane of the B-cells. Differences in signal patches or signal peptides would allow for changes in the targeting of these antibodies.

8. (D)

The alpha-helical structure of membrane-spanning regions is useful to shield the polar amino acid side chains from the hydrophobic interior of the cell membrane. Pleated sheets or other conformations would simply not be as stable as they passed through the membrane. However, there is no evidence presented that the alpha-helix is always the most stable structure, so eliminate (A) for being too extreme an answer choice; nor is there any suggestion that beta-pleated sheets have anything to do with cholesterol, so (B) is out as well. There is no polarity of the space between the lipid bilayers, (C), because that space is hydrophobic and nonpolar, and alpha-helices do not resemble the structure of phospholipids, choice (E)—also, phospholipids do not have alternating polar and nonpolar regions as suggested.

9. (B)

Gap junctions are connections between cells that enable ions and other small material to move between adjacent cells. They are essential for rapid electrical conduction across large tissues or organs, such as in the heart. Plasmodesmata, choice (A), occur only in plants, and lamella (choice D) are inner sections of plant cell wall. Try to eliminate answer choices quickly if they are not relevant to the information presented in the question stem—here, the stem mentions "animal cells."

10. (A)

Removal of cyclin D inhibition would allow cells to progress unhindered from G_1 into S phase. This could cause, in the absence of other controls, a more rapid onset of mitosis and uncontrolled cell division.

11. (E)

This definition is that of glycolysis, a process that degrades sugars in order to recycle ATP. If you have trouble distinguishing between all the *gly-* words here, try to remember that the word glycolysis is really made of two parts: *glyco-* for "sugar" and *-lysis* meaning "to break"—therefore, glycolysis is the breakdown of sugar.

12. (A)

The building blocks of lipids include glycerol and fatty acids. Glycerol is a three-carbon molecule onto which fatty acids can be attached in a condensation reaction (otherwise known as a dehydration synthesis reaction).

13. (D)

Whereas plants store excess sugars as starch, animals store sugars as glycogen.

14. (D)

Glycogen is stored mainly in the liver and muscle for breakdown in times of need. Breakdown of glycogen is called glycogenolysis (to add another *gly-* word to your repertoire!) and the process supplies fresh glucose to cells for ATP production via glycolysis and cellular respiration.

Part III

ORGANISMAL BIOLOGY

Introduction to Organismal Biology

Your knowledge of the biomolecules that make up the molecular foundation of life on Earth will serve you well in this section, which looks at life on a more macro level. Here, we will study the tissues and organs that build organisms and the interrelatedness between organisms on Earth. The GRE will test your knowledge of how various systems work and the similarities of these systems across phyla (e.g., digestive systems in earthworms, jellyfish, and mammals). You will also need to be familiar with plant structure and growth as well as with the major kingdoms of life on Earth.

A major area covered here is comparative anatomy and physiology, where the GRE will test you on **hormonal regulation** of body systems, **homeostatic mechanisms** of life, and key evolutionary adaptations across species. A sampling of digestive, nervous, excretory, and gas exchange systems are looked at in depth as well as why various organisms are classified into their particular kingdoms. This section cannot present every aspect of organismal biology, but it will give you an overview of the areas you should be familiar with for the GRE.

On the GRE exam, this section will cover three main topic areas: Animal Structure, Function, and Organization; Plant Structure, Function, and Organization; and the Diversity of Life. As such, this section of the book will address each of these subsections.

CHAPTER TWELVE

Digestion and Nutrition

Digestion involves the degradation of large molecules into smaller molecules that are then absorbed and used directly by cells. In simpler animals, digestion can occur in single cavities, from which biomolecules can diffuse into every cell of the organism (see *hydra* below). In more complex animals, branching digestive canals and absorption into the bloodstream are necessary for the effective breakdown of nutrients and their transport to all cells of the body. Mammalian digestive tracts, which are the focus of this chapter, are organized into regions specialized for the digestion and absorption of specific nutrients.

COMPARATIVE ANATOMY AND PHYSIOLOGY OF DIGESTIVE SYSTEMS

Let's take a look at how different organisms ingest and digest their food.

Protozoans utilize intracellular digestion. In amoebas, pseudopods surround and engulf food (via phagocytosis) and enclose it in food vacuoles. Lysosomes containing digestive enzymes fuse with the food vacuole and release their digestive enzymes, which act upon the nutrients, breaking down macromolecules like proteins, nucleic acids, and polysaccharides. The resulting simpler molecules then diffuse into the cytoplasm. The unusable end products are eliminated from the vacuoles.

In the paramecium, cilia sweep microscopic food such as yeast cells into the oral groove where a food vacuole forms around food. Eventually, the vacuole breaks off into the cytoplasm and progresses toward the anterior end of the cell. Enzymes are secreted into the vacuole and the products diffuse into the cytoplasm. Solid wastes are expelled at the anal pore.

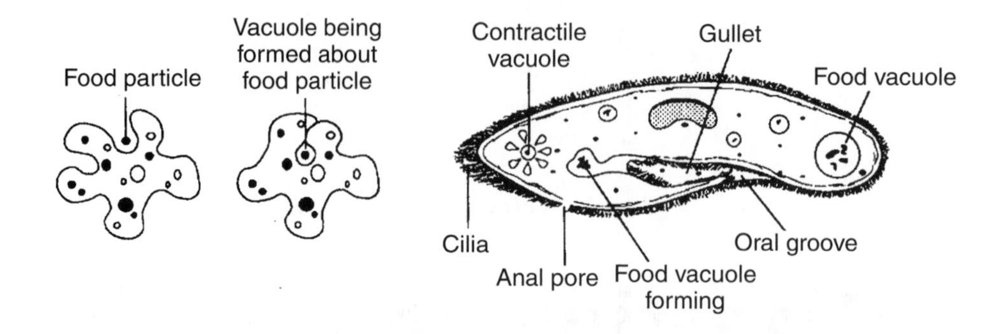

Hydra (phylum Cnidaria) employ both intracellular and extracellular digestion. Tentacles bring food to the mouth (ingestion) and release the particles into a cuplike sac called the gastrovascular cavity. The endodermal cells lining this gastrovascular cavity secrete enzymes.

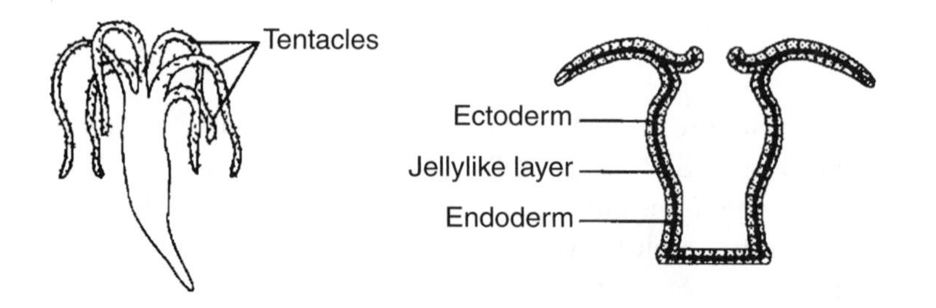

Thus, digestion principally occurs outside the cells (extracellularly). However, once the food is reduced to small fragments, the gastrodermal cells engulf the nutrients and digestion is completed intracellularly. Undigested food is expelled through the mouth. Every cell is exposed to the external environment, thereby facilitating intracellular digestion.

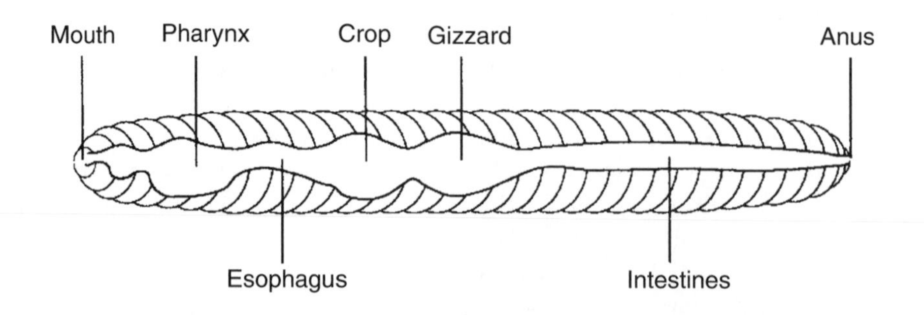

Since the earthworm's body is many cells thick, only the outside skin layer contacts the external environment (see figure). For this reason, this species requires a more advanced digestive system and circulatory system. Like higher animals, earthworms have a complete one-way, two-opening digestive tract. This enables specialization of different parts of the tract for mechanical and chemical digestive processes and absorption of the food that has been ingested. These parts include the mouth, pharynx, esophagus, crop (to store the food), gizzard (to grind the food), intestine (which contains a large dorsal fold that provides increased surface area for digestion and absorption), and anus (where undigested food is released). The digestive tract of arthropods is similar, with the addition of complex feeding appendages and salivary glands (see figure).

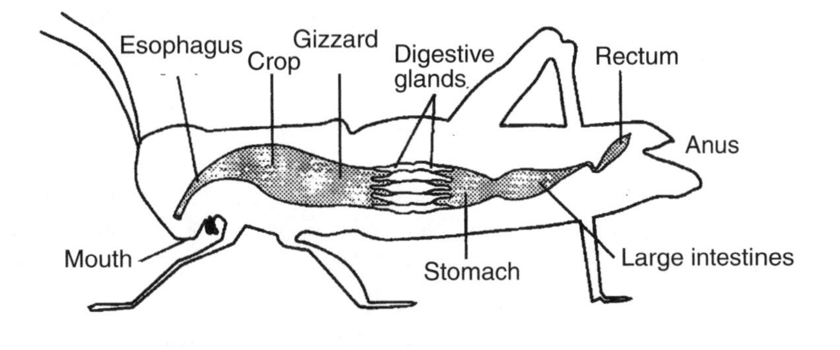

MAMMALIAN DIGESTION: STRUCTURES, HORMONES, AND ENZYMES

Most animal systems involve the regulation of some physical parameter that maintains normal cell **homeostasis**—usually a **negative feedback** regulatory loop. The distinguishing feature of negative feedback is that the control signals fed into certain cells cause the measured parameter to change in the opposite direction from that which caused the initial error signal. For example, blood glucose concentration is constantly regulated to provide a ready source of glucose to cells of the body. If blood glucose increases from the normal resting level, insulin gets released from the pancreas to stimulate cells to uptake more glucose. This decreases blood glucose levels; hence, this is negative feedback because an increase in blood glucose results in an action to decrease blood glucose in order to regulate levels back to normal resting concentrations.

Our digestive systems break down and absorb essentially anything digestible that enters our system. Generally, digestion and absorption continue regardless of whether the individual needs certain nutrients. The coordinated movement of food through the digestive system facilitates digestion, which involves a variety of negative-feedback loops.

Overview of Digestion

Ingested food is first subjected to mechanical shearing and grinding in the mouth in order to reduce the size of food particles. Saliva begins digestion of carbohydrates and lubricates the ball of food that is known as a "**bolus.**" Stretch receptors in the stomach cause chief cells and parietal cells to secrete gastric "juices," a highly acidic salt solution that denatures proteins by disrupting hydrogen bonds and secondary structure. Gastric juice thereby causes connective tissues of animal meat to break apart. Unfolding of proteins provides increased surface area for proteases active **only** at low pH that are secreted by stomach cells.

As the food mass breaks up, it stratifies in the stomach according to size, density, and water solubility. Fat cells and lipids float to the top of the stomach, carbohydrates and smaller proteins remain in the middle, and the larger proteins sink to the bottom. This is very important for the next step, because the contractions of the stomach force proteins and carbohydrates out first and the lipids next. Evolutionarily speaking, this design favors early digestion and absorption of the most readily available sources of chemical energy (polysaccharides); protein and fat digestion come afterward.

The partially digested mass of food, known as **chyme**, enters the small intestine. Pancreatic fluid, rich in bicarbonate ions to neutralize stomach acid and full of enzymes to break down proteins and lipids, is secreted. Bile is secreted by the liver to help fat absorption by emulsifying them. The proper mixture (depending on what's in the meal that was eaten) of chyme, pancreatic fluid, and bile is obtained by adjusting the rates of gastric emptying, pancreatic secretion, bile secretion, and intestinal motility (muscle movement). The smooth muscle of the small intestine, like an elevator, continuously moves the chyme down the intestine. Chemical breakdown continues and products are absorbed by cells lining the inside if the small intestine. By the time the chyme gets to the end of the small intestine—the terminal **ileum**—all of the digestible food will have been digested and absorbed. **Bile salts** secreted by the liver are reabsorbed here and recycled back to the liver to be used again for the next meal.

Diseases like cholera cause death through rapid dehydration. Bacterial toxins that cause the disease make large intestinal cells so leaky that the body loses more water through diarrhea than it can absorb. This is why the reabsorption of water and salt is so important in the large intestine.

Unabsorbed material is pushed into the large intestine where the remaining water and salt are reabsorbed. The leftover material is the **feces** to be stored in the terminal colon (large intestine) until enough material accumulates that the defecation reflex is activated.

Overall, you should consider that the digestive process involves the secretion of over 7 liters of fluids and enzymes *per day* to the inside of the digestive tract (which is technically outside of the body). Humans only have 5 or 6 L of blood in them; thus, an enormous amount of fluid is used to digest, and the fluid needs to be restored quickly if dehydration and eventual death are to be avoided.

The secreted fluids and enzymes used in digestion are ultimately reabsorbed back into the body: The process in mammals is not all that different from the way a fly digests and absorbs its food—just more discrete!

Structures, Hormones, and Enzymes

The following is a list of the various organs involved in mammalian digestion along with the key processes that take place. Be familiar with these structures for the GRE exam.

Human Digestive Tract

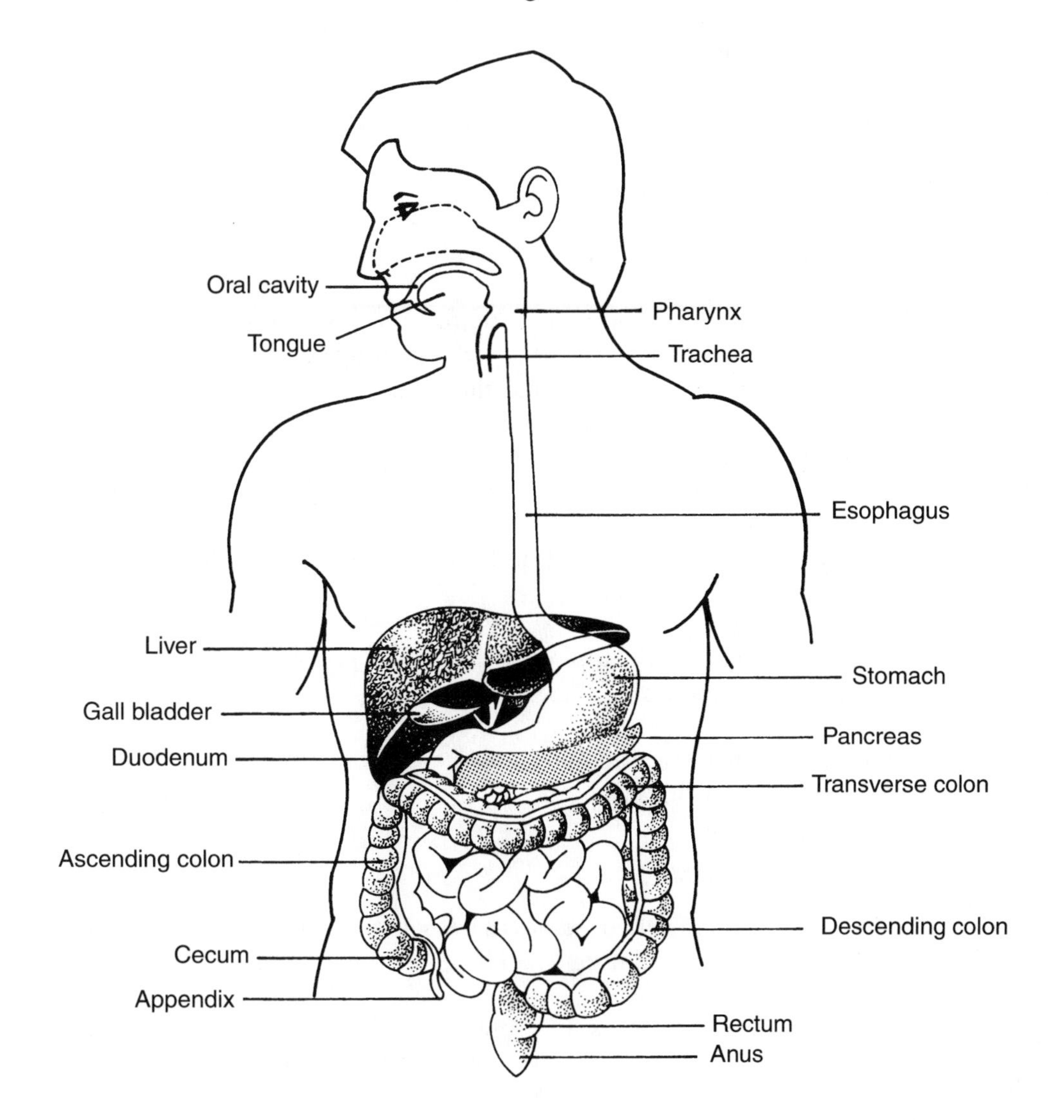

MOUTH

- **Saliva**, containing **salivary amylase**, is secreted from salivary glands to begin starch digestion.
- The amylase is inactivated by the low pH in the stomach, but amylase in the center of the food bolus can remain active for a long time and helps to digest up to 50 percent of the starch in a normal meal.
- Both **chemical** and **mechanical** digestion take place here.

ESOPHAGUS

- This is the passageway for food from the mouth to the stomach.
- The UES (**upper esophageal sphincter**) and the LES (**lower esophageal sphincter**) are bands of muscle at the bottom and top, respectively, of the muscular esophagus that regulate the passage of food down to the stomach.
- The UES prevents air from entering the GI system.
- The LES prevents food from backing up again into the esophagus when the pressure in the chest cavity decreases—as lungs inflate. When food gets back past the LES, you get heartburn and indigestion.
- Skeletal (striated) muscle at the top of the esophagus gradually transitions into smooth muscle by the bottom of the esophagus.
- **Peristalsis** is the name of the muscular waves created by smooth muscle that move food down the length of this passage.

Calcium is needed for H^+ secretion from cells in stomach lining. The more calcium that is around, the more H^+ secreted. So antacids like Tums that boast a high level of calcium to supplement your diet often increase the amount of calcium, and therefore acid, in your stomach!

STOMACH

- This organ serves as a reservoir that allows for the **ingestion** of food faster than it can be **digested** and absorbed.
- The process of protein digestion begins here by exposing food to a low pH environment and to the enzyme pepsin.
- The stomach delivers chyme to the **duodenum** (beginning of the small intestine) at a rate compatible with the rate of breakdown and absorption taking place in the small intestine.

- Secretory cells that produce HCl, **pepsinogen** (the inactive form of pepsin), and mucous are inside the gastric pits of the mucosal layer of the stomach. The HCl in the stomach activates pepsinogen into pepsin. Why do you think it is advantageous to keep enzymes such as pepsin in their inactive form before they are secreted and activated by stomach acid? *Because they can cause a great deal of damage to the cells that produce them (pepsin shreds proteins apart).*

- The stomach normally secretes about 2000 mL of concentrated HCl/day and the pH of the stomach is usually around 2. This represents a 3,000,000-fold increase in the H^+ ion concentration compared to blood!

Too much H^+ secretion, compounded by infection with Helicobacter pylori bacteria that infect the gastric mucosal cells, causes stomach ulcers. When infected, these mucosal cells do not produce enough mucous to protect the stomach lining against the strong acid.

SMALL INTESTINE

- There are three main areas of this organ: duodenum (1st segment), jejunum (2nd), and ileum (3rd). The vast majority of chemical digestion and absorption of biomolecules takes place in these sections.

- The small intestine is more than six meters long in humans! (This is discussed more in the section on evolutionary adaptations.)

- At the end of the stomach (the pylorus), there is a thick band of muscle called the **pyloric sphincter**. This muscle regulates passage of chyme into the small intestine, letting it squirt into the intestine a bit at a time. Meals take from 2-6 hours to clear the stomach.

- Complex feedback loops regulate how fast the stomach empties its contents into the duodenum. If gastric emptying occurs too fast, malabsorption takes place and some food is never fully broken down. You should be aware of these **feedback loops** and the hormones that control them for the GRE:

    1. **CCK (cholecystikinin): A hormone released by small intestine cells in the duodenum.** This hormone stimulates the release of amylase (for starch digestion), lipase (for fat digestion), and bile (for fatty acid emulsification). It also stimulates contraction of the pyloric sphincter to slow down chyme coming through from the stomach to the small intestine. *Because CCK is released due to fat-digestion products (such as fatty acids), it gives the fat products time to be further digested by CCK-induced pancreatic lipases while the hormone shuts down the pyloric sphincter.* CCK controls the negative feedback loop for fat breakdown.

2. **Secretin: A hormone that stimulates the release of bicarbonate ions from the pancreas if too much chyme enters the duodenum at one time.** Secretin also shuts down (causes the contraction of) the pyloric sphincter, again to control how fast stomach contents move into the small intestine (the faster the movement of chyme into the duodenum, the faster the pH of the small intestine drops).

   Secretin controls the negative feedback loop for pH control of chyme.

3. **Gastrin: A hormone that stimulates the release of HCl and pepsinogen to further break down proteins and amino acid chains in the stomach before they enter the duodenum.** Elevated concentrations of peptides and amino acids in the duodenum signal that the stomach is emptying too fast for pancreatic juices to handle. Gastrin gets released by cells in the duodenum and makes the pyloric sphincter contract. Gastrin controls the negative feedback loop for protein breakdown.

Digestion of products one eats is necessary for proper absorption. The key final step in digestion comes when food molecules are absorbed across the intestinal membranes into the nearby bloodstream so that biomolecules can be transported to cells around the body.

The small intestine comprises a huge amount of surface area. Large, circular folds called **villi** (singular: villus) protrude into the interior of the intestine. These villi are made of epithelial cells that have **microvilli** to further increase surface area for absorption. These microvilli make up the "**brush border**," so-called because the microvilli look like fine bristles across the top of the cells. In the core of each villus lie blood capillaries that drain from the villus into the hepatic portal vein, which pushes blood rapidly through the liver so that biomolecules can be further modified, packaged, or stored.

WHAT IS PANCREATIC JUICE?

The pancreas is both an exocrine organ (secreting into the intestine) and an endocrine organ (secreting into the bloodstream). In its exocrine function, the pancreas releases bicarbonate ions to neutralize the pH of chyme in the small intestine, proteases to break down proteins, lipases to break down lipids, and amylases to break down starch). The intestinal cells themselves can produce their own peptidases (to break down small polypeptides), lipases, and disaccharidases (to break down sugars such as maltose, lactose, and sucrose). Exocrine pancreatic secretions enter the small intestine through the pancreatic duct.

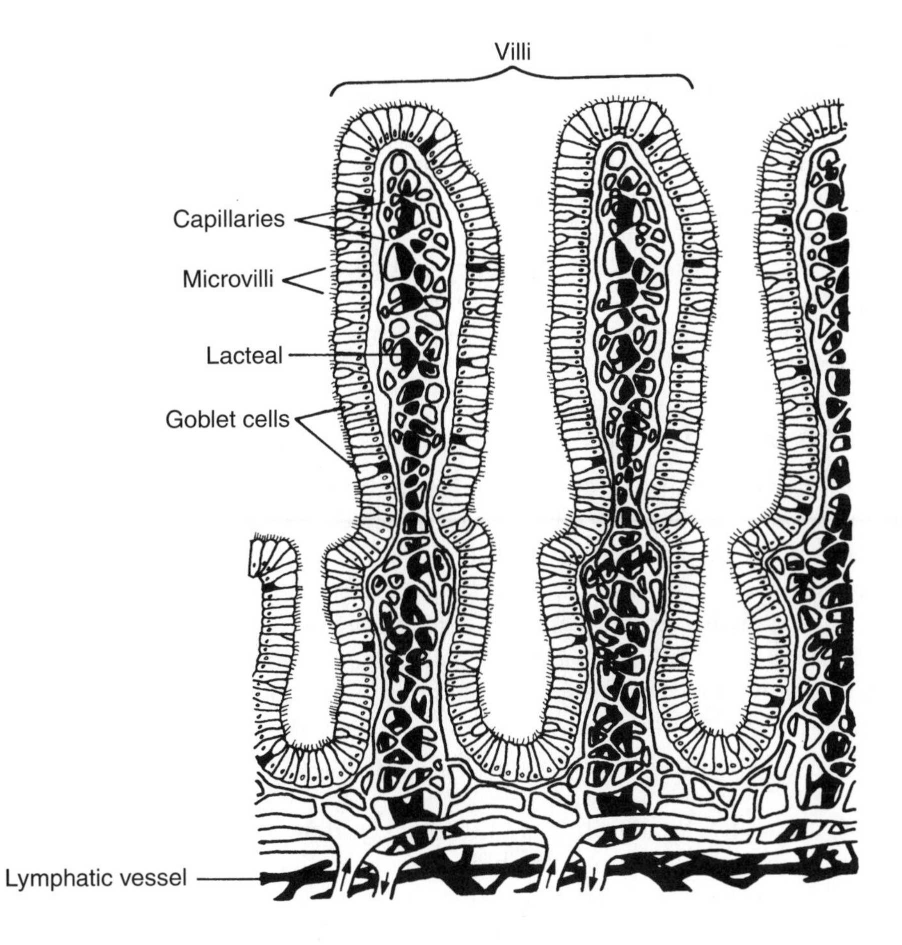

LARGE INTESTINE

- Also known as the colon, the large intestine's main job is to reabsorb water and salts.
- A variety of bacteria reside in the large intestine, digesting certain material that would not otherwise get broken down by our own enzymes. Some of these bacteria make useful products, like Vitamin K, which is needed for proper blood clotting.
- Cellulose fibers in plant material (salads, fruits, etc.) are broken down by bacteria as well, and help to maintain proper consistency of stool.

Fat products are usually transported across cell membranes in protein-coated vesicles (chylomicrons are one example). The reason for this is that fats can easily remain dissolved within the fatty bilayer of a cell membrane unless surrounded by proteins as they are transferred across the membrane.

Absorption of Amino Acids and Proteins

By the time proteins reach the small intestine, HCl and pepsin in the stomach have denatured them and have broken the hydrogen bonds within their secondary structure. Proteases such as trypsin and chymotrypsin, secreted by the pancreas further break polypeptide chains into individual amino acids in the small intestine. Neutral (uncharged) amino acids can diffuse right across the "brush border" of small intestinal cells. Yet, charged amino acids are carried across by active transport using membrane spanning proteins or via facilitated diffusion. Small polypeptides can often enter the bloodstream from the intestine by endocytosis into intestinal cells. These polypeptides are broken into amino acids within the cells that absorb them.

Absorption of Fatty Acids and Lipids

Bile salts, secreted by the liver after being stored in the gall bladder, form **micelles** around fat droplets in the intestine. Bile salts are *amphipathic* (like detergents)—polar on one side of the molecule and nonpolar on the other—and will stabilize fat droplets in aqueous solution. Because they surround and penetrate into larger fat droplets, they allow large fat droplets to be broken up into smaller droplets, thereby permitting fats to pass directly into intestinal cells and into the bloodstream. This process is called **emulsification**. The micelles that surround fats in the intestine do not themselves enter the intestinal cells, but rather exchange the emulsified lipids from within the micelles to the inside of the cells, leaving behind the bile salts, which get reabsorbed and recycled in the large intestine.

The lipids that have left the intestine are then packaged into larger fat droplets again, called **chylomicrons**. These droplets enter the bloodstream by first being absorbed into lymphatic sacs called **lacteals** that are present in each intestinal villus. The lymphatic system will eventually dump these chylomicrons into the general circulation and they will be brought to the liver, which distributes them throughout the body using special carrier molecules to bring them to cells. Cholesterol is handled in a similar fashion.

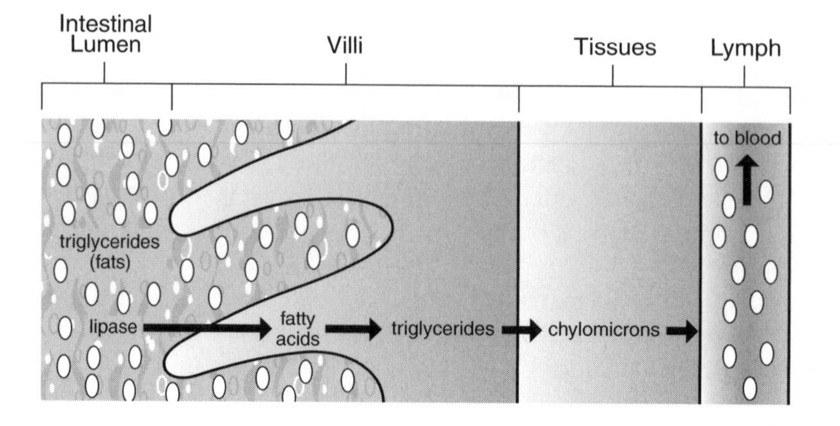

EVOLUTIONARY ADAPTATIONS AND ESSENTIAL NUTRITION

Evolutionary Adaptations

The variations seen among vertebrate digestive systems are a result of dietary differences. These variations involve the shape of teeth, the length of the alimentary canal (how long the digestive tract is), and the presence of enlarged, multichambered stomachs as seen in mammals called **ruminants.** Whereas carnivores possess sharp and pointed incisors to kill prey and tear flesh, herbivores have round, flat teeth for the crushing and grinding of stems and leaves. In addition, herbivores and animals that eat mainly vegetation have much longer digestive systems than those that are primarily meat-eaters. The reason for this is that longer periods of digestion are needed to break down and absorb the biomolecules found in vegetation, especially cellulose which must be broken down by symbiotic bacteria living within the gut. Animals such as cattle and horses have enlarged cecums, a pouch where the small intestine joins the large intestine. It is here that hordes of bacteria sit and digest cellulose present in the plant material that is consumed in large quantities. In fact, cattle and sheep also have multichambered stomachs, divided into three or four sections (chambers) which allow slow and repeated digestion of food before the food actually proceeds into the small intestine for absorption.

Essential Nutrition

Many essential **vitamins** are absorbed in the intestine as well. These vitamins often act as cofactors for enzymes or other proteins. Vitamin A is a precursor used to synthesize an essential cofactor for rhodopsin, a pigment used in the eye for proper vision. Niacin (Vitamin B3) is used to build NAD^+, the electron carrier used in cell respiration. Vitamin C is used to properly synthesize collagen, a protein that supports and builds up tissues. **Minerals** are also important: Iodine is required for the building of thyroid hormone, used to regulate metabolism, and iron is used mainly in blood to build hemoglobin, calcium, and phosphorus.

Of the 20 amino acids needed to build proteins, humans can synthesize 12. The remaining eight must be acquired from food. Those amino acids that must be eaten are known as **essential amino acids** because we cannot synthesize them. Deficiencies in any one of them can cause protein deficits which result in wasting syndromes or physical and mental handicaps. There are also essential fatty acids, though our bodies can make most of the fatty acids we need on a daily basis.

Vitamins A, D, E, and K are fat-soluble vitamins, meaning that they dissolve into cell membranes and other fatty tissues. They are not excreted in the urine when taken in excess like water-soluble vitamins. Therefore, people can overdose on A, D, E, and K.

CHAPTER THIRTEEN

Gas Exchange and Transport

Gas exchange and transport depend upon the movement of gases or fluids around an organism's body and the passage of these gases or material contained within the fluids into and out of cells. All gas exchange is based upon simple diffusion across cell membranes, and the key characteristic that has evolved across almost all species for maximal gas exchange is *moist membranes that cover lots of surface area.* This chapter compares cardiovascular and respiratory systems across animals (and protists) and looks at the maintenance of blood pressure, extremity temperature, and proper oxygenation.

COMPARATIVE ANATOMY AND PHYSIOLOGY OF GAS EXCHANGE SYSTEMS

The simplest form of gas exchange in living organism is the movement of oxygen or carbon dioxide into or out of a single cell, as is the case with bacteria or protists. Because these gases are small, they are readily able to diffuse directly through the lipid bilayer of all cells according to the concentration gradient that is present.

Gas Exchange in Simple Animals

The next step up in complexity would be the exchange of gases from a fluid in a branching or simple **gastrovascular cavity** into surrounding cells. Almost all cells are in close contact with the external environment. This can be seen in cnidarians such as the hydra and jellyfish shown below:

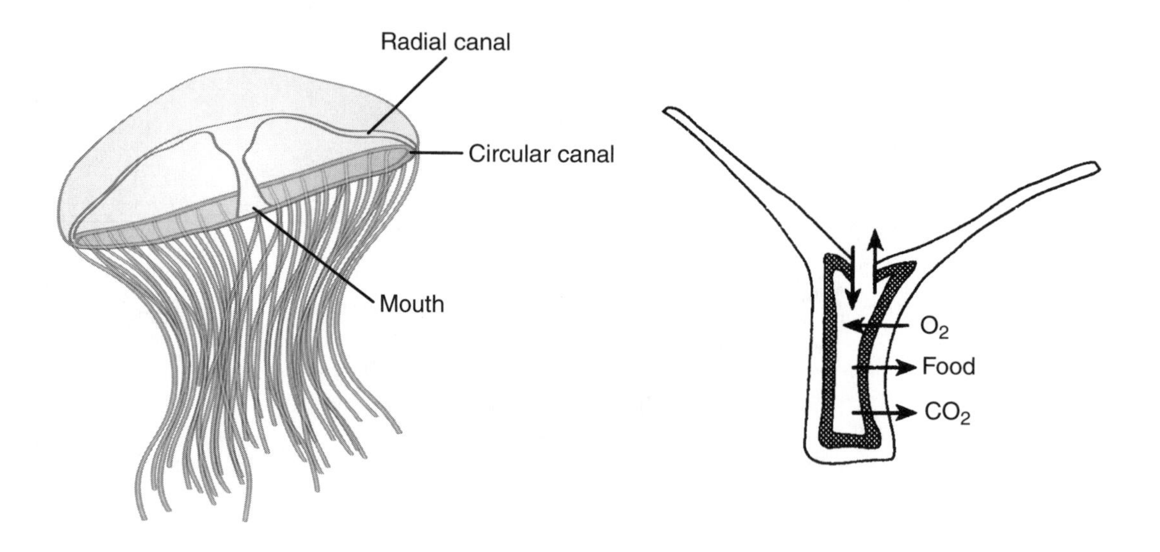

Still more complex evolutionarily is the exchange of gases through the **skin**, which is common in smaller organisms that are long and thin, such as the annelid worms and tapeworms. Annelids, which include the earthworms, have a great deal of surface area for a relatively small amount of internal volume, which makes breathing through the skin an adequate way to move sufficient oxygen in and release sufficient carbon dioxide as waste. Worms and other small animals that breathe through their skin keep their outer membranes moist by the secretion of copious amounts of *mucous*.

The arthropod respiratory system consists of a series of respiratory tubules called *tracheae*. These tubules open to the outside in the form of pairs of openings called *spiracles*. Inside the body, the tracheae subdivide into smaller and smaller branches, enabling them to achieve close contact with most cells. In this way, this system permits the direct intake, distribution, and removal of respiratory gases between the air and the body cells. No oxygen carrier is needed and specialized cells for this purpose are not found. Since a blood system does not intervene in the transport of gases to the body's tissues, this system is very efficient and rapid, enabling most arthropods to produce large amounts of energy relative to their weights. The direct diffusion of air through trachea is one factor that limits body size in arthropods.

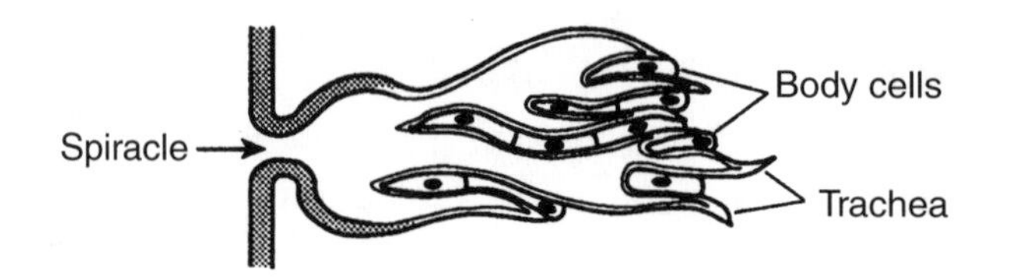

Gas Exchange in Amphibians, Reptiles, Birds, and Fishes

Most amphibians (frogs, toads, salamanders) breathe mainly through their skin, despite having small, functional lungs. As with the annelids, the skin must be kept moist at all times in order for effective gas exchange to take place. For larger organisms, however, the solution to effective gas exchange is to create a surface area that is effectively larger then the skin's surface area. This is accomplished by extensive branching or folding. In addition, reptiles and birds exchange gases internally because external membranes would dry out on land.

All reptiles, unable to breathe through scaly, waterproof skin, use their lungs for gas exchange. The lungs have multiple branching structures and inflate and deflate due to muscular movements that expand and contract the surrounding rib cage. In addition, some reptiles breathe through their cloacae. The **cloaca** (pronounced clo-ay-cah) is an opening found at the tail end of reptiles and used for excretion. Similar to the anus in mammals and covered in moist mucous membranes, the cloaca can be used as a secondary organ of gas exchange in reptiles.

Unlike reptiles and amphibians, birds are **homeotherms**, and they maintain a nearly constant body temperature even as their surrounding temperature changes. They use the heat produced by muscular activity to maintain their body temperature, and this muscle activity depends on a high rate of cellular respiration (ATP production). Although the ventilation of bird lungs is similar to that of reptiles, the presence of air sacs next to the lungs increases their effectiveness. Though no gas exchange occurs in these **air sacs**, they allow fresh air to continue flowing through the lungs even during exhalation. This is because the sacs store large quantities of air beyond that which fills the lungs during a single inhalation. Air flows through the lungs in one direction only, unlike the airflow in mammals that ends in alveoli and then reverses direction. In many birds, some bones are hollowed out and penetrated by air sacs as well, decreasing the overall weight of the bird for better takeoff and flight.

Gills, such as those found in the bony fishes, are a step up in complexity from those organisms that breathe in the water through branching gastrovascular cavities. These are often simple, tubular projections from the body with some branching structures for added surface area. Gills can be found in many organisms other than fishes, such as electric eels, tadpoles, and mollusks; however, a study of gills in fishes will provide an understanding of their usefulness in an aquatic environment. In water, fishes obviously need expend no energy in order to keep the gill surfaces moist for diffusion. At the same time, however, the gills must be very efficient at extracting oxygen from the water since water holds only a fraction of the oxygen available in the air.

When aquatic organisms "pull" oxygen out of the water in order to breathe, they are *not* breaking down H_2O to get the oxygen from water molecules! The oxygen they breathe is O_2 gas dissolved among the water molecules.

The respiratory surface of the fish contained within the relatively small gill area near the head is, in fact, greater than the surface area of the entire surface of the fish. The highly branched gills are filled with tiny blood vessels, most so thin that blood cells can pass through only in single file. Four gill arches can be found on either side of the head and thin filaments jut out from each of these bony arches. Each filament bears hundreds of platelike structures called **lamellae** where the actual diffusion of oxygen from the water into the fish's bloodstream takes place.

The key aspect of the gas exchange system in fishes is that as blood is flowing toward the head of the fish, water is flowing toward the tail (as the fish swims). This setup favors the exchange of oxygen from water to blood across the entire gill surface, whereas the blood would already have been saturated with oxygen halfway through the lamellae if water and blood were flowing in the same direction. This flow of blood in a direction opposite the source of oxygen is called **countercurrent exchange**. The movements of a fish's mouth as it swims serve to open up and close the gill coverings, called **opercula**, so that water is continuously flowing over the lamellae.

HOW DOES COUNTERCURRENT EXCHANGE WORK TO IMPROVE THE AMOUNT OF OXYGEN PICKED UP BY BLOOD CELLS?

Think about concentration gradients here: If blood and water flow in the same direction, when the blood has picked up 50 percent of the available oxygen from the water, the water has lost 50 percent of its available oxygen and there is no more gradient! If blood constantly flows toward "fresh" water as it does in the fish, when the blood has picked up 50 percent of the available oxygen, there is still more highly oxygenated water flowing toward it and the blood can actually become saturated with closer to 80% of its oxygen capacity.

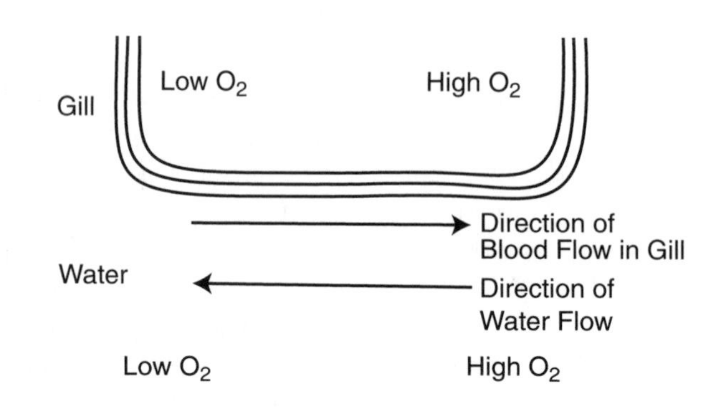

Mammalian Gas Exchange

While organisms such as amphibians, reptiles and even some fishes (e.g., lungfish) may possess **lungs** to aid in breathing and gas exchange, these organisms all use other structures as their primary means of oxygen and carbon dioxide exchange. In mammals, however, the lungs are the primary structure. Because the lungs are restricted to one location in the body, a circulatory system is needed to bridge the gap between the lungs and the cells around the organism's body. All mammalian lungs end in blind sacs called **alveoli** that are surrounded by a dense net of capillaries into which oxygen diffuses and from which carbon dioxide is excreted.

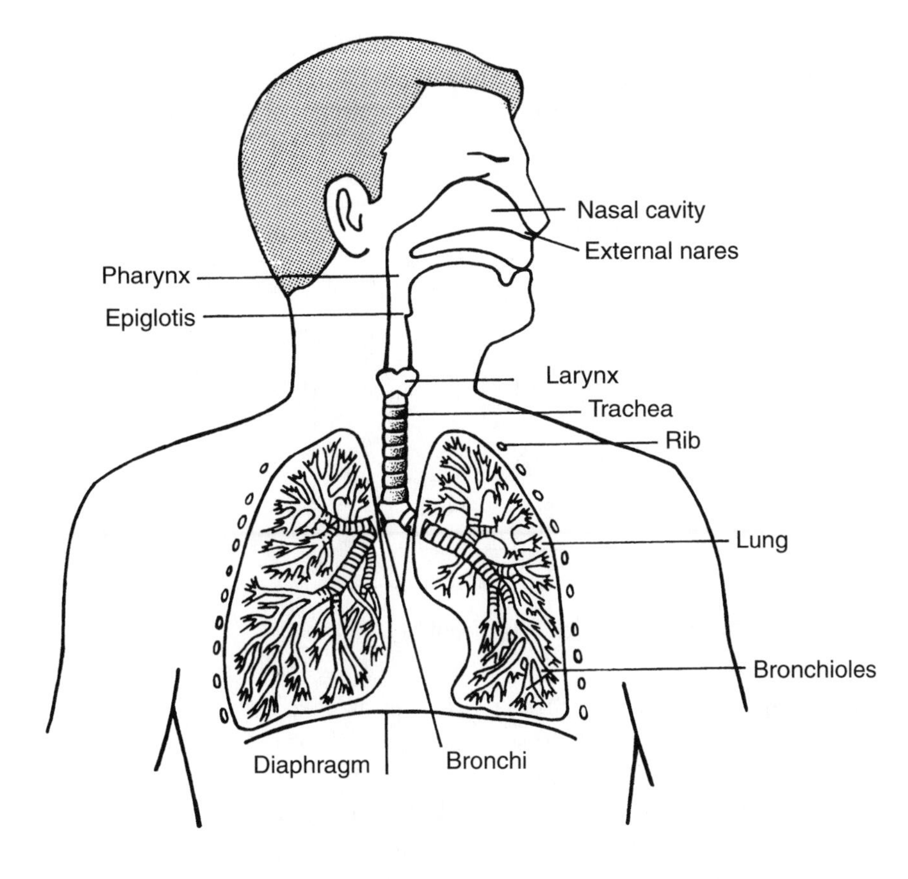

In mammals, the path that air takes once inhaled is as follows: once brought into the nose or mouth, air enters the **pharynx** (throat) and passes across the **larynx** (voicebox) that lies within the **trachea** (windpipe). The trachea branches into two **bronchi** (singular: bronchus), which further branch into **bronchioles** and then alveoli. Cells that line the bronchioles and alveoli keep the lungs moist, secreting mucous and other watery fluids into the lung openings. There are also a great many cells covered with short, hairlike cilia to keep mucous and other particles flowing across the inside epithelial surface. The millions of alveoli where the gas exchange actually takes place are separated from capillaries by only a thin layer or two of cells.

Deoxygenated blood enters the pulmonary (lung) capillaries from the systemic circulation having a *low partial pressure of oxygen* (it has lost its store of O_2 at the tissues). Inhaled air in the alveoli has a much higher partial pressure (there is much more oxygen relative to other gases in the alveoli compared to within the capillaries). Therefore, oxygen diffuses down its concentration gradient into the capillaries, where it binds to

WHAT IS THE DIFFERENCE BETWEEN POSITIVE AND NEGATIVE PRESSURE BREATHING?

Frogs and other amphibians can ventilate themselves literally by pushing air down their windpipes. They open and lower the floor of their mouths and gulp in air. This is positive pressure breathing. Mammals have a muscular band of tissue called the **diaphragm** that separates the chest cavity from the abdominal cavity. As this muscle contracts and pulls down, pressure in the chest cavity decreases, sucking air in from outside. This drop in air pressure from within, called negative pressure breathing, allows breathing to take place.

hemoglobin molecules in red blood cells and returns to the heart to be pumped out to the body. In contrast, the partial pressure of CO_2 in the capillaries is greater than that of the inhaled alveolar air; thus, CO_2 diffuses from the capillaries into the alveoli, where it is exhaled.

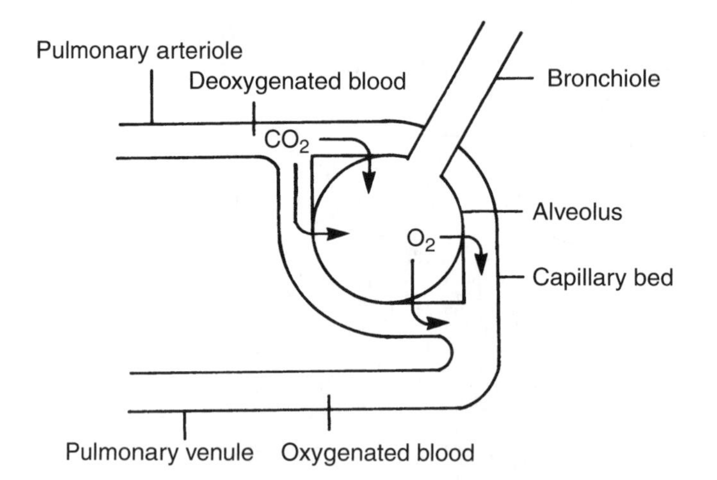

At high altitudes, the partial pressure of O_2 in the atmosphere declines, making it more difficult to get sufficient oxygen to diffuse into the capillaries. The body often compensates for this by increasing the rate of breathing (hyperventilation) and by increasing the number of red blood cells available to carry oxygen. **Erythropoetin**, a hormone released by the kidneys, is involved in the increased production of red blood cells: It travels to the bone marrow to stimulate red blood cell production there.

Within each new erythrocyte (red blood cell), there are approximately one million **hemoglobin** (**Hb**) molecules. Each hemoglobin molecule is made of four subunits capable of binding to four different oxygen molecules. The binding of the first oxygen molecule *induces a conformational change in the hemoglobin* molecule as a whole so that the binding of the other three molecules becomes progressively easier. Similarly, the unloading of one oxygen facilitates the unloading of the other oxygens. This allosteric effect is reflected in the S-shaped oxygen dissociation curve for hemoglobin (see diagram below). You will notice that hemoglobin is able to hold onto its oxygen molecules with great affinity until partial pressures reach about 40 mm Hg (hemoglobin's oxygen saturation remains over 70 percent until that point). Then, it quickly dumps off its oxygen. This is quite useful, for it makes sure that the oxygen in the red blood cells is not completely grabbed up by other cells until the red blood cells reach the farthest extremities of the body. As you will also notice in the diagram below, active tissues use up oxygen much faster and, therefore, have a lower partial pressure of O_2. Yet, the Hb molecules still hold onto enough oxygen to supply these needy tissues with some oxygen.

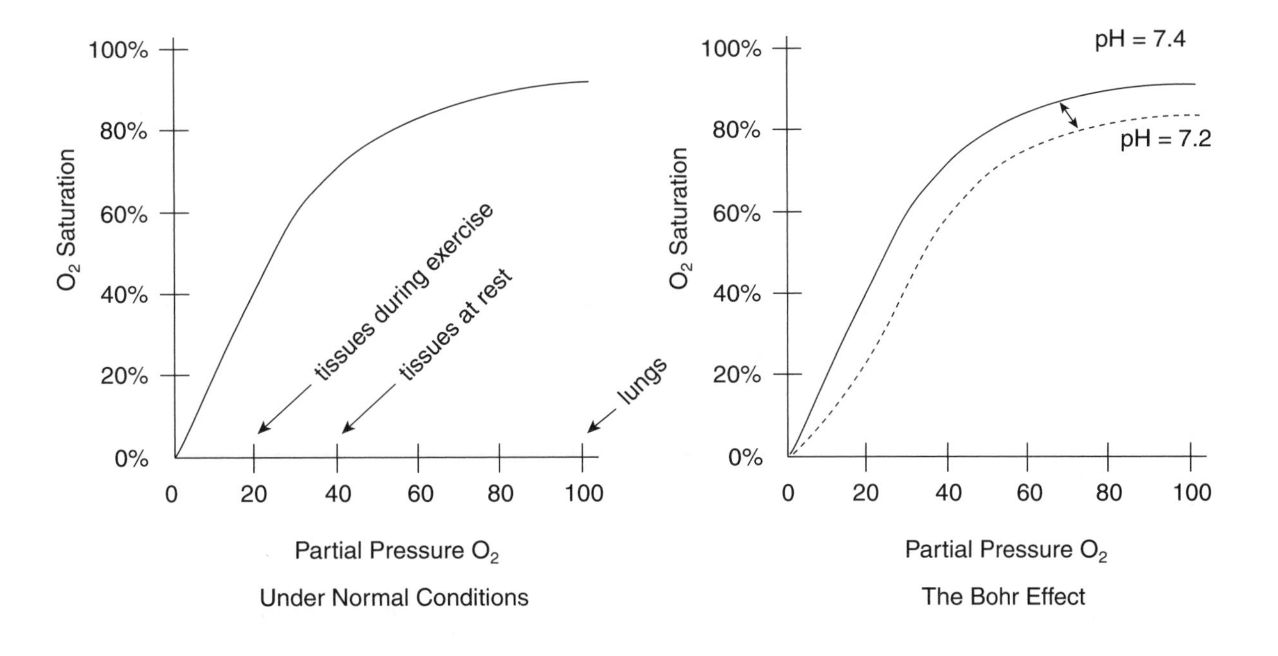

In the diagram above, notice how the oxygen-dissociation curve of hemoglobin is shifted to the right as pH drops. This evolutionary adaptation, known as the **Bohr Effect**, means that hemoglobin ends up losing oxygen much more readily in acidic environments than it does normally, which is extremely helpful for getting oxygen into active cells. The question, of course, is how do active cells make their surrounding more acidic? Carbon dioxide produced by the Krebs Cycle forms *carbonic acid* in watery environments, which explains the more acidic pH found in active tissues.

Another respiratory pigment, **myoglobin**, is present in muscle cells and in other places where tissues use up large amounts of oxygen rapidly. Myoglobin's affinity for oxygen is much higher than hemoglobin's, and it can strip red blood cells of their oxygen as they pass through active muscle tissue, holding onto this oxygen until the muscle cells are very depleted in oxygen. In other words, myoglobin's oxygen-dissociation curve would be shifted to the left of hemoglobin's, as it remains saturated with oxygen at far lower partial pressures than does hemoglobin. Diving mammals, such as whales, have enormous stores of myoglobin in their tissues, which explains why they can stay underwater for such long periods of time.

COMPARATIVE ANATOMY AND PHYSIOLOGY OF CARDIOVASCULAR AND TRANSPORT SYSTEMS

For the smallest organisms, gas exchange and circulation are one and the same. In single-celled organisms such as bacteria or protists, gas exchange through the cell membrane supplies all the oxygen that the organism needs. As organisms get larger and the distance increases from the oxygen supply to the cells requiring the oxygen, the oxygen is distributed to regions far from the oxygen source. One system for doing this is for the organism to develop a deep pocket, or gastrovascular cavity, as we have already seen in the hydra. Other cnidarians, such as the jellyfish, exhibit a branching central cavity for better circulation to all cells.

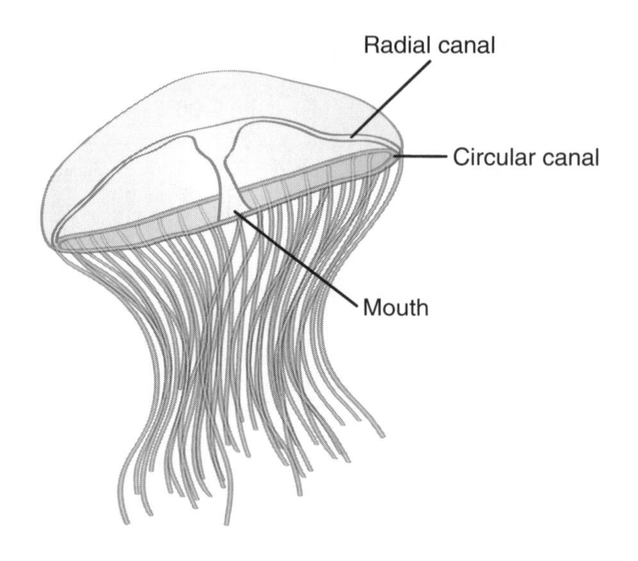

HOW IS CARBON DIOXIDE TRANSPORTED TO THE LUNGS FROM THE TISSUES?

The majority of CO_2 that leaves the cells combines with water in the bloodstream and forms carbonic acid, $H_2CO_3^-$ This immediately dissociates into H^+ (hydrogen) ions and HCO_3^- (**bicarbonate**) ions. The hydrogen ions are picked up by hemoglobin, as is some of the CO_2, and the bicarbonate makes its way to the lungs. Therefore, most CO_2 is transported through the blood as bicarbonate ion. The process is reversed at the lungs, where H^+ released by hemoglobin recombines with the bicarbonate and CO_2 is split out and exhaled.

Throughout the branching jellyfish circulation exist cells with cilia that beat back and forth to keep the oxygen and nutrient-filled water flowing. Only these ciliated cells lining the circular and radial canals have direct access to the oxygen and nutrients, but it is not far from these cells to the other jellyfish cells and simple diffusion can handle the rest of the distribution.

In insects and other arthropods, an *open circulatory system* exists, where there is no distinction between the blood and the fluid that bathes the cells of the organisms (interstitial fluid). This fluid, called **hemolymph**, is pumped into a connected system of tubes, along which lie many small bands of muscle cells, sometimes called hearts. These hearts push the nutrient-carrying hemolymph around the insect's body to nourish the cells that bathe in the fluid. The ventilatory system in insects and many other arthropods, however, is *completely distinct* and separated from this cardiovascular system. Recall that insects get air to their cells through a complex system of tracheal tubes that bring oxygen directly throughout the insect's body. There is no need for the hemolymph to pick up oxygen at all in most arthropod systems.

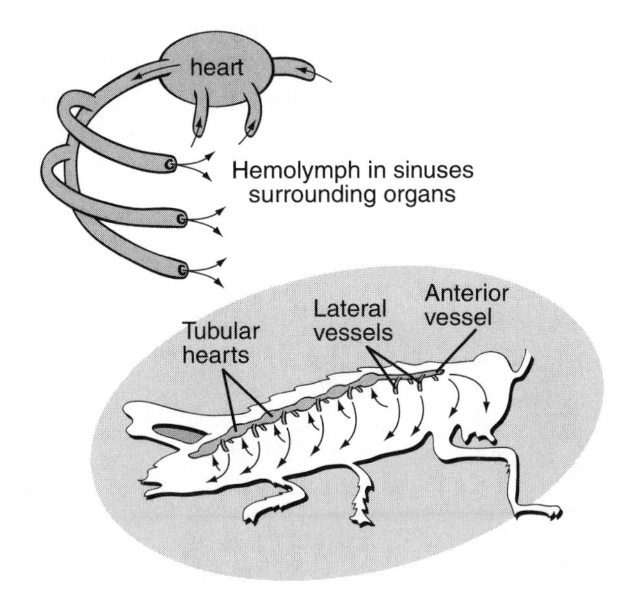

In annelid worms, such as earthworms, we see the first development of an entirely *closed circulatory system*. Here, diffusion of gases and nutrients occurs into and out of blood that is contained within vessels that do not allow blood to mix with the watery fluid (**interstitial fluid**) bathing the cells of the organism. Oxygen and nutrients diffuse out of the bloodstream into the interstitial fluid and then into cells. Similar to insects, annelids have many hearts, but the hearts form closed loops between the dorsal and ventral blood vessels. Notice as well that the blood circulation is almost entirely along the outside of the earthworm, close to the skin where gas exchange occurs.

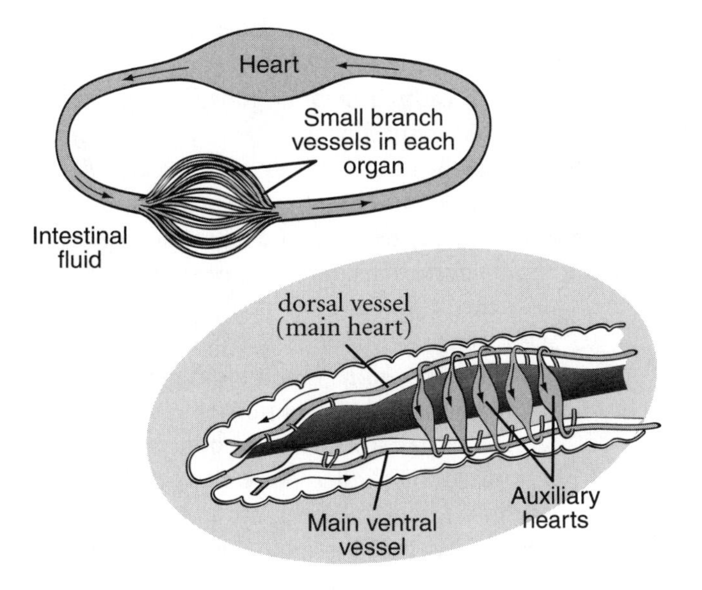

The diagrams below compare the evolutionary changes in the closed circulatory systems of vertebrates from the basic two-chambered heart found in fishes, to the three-chambered heart found in reptiles and amphibians, and finally to the four-chambered heart found in mammals and birds. We will look at each of these organisms in turn.

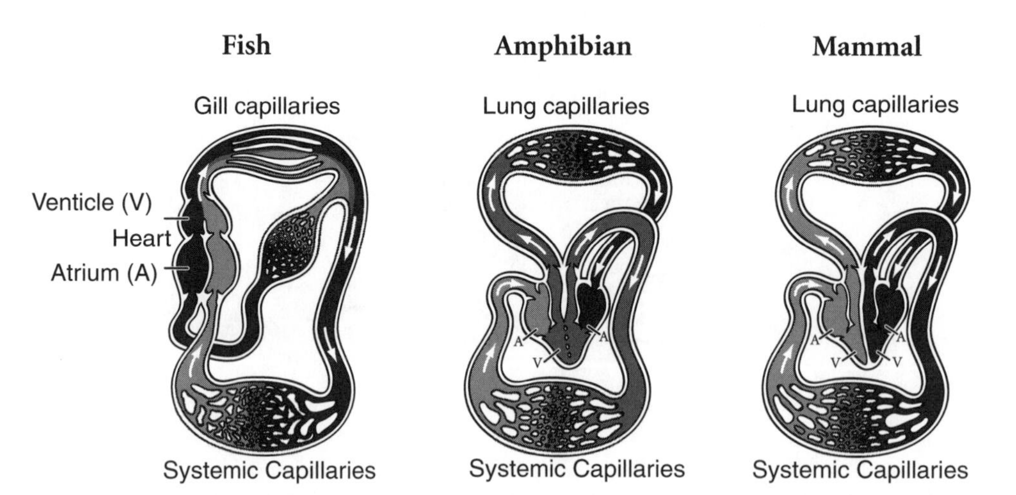

In fishes, the **two-chambered heart** is comprised of one **atrium**, where blood is received, and one **ventricle** for pumping blood back out to the gills. No oxygenated blood flows through the heart of a fish, since blood is oxygenated only at the gills, and the heart is located ventrally (in the belly of the fish). Because blood does not return to the heart after being oxygenated, the fish depends upon swimming movements of its body to help push the blood through its systemic circulation. Notice as well that there is *one circuit* through the body—out to the gills, through the body, and back to the heart.

In reptiles and amphibians, the **three-chambered heart** is comprised of two separate atria and one larger ventricle. Evolutionarily, breathing through lungs rather than through skin or gills may have forced the change from one circuit through the body to two circuits (out to the lungs first, then out to the body, stopping at the heart between each circuit). A three-chambered heart with *two circuits* is more efficient than a two-chambered heart, because blood can be delivered with high velocity both to the lungs for oxygenation and then again to the cells around the body after getting a boost from the heart before each circuit. Three chambers, however, is less efficient than the four chambers we see in mammals and birds, since the single, large ventricle mixes both deoxygenated blood returning from the systemic (body) circulation and oxygenated blood returning from the pulmonary (lung) circulation. Many active reptiles, especially the crocodilians, have evolved an elongated **septum**, or division, that nearly cuts the ventricle into two halves, one half for each of the atria. This progression toward a **four-chambered heart** (which has a completed septum down the middle of the ventricle) allows for more separation between oxygenated and deoxygenated blood.

Three chambers are not better than two chambers! Two chambers may work very well for different kinds of organisms living in different kinds of places. Fishes might well be at a disadvantage carrying around a bulkier three-chambered heart that weighs more.

In mammals and birds, blood follows a double circuit as well. From the diagram below you'll see that the right and left sides of the heart are each their own separate pumps: the right side pumps deoxygenated blood to the lungs, while the left side pumps oxygenated blood to the aorta and out to the body. The two upper (receiving) chambers are atria, and the two lower chambers are ventricles. While the atria are fairly thin-walled and not very muscular, the ventricles, particularly the left ventricle, are thick with muscle tissue for sustained, forceful pumping.

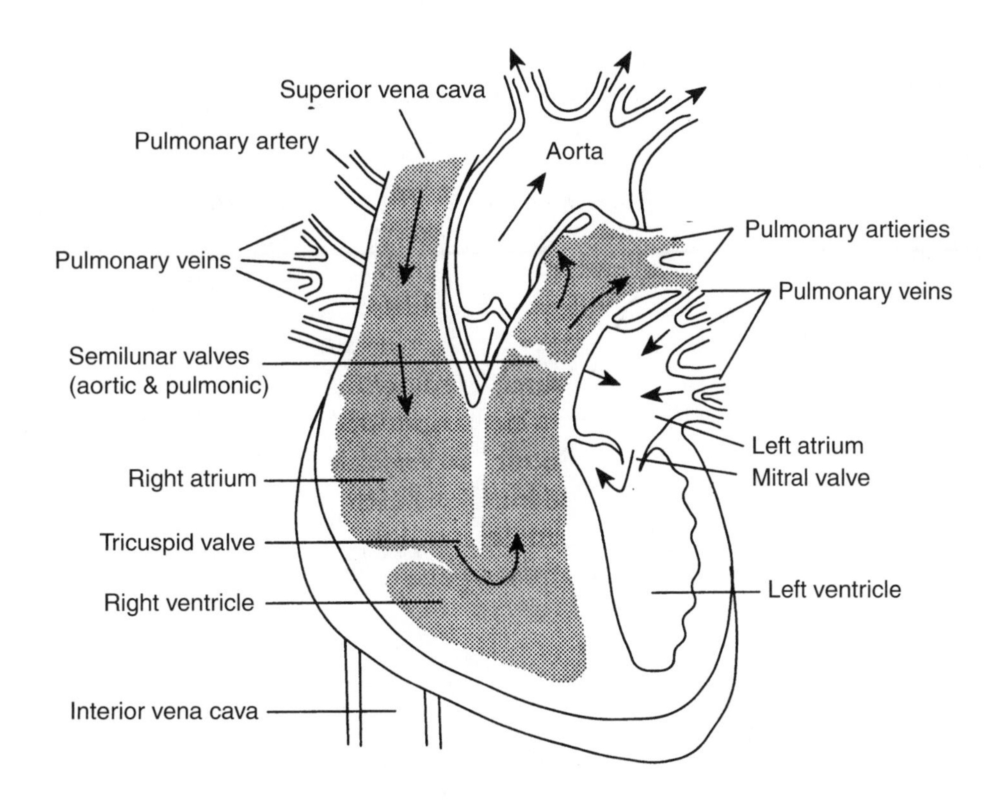

The atrioventricular (AV) valves, located between the atria and ventricles on either side of the heart, prevent backflow of blood into the atria when the ventricles are contracting. The semilunar valves in the pulmonary artery and the aorta serve a similar function: to prevent the backflow of blood into the ventricles from the lungs or aorta once the blood has been pumped out of the heart. The heart's pumping cycle is divided into two alternating phases: **systole** and **diastole**, which together make the heartbeat. Systole is the period during which the ventricles contract, forcing blood into the lungs and aorta, and diastole is when the ventricles relax and fill with blood.

BLOOD PRESSURE AND TEMPERATURE REGULATION

Blood pressure is the force per area that blood exerts on the walls of blood vessels. It is expressed as a systolic number over a diastolic number. Blood pressure drops as blood moves farther away from the heart, partly because the distance from the pump has increased and partly because of the muscular construction of arteries versus veins.

> When you have your blood pressure taken, you'll get a number such as 120/80. The top number, the larger one, is your systolic blood pressure and reflects how much pressure is in your arteries when your heart contracts. The bottom number is your diastolic pressure and reflects the pressure in your arteries as the heart relaxes. As your arteries fill with fatty plaques and harden with age, the top number typically goes up.

Arteries are thick-walled, muscular, elastic vessels that transport oxygenated blood away from the heart, while veins are relatively thin-walled, inelastic vessels that bring deoxygenated blood black to the heart. When veins are not filled with blood, they often collapse on themselves, only to swell open when blood is pushed through. In fact, because veins have little contractile ability of their own, blood is often pushed back to the heart from the extremities when the skeletal muscles near the veins in those areas contract. As you walk, for instance, contractions of your leg muscles propel blood sitting in leg veins back toward the heart. In contrast, arteries maintain an open **lumen** at all times and can contract against the blood within, helping to propel it away from the heart. Capillaries are the tiniest blood vessels in organisms, usually wide enough to allow blood cells to pass through only in single file. It is here, and not in arteries or veins, that diffusion of gases between the blood cells and tissue cells takes place. Keep in mind that the *predominant muscle type found in all blood vessels is smooth muscle*, under autonomic nervous system and hormonal control.

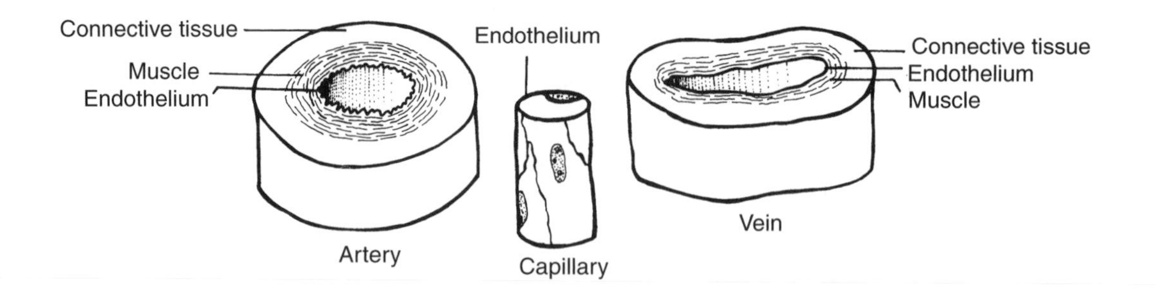

The path that blood takes from the heart to the rest of the body begins in the aorta, the major artery leading away from the left ventricle. The aorta splits into other arteries, smaller arterioles, and finally capillaries. As you can see from the diagram below, the increase in *overall surface area* from arteries to capillaries is huge, and this surface area is immediately decreased as capillaries coalesce into venules and then veins for the trip back to the heart.

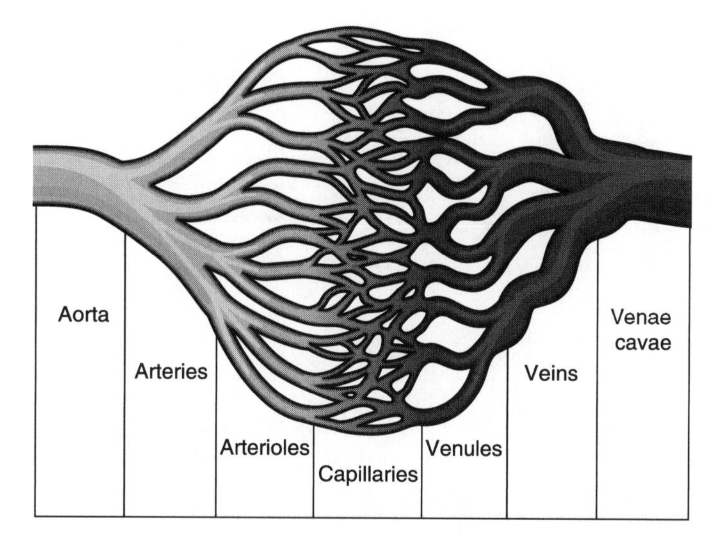

Because of this increase in surface area and friction between the blood cells and the blood vessel walls, average blood pressure decreases the farther away blood travels from the heart. In the veins, there is virtually no blood pressure at all.

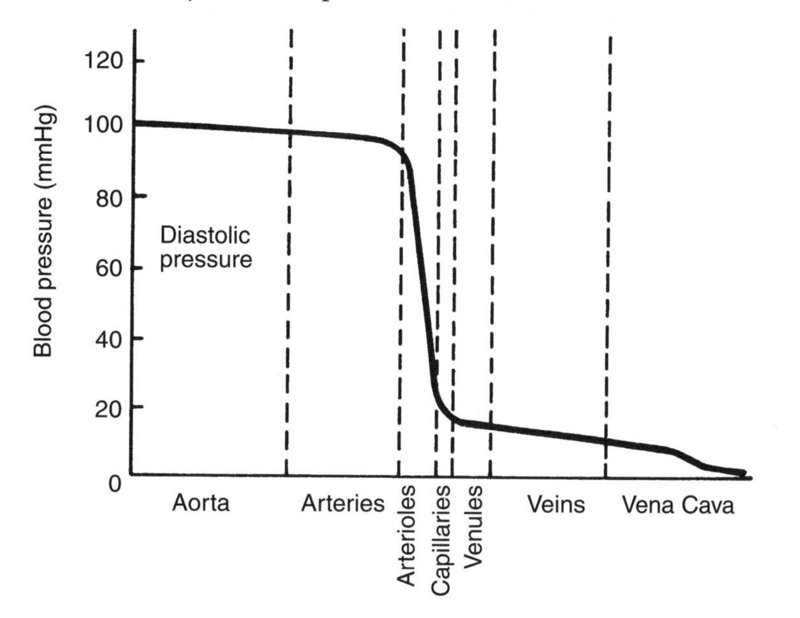

If all the capillaries of the body had blood flowing through them at all times, your overall blood pressure would be close to zero! In other words, you do not have enough blood in your body to supply every capillary with blood all the time. Thus, the body "shunts" this blood from place to place, closing off some capillaries while opening others. This selective dispersal of blood is accomplished by using small bands of muscle, called **precapillary sphincters**, which exist at the base of capillaries as they branch off from nearby arterioles. When these sphincters are closed, the blood will simply bypass the capillary bed, moving from the arteriole directly into the venule on the other side of the bed. When open, these sphincters allow blood to pass into the capillary bed to nourish the tissue cells there.

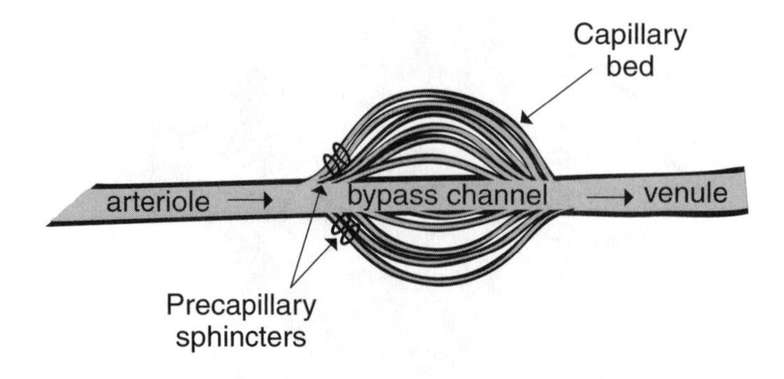

Another consequence of regulating blood flow using precapillary sphincters is temperature regulation. Have you ever noticed how your skin turns whitish and pale when it is cold outside, and redder when it is hot? Exercising also causes the skin to become hotter and redder in color. The reason for this is that when it is cold outside, autonomic nerves in your skin cause as many precapillary sphincters as possible to close, in order to conserve heat (the more blood near the surface of your body, the faster you lose heat). When it's hot or when you're exercising, many sphincters near the surface of the skin will be open, allowing more efficient cooling of the body. **Vasodilation**, or keeping open lots of blood vessels in the extremities, and sweating are the two main mechanisms the body uses to cool itself.

CHAPTER FOURTEEN

Excretion and Metabolism

Excretion refers to the removal of **metabolic wastes** produced in the body. It is distinguished from elimination, the removal of indigestible material. Most of the body's activities produce metabolic wastes that must be removed. Aerobic respiration leads to the production of carbon dioxide waste. **Deamination** of amino acids leads to the production of nitrogenous wastes, such as urea or ammonia. All metabolic processes lead to the production of mineral salts, which must also be excreted.

COMPARATIVE ANATOMY AND PHYSIOLOGY OF OSMOREGULATION

Just like digestion, respiration, and transport, excretion is built upon a foundation of *moist membranes for diffusion and lots of surface area.* The most basic form of excretion takes place in single-celled organisms, such as the bacteria and protists, whose cells are all in contact with the external, aqueous environment. Water-soluble wastes, such as **ammonia** and **carbon dioxide**, can exit the cells via diffusion through the cell membrane. This type of excretion is completely passive and is based upon a concentration gradient that works because wastes build up inside the cell. Wastes are in higher concentration inside than outside. Some freshwater protozoans, such as the paramecia, also possess **contractile vacuoles**, organelles that can actively pump excess water to the outside of the cell, much like bailing out a sinking boat. This allows the paramecia to maintain appropriate water volume and osmotic pressure even in the face of a low salt environment around it that pushes water into the cell.

In animals such as the hydra, much of the same kind of excretion exists as in the protists and bacteria. Simple diffusion into the gastrovascular cavity effectively removes ammonia and carbon dioxide wastes that build up. In flatworms, the excretory system transports waste throughout a series of tubes, **protonephridia**, that end in hollow bulbs known as

flame cells, so named for the cilia that flicker back and forth within them to keep the fluid moving. Excess water and salts enter the flame bulbs and are pushed into the protonephridia. Salts are reabsorbed into the flatworm, as the excess water is eliminated in a dilute urine. In more complex animals, such as the annelid worms (e.g., earthworms), carbon dioxide excretion takes place directly through the moist skin, but specialized tubes known as **nephridia** are used to excrete mineral salts and the nitrogenous waste, urea. Each segment of an earthworm has a pair of nephridia (also known as **metanephridia**), which pull fluid from the adjacent section.

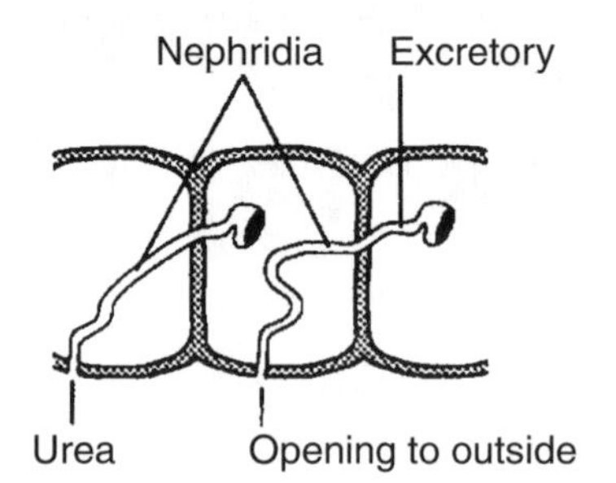

These tubes allow the reabsorption of most salts back into the internal cavity, while excreting a very dilute urine. Because moist skin of the earthworms allows for the absorption of lots of water from the surrounding soil, the nephridia help in osmoregulation, or water balance. In insects, **Malphigian tubules** are outfoldings of the digestive tract in the midgut of the organism. These tubes absorb water, nitrogenous wastes, and salts from the **hemolymph**, the circulatory fluid in the body of the insect. At the rectum, water and salts are mostly reabsorbed, allowing the excretion of a fairly concentrated, almost solid urine, helping to conserve water on land. Nitrogenous wastes excreted in solid form are usually **uric acid** crystals. Carbon dioxide is released back into **tracheae**, into which oxygen had been drawn for respiration.

MAMMALIAN KIDNEYS AND HORMONAL REGULATION

The principle organs of excretion in humans are the lungs, liver, skin, and kidneys. In the lungs, carbon dioxide and water vapor diffuse from the blood and are continually exhaled. Sweat glands in the skin excrete water and dissolved salts (and small quantities of urea). Perspiration serves to regulate body temperature, since the evaporation of sweat produces a cooling effect. The liver processes blood pigment wastes and other chemicals for excretion. **Urea** is produced by the deamination of amino acids in the liver, and it diffuses into the blood for excretion in the kidneys. The kidneys function to maintain the osmolarity of the blood and to conserve glucose, salts, and water.

On an average day-to-day basis, an adult human maintains more or less the same weight, a relatively constant blood pressure, and a certain amount of salts, metabolic wastes, and water volume in his tissues. This is due to the fact that **renal** (kidney) **output** varies around a set point for each individual, whereby renal output of salt, water, and urea goes up as dietary intake increases, and goes down as dietary intake decreases. The mechanism by which the kidney achieves this is through filtration of the blood through many small capillary bundles called **glomeruli** (singular: **glomerulus**) and reabsorption or rejection of this filtrate as it passes down long, twisty tubes known as **nephrons.** Within each kidney, there are perhaps 1,000,000 glomeruli, each attached to a nephron tube. The **filtrate** that makes it through the nephron tubes and is reabsorbed back into the bloodstream is called the final filtrate, or **urine.** It empties from the ends of the nephrons into the renal pelvis and is then carried to the bladder for storage an eventual release during urination.

Accumulation of nitrogenous waste is dangerous; it can cause changes in mental alertness and bleeding, particularly in the stomach and small intestine.

The Structure of the Kidney

The kidney is divided into three regions: the outer **cortex**, the inner **medulla**, and the **renal pelvis.**

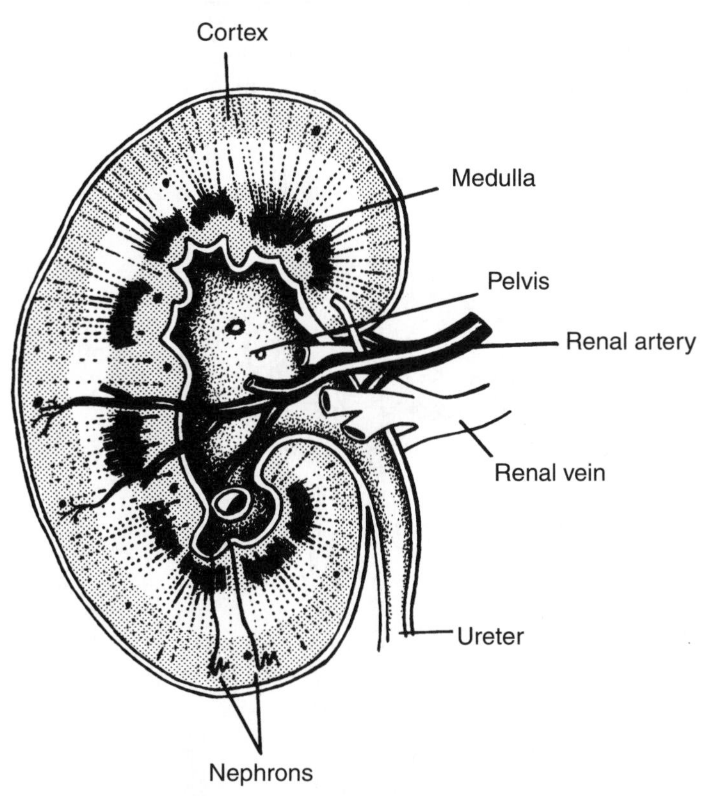

Approximately 200–300 liters of blood pass through the two kidneys every day, entering via the **renal artery** and exiting via the **renal vein.** The average human has only 6 or 7 liters of blood in his body, which means that the *entire blood volume is swept through the kidneys at least 20 times every day.* As blood enters, it is quickly disseminated into tiny capillaries that end at a glomerulus, a tiny ball of capillaries contained within a pouch called **Bowman's capsule.** Bowman's capsule leads into a long and coiled nephron that is divided into functionally separate units: the **proximal convoluted tubule**, the **loop of Henle**, the **distal**

convoluted tubule, and the **collecting duct**. The loop of Henle of most nephrons dips down through the medulla of the kidney, while the glomeruli and convoluted tubules remain positioned in the cortex.

Filtration and Urine Production

In the glomerulus, blood is filtered out of capillaries into Bowman's capsule and the nephron. Many substances do not pass through the glomerulus into the nephrons; instead, they simply exit the kidneys through the renal vein without ever passing through the nephron tubes. Large proteins, such as **albumin**, and proteins with lots of negative charge, are repelled due to the structure of the glomerulus, with its triple-layered capillary cells. Cells remain within the bloodstream. Water, salts, sugars, and urea, however, do enter the nephron, and this fluid entering the nephron is called the filtrate. Notice in the diagram below how the bloodstream drops off material at the glomerulus and then doubles back to wind very closely around the proximal and distal tubules and the loop of Henle. The closeness of the bloodstream to the filtrate in the nephron tube sets up a concentration gradient whereby much of the material in the filtrate will diffuse back into the nearby blood, or be moved there by active transport. After filtration, the next step is the selective reabsorption of most of the filtrate.

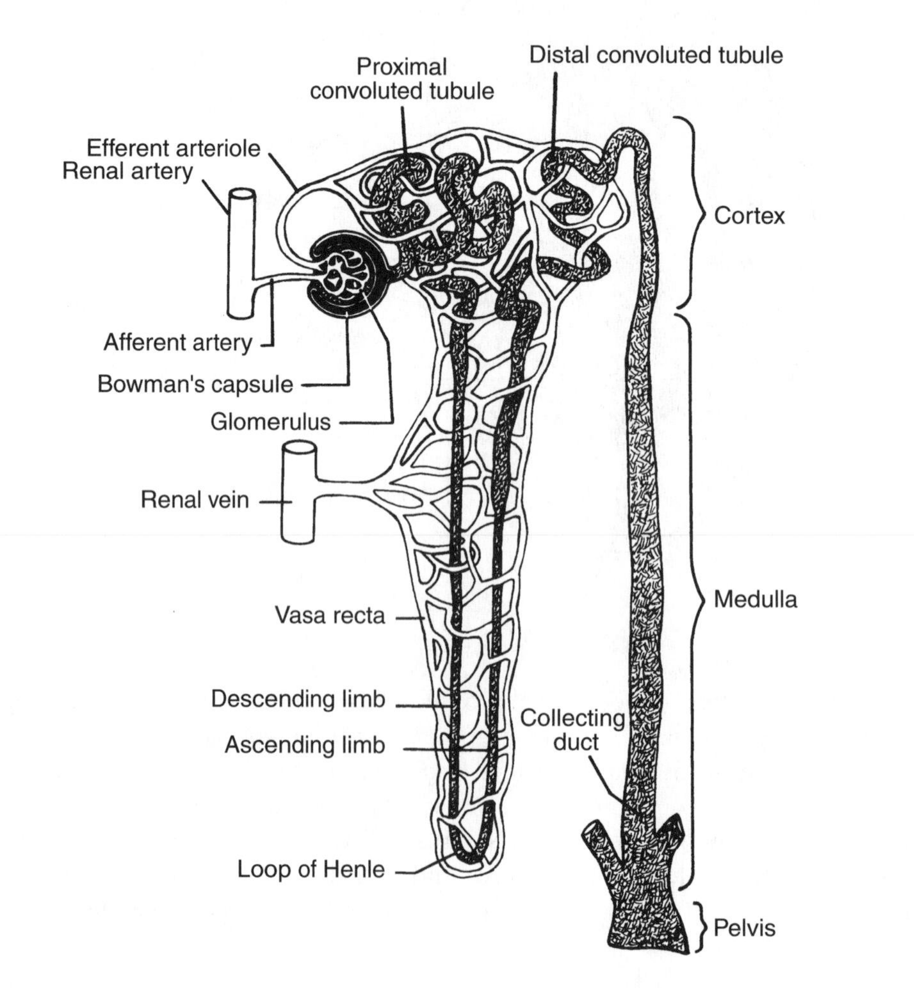

The filtrate finds itself within the proximal tubule. On the surface of the cells that line this tubule are Na^+/H^+ ion exchange pumps. While kicking out hydrogen ions into the tube, these membrane proteins absorb sodium from the filtrate so that it travels into the cells that line the tubule. This sodium moves through to the opposing end of the cells where it is extruded into the bloodstream by a Na^+/K^+ ATPase pump. The end result is the movement of sodium ions from the filtrate into the bloodstream. Water naturally follows due to the hypertonic conditions that occur as salt concentration goes up in the nearby bloodstream, and it is here in the proximal tubule that most of the water and salt that were originally filtered out of the blood are returned back into the blood.

Of the 200 liters of fluid that pass through the kidneys each day, total urine volume excreted is only 1 or 2 liters. That means 99 percent of what passes into the kidneys is reabsorbed back into the bloodstream.

Proximal Tubule

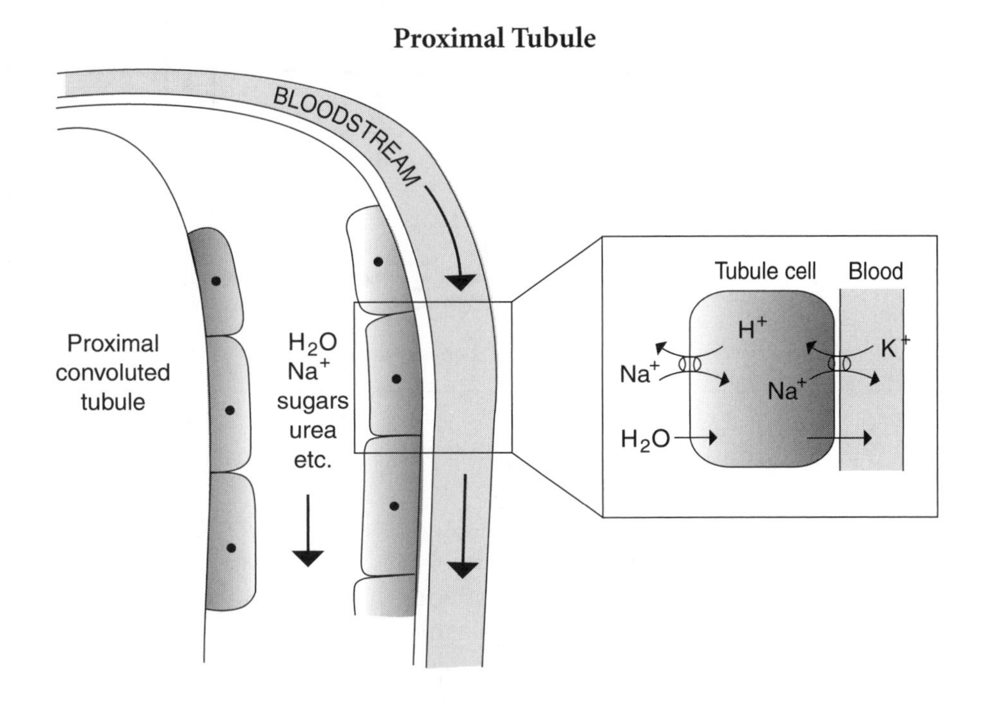

The proximal tubule is also responsible for the reabsorption of glucose using active transport. In a healthy individual, there is no glucose excreted in the urine because all glucose is efficiently reabsorbed in the proximal tubule. In the Loop of Henle, either water or salt is absorbed independently. The descending part of the loop absorbs water only, and the ascending loop absorbs salt ions only. More water and salt reabsorption takes place in the distal tubule, and in the collecting duct, certain hormones regulate the concentration of the urine.

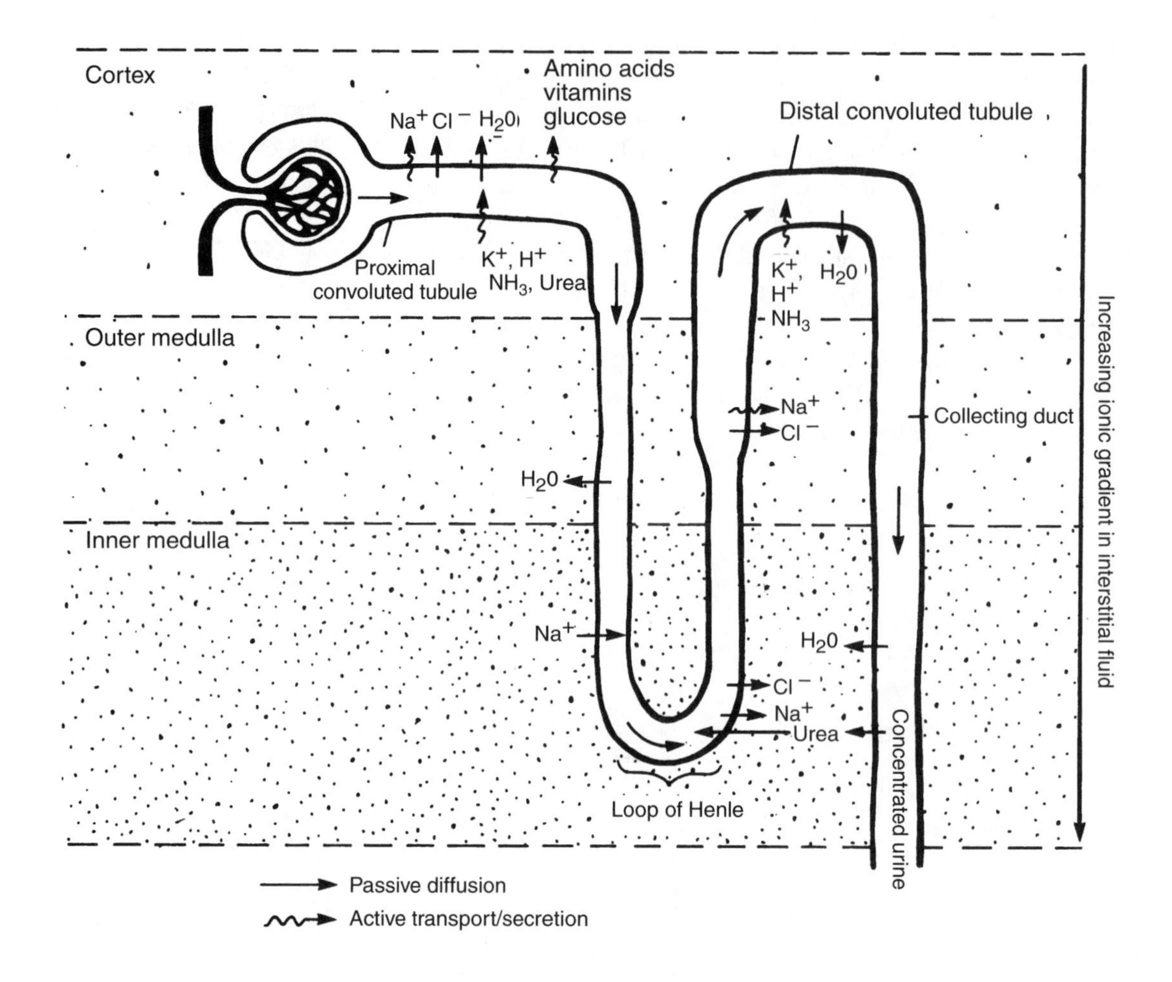

The selective permeability of the tubules establishes an **osmolarity gradient** in the surrounding interstitial fluid. By exiting and reentering at different segments of the nephron tube, solutes create a situation in which tissue osmolarity increases from the cortex into the inner medulla. While most of the urea that ends up in the collecting duct is excreted, some diffuses into the nearby interstitial fluid and reenters the ascending loop of Henle. Overall, the gradient that is established by the concentration of solutes in the kidney medulla helps to maximize water conservation and make the urine as hypertonic (concentrated) as possible.

Hormonal Regulation

Cells in the collecting duct called **principle cells** respond to various hormones, such as **aldosterone**, to absorb more water or salt depending on how concentrated the filtrate is at that point. Aldosterone is produced in the adrenal cortex and stimulates the reabsorption of Na^+ from the collecting duct. Na^+ stimulates water reabsorption as well, so aldosterone is released when blood pressure falls in the body. People with Addison's disease produce insufficient aldosterone and have urine that is too high in Na^+ ion concentration. This causes dehydration and a drop in blood pressure, as water is pulled out of the bloodstream to follow the excess salt in the urine. Another hormone, **angiotensin**, is also produced when blood volume falls, and it works in a similar fashion to aldosterone: retaining water in order to increase blood volume.

ADH, or antidiuretic hormone, also controls the concentration of the urine and acts on cells lining the loop of Henle collecting duct of the nephron. As its name implies, release of this hormone, which takes place in the posterior pituitary, causes water reabsorption and a decrease in urine output. ADH is also known as **vasopressin**, and it works mainly by opening up active transport water channels called **aquaporins** on the luminal surfaces of principle cells in the collecting duct. This causes water to leave the filtrate and move into the medulla, where it is eventually reabsorbed into the bloodstream, leaving the filtrate more concentrated and the eventual urine much less dilute.

WHAT EVENTS WOULD MAKE BLOOD PRESSURE FALL?

Blood is mostly water, so anything that causes water loss, such as vomiting, diarrhea, blood loss, or heavy sweating, can cause a drop in blood pressure. The body responds by releasing aldosterone and angiotensin, which allow the cells in the collecting duct to reabsorb more salt and water. This extra fluid adds volume, and therefore pressure, back to the blood.

THERMOREGULATION AND HOMEOSTASIS

Homeostasis is the process by which a stable internal environment is maintained within an organism. Important homeostatic mechanisms include the maintenance of a water and solute balance (osmoregulation), which you have just learned about, regulation of blood glucose levels, and the maintenance of a constant body temperature. In mammals, the primary homeostatic organs are the kidneys, the liver, the large intestine, and the skin.

Thermoregulation

The hypothalamus is the brain region that acts to control the body temperature of an organism that is able to set its internal temperature. Mammals fall into this category. Hormones such as **epinephrine** or thyroid hormone, released by the adrenal glands, can increase one's metabolic rate and, therefore, one's heat production. Muscles can generate heat by contracting rapidly (**shivering**). Heat loss is regulated through the contraction or relaxation of precapillary sphincters, as explained in the last chapter. Alternative mechanisms are used by some mammals to regulate body temperature. **Panting** is a cooling mechanism that results in the evaporation of water from the respiratory passages. **Sweating** is also important in increasing evaporative heat loss, so that the body cools. Fur is used to trap heat, and hibernation during the winter conserves energy by decreasing heart rate, breathing, and metabolism. Animals able to regulate their internal temperature even in the face of a changing external temperature are called **endotherms** or homeotherms. Mammals and birds are capable of this type of regulation, yet reptiles, amphibians, and most other animals are not and are known to be "cold-blooded," or **ectotherms**.

There are two laws you should know for the GRE that relate heat and body size:

1. Bigger bodies produce less body heat per pound per hour.
2. Bigger bodies lose less body heat per pound per hour.

Heat regulation is found even among some plants, such as the skunk cabbage and the Lotus. Lotus plants can maintain their internal temperatures at about 85 degrees Fahrenheit, even as the temperature around them drops to near 50 degrees. This seems to play an important role in pollination and reproduction.

Metabolic heat production drops in a very specific manner as body size increases. Compared to an elephant that might weigh 10,000 pounds, a small mammal weighing only one pound produces about 10 times more heat per pound than the elephant does. Yet of the two animals, the elephant stays warmer since it has much less overall surface area compared to internal volume than does the small mammal. In other words, the small mammal gives off much more heat to its surroundings.

CHAPTER FIFTEEN

Nervous/Endocrine Systems and Behavior

Even the smallest, single-celled organisms on Earth respond to their environment. They can respond to light or to chemicals in their environment, and these stimuli cause certain behavioral responses, such as orienting themselves toward or away from a particular stimulus. In fact, many of the receptors and signaling molecules used by single-celled organisms are used in the same way by more complex, multicellular life. Bacteria, for example, can change the direction of their flagella according to the presence of certain chemicals around them. These chemicals bind to cell-surface receptor proteins, which initiate an intracellular signal ending with the movement of flagella in a certain direction. Even light can open ion channels in the membranes of many different organisms, from bacteria to plants. In more complex organisms, chemical signals are balanced with electrical signals in order to achieve a balance of short-term and long-term behavioral responses. This chapter will look at the nervous and endocrine systems, and their effects on animal behavior.

THE MAMMALIAN NERVOUS SYSTEM AND BRIEF COMPARATIVE ANATOMY

Simple nervous systems in animals can be found in creatures such as the hydra or in other cnidarians. While the hydra lacks a central nervous system, it does possess a very basic **nerve net** made of modified neurons that allows impulses to travel in either direction up or down the body to coordinate simple movements. The annelid worms (e.g., earthworms) and other worms often have primitive brains made of **ganglia**, fused groups of neuron cell bodies that extend their axons down the length of the worm. There is a ventral nerve cord from which arise peripheral nerves that extend across the width of the worm.

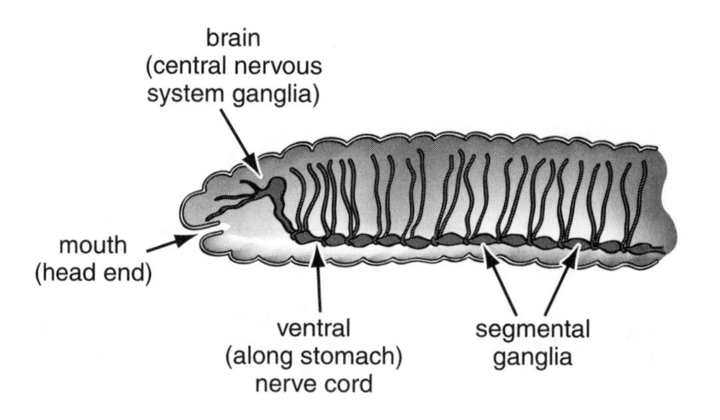

Insects and other arthropods have a nervous system similar to that of the annelid worm above, yet have taken this a step further in their development of specialized sensory structures such as antennae and eyes. Keep in mind that all animals, except the simplest invertebrates, have *three main components* to their nervous systems: complex nerve clusters (brain) at their head end, a central nerve cord (either dorsal or ventral), and peripheral (side) nerves that come off this central nerve cord. These components are responsible for the *three main functions* of a nervous system: sensory input, motor output, and integration, which will be discussed as the chapter progresses.

The Mammalian Nervous System

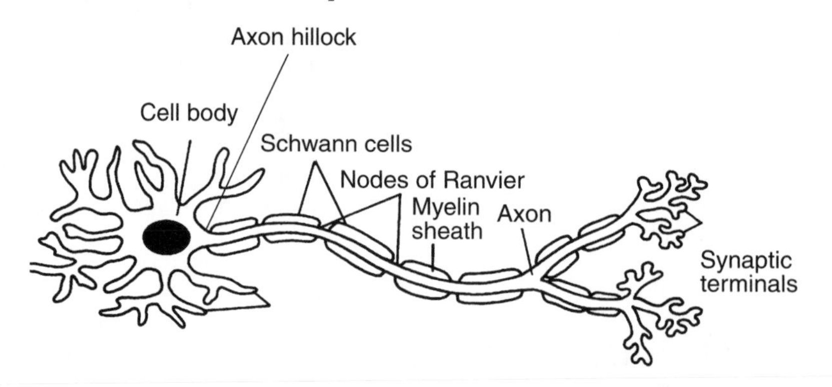

The nerve cell, or neuron, is the basic functional unit of the mammalian nervous system. Each neuron has the following components:

Cell body: contains the nucleus and organelles; site of protein synthesis

Dendrites: extensions of cytoplasm from the cell body that receive chemical signals from nearby neurons and can initiate a new electrical signal

Schwann cells: separate cells from the neurons; secrete myelin, an insulation for nerve cells that helps make signals move faster down the axon

Nodes of Ranvier: spaces along myelinated axons where Schwann cells have not laid down myelin covering

Axon: the main elongated extension of the cell body through which the electrical signal travels; signal in one direction only from cell body toward synaptic knobs

Synaptic knobs: release neurotransmitters, chemicals that communicate with surrounding nerve cells

Dendrites are highly branched and can receive signals from many different neurons at the same time. In fact, some specialized neurons such as the Purkinje cells of the brain can receive signals from over 1,000 different neurons at once into a single cell body. The amount of integration that takes place in a cell such as this, filtering out and summing together all of this input, is enormous. Most neurons have a single, long axon enclosed in layers of **myelin.** Myelin is a complex polysaccharide and lipid membrane, which is actually a continuation of the membranes of nearby Schwann cells that throw off myelin like blankets in order to cover nearby axons. Myelin insulates the axon, much like electrical tape insulates bare wire, and neurons that lose their myelin sheaths cannot transmit signals fast enough to give appropriate stimuli to muscles or organs. Each Schwann cell can contribute a single internodal segment of myelin (notice in the diagram above how the myelin wrapping occurs in segments, called nodes). Therefore, many Schwann cells are needed to coat a nearby axon with myelin. In the central nervous system (*brain and spinal cord*), the equivalent cell to a Schwann cell is known as an **oligodendrocyte.** Whereas a single Schwann cell can myelinate only a single axon, oligodendrocytes can send off myelin sheaths in several directions at once.

Multiple sclerosis is an autoimmune disease in which the myelin sheaths on certain axons are degraded and destroyed. Patients gradually lose their ability to control muscles and glands, since the myelination is necessary for accurate and timely transport of signals.

Schwann cells and oligodendrocytes are not the only "support cells" present around neurons. **Astrocytes** in the central nervous system (CNS) far outnumber the neurons of the CNS. The extensions of astrocytes stick to various parts of neurons and help to break down and remove certain neurotransmitter chemicals as well as engulf debris. **Ependymal cells** line the fluid-filled cavities of the brain and spinal cord and secrete cerebrospinal fluid (CSF) that helps cushion the CNS.

Resting and Action Potentials

Signals along the axon are electrical in nature and depend upon the flow of ions across the axon membrane and cell body membrane. How does electrical voltage arise in a cell and how does a signal arise from ion flow? All living cells have an electrical charge difference across their membranes, the inside of the cells being more negatively charged than the outside. The reasons for this are:

1. DNA is a negatively charged molecule due to lots of negative phosphate groups (remember the basic units of nucleotides).

2. Many proteins (amino acid side chains) in the cell are negatively charged.

3. The Na/K (sodium/potassium) pumps in the cell membrane kick out three positive sodium (Na^+) ions for every two positive potassium (K^+) ions they move into the cell. Overall, that means that one positive charge is leaving the cells every time one Na/K pump "turns."

4. While sodium is kicked out of the cell and potassium is brought in due to the action of the Na^+-K^+ ATPase pump (ATPase means that active transport is involved here—the pump uses ATP to work), some potassium leaks out. This passive diffusion of potassium through "leakage" channels works along with the sodium-potassium pump to create an overall charge difference across the axon or cell body membrane.

Resting Potential of a Neuron

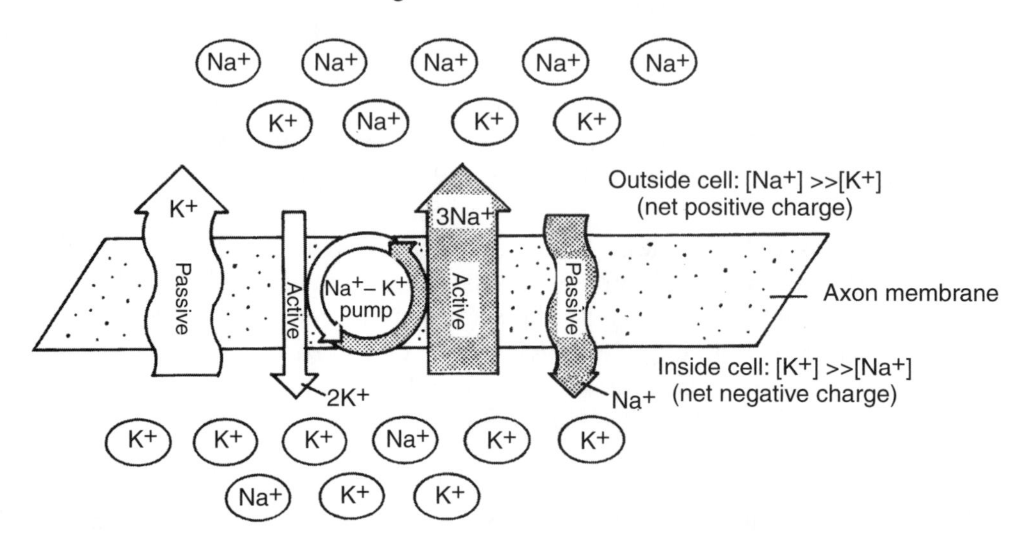

This voltage difference across the membrane can be measured with ultrafine electrodes stuck into the cell membrane, and the charge difference is called a **membrane potential**. For almost all neurons at rest, this membrane is about –70 millivolts (mV). The outside of the cell is taken to be at 0 mV, which means that the inside is –70 mV with respect to the outside. The principle cation (positively charged ion) inside neurons is K^+, while the principle cation outside is Na^+. The principle anions (negatively charged ions) inside neurons are the proteins, amino acids, and phosphate groups, while the principle anion outside is chlorine (Cl^-).

Ions do *not* readily travel across cell membranes because of their charges, and they must be carried across either by membrane-spanning proteins or specific protein channels that let only certain types of ions through. Some channels let only Na^+ through (sodium channels), some only K^+, some only Cl^-, etc.

Although all cells have membrane potentials (essential for proper osmotic balance), only certain ones such as neurons can generate changes in this potential. As explained before, the resting membrane potential of a neuron is approximately -70 mV. In response to particular stimuli, special "gated" channels in the membrane will open. As these voltage-gated channels open, a nerve response can be initiated if the membrane depolarizes above threshold. The first channels to open are voltage-gated

Na^+ channels, which allow the rush of sodium ions into the cell due to the concentration gradient that had been established by the sodium-potassium pumps. At that point, the inside of the neuron becomes very positively charged where sodium channels have opened, due to the rush of these positive ions inward. Shortly after this **depolarization**, potassium channels open, allowing the rush of K^+ ions out of the cell given the established concentration gradient. This outflow of positive charge, known as **repolarization**, restores the membrane potential back toward the resting potential of –70 mV. Occasionally, so many K^+ ions flow out of the cell that the membrane potential undershoots the resting charge of –70 mV, a phenomenon called **hyperpolarization**. The graph below sums up this "action potential" or nerve impulse:

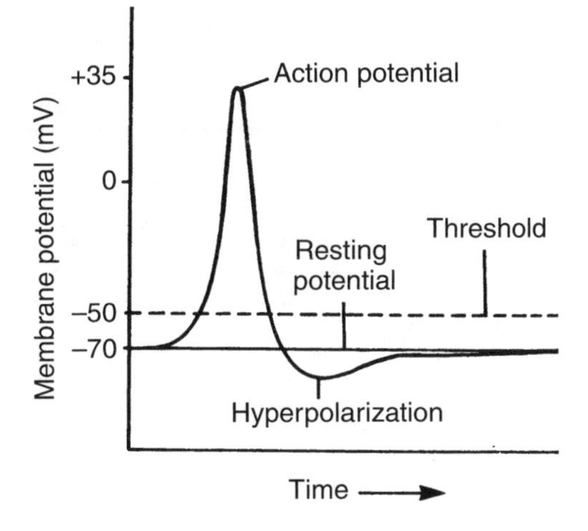

Sodium and potassium channels are open for a very short time (milliseconds) only, yet the rush of positive charge into one region of a neuron can set off a cascade of channel opening along the entire length of the axon. Voltage-gated sodium channels adjacent to where the axon first depolarizes (usually at the axon hillock) will open, causing sodium channels further down the axon membrane to do the same. In such a way, the action potential is propagated down the entire length of the axon. Once opened, sodium channels inactivate for a brief period of time, which means that the action potential (flow of positive charge down the axon) occurs in one direction only, down toward the axon terminal. The inability of channels to open again creates a refractory period in which another impulse cannot travel along the axon. This period limits the number of signals that can travel through the axon in a given period of time.

Nerve cells respond to a wide variety of stimuli, including light, temperature, pressure, taste, and touch by opening up sodium channels.

Action potentials are an all-or-none response. Every action potential is of the same magnitude. Larger signals are transmitted by the firing of a larger number of neurons or by a greater action potential frequency, NOT by larger action potentials.

Propagation of an Action Potential

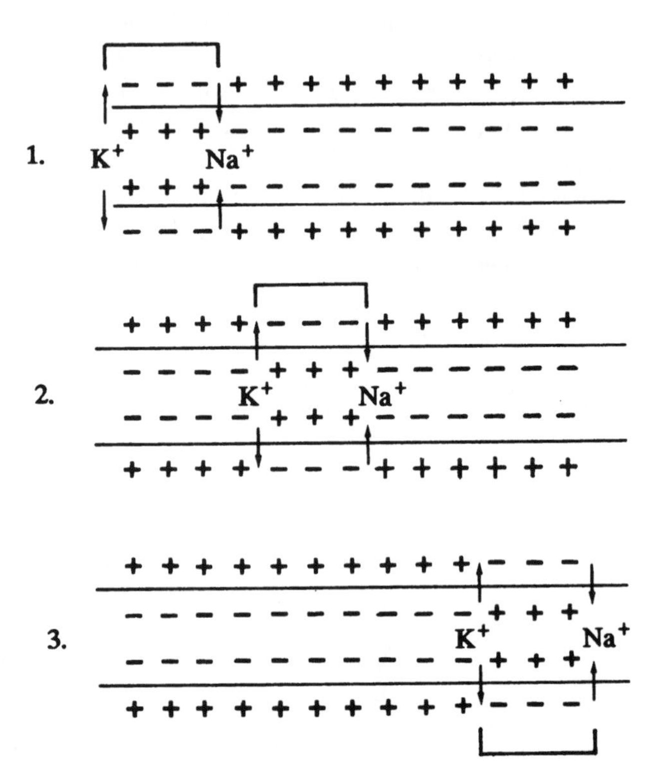

Different kinds of axons will propagate the action potential at different speeds. The *larger diameter axons have faster transmission than small diameter ones* (Recall your physics knowledge: Resistance to the flow of electrical current is inversely proportional to the cross-sectional area of the "wire" carrying the current). The membrane of the neuron is not a perfect conductor of electricity, and the current of the action potential will diminish over time as the signal travels down the axon if it is not replenished. That is why successive depolarizations down the axon membrane succeed in bringing the charge all the way from the cell body of the neuron to the axon terminal. Myelin increases the speed at which the signal travels, as well as holds the temporary influx of positive charge inside the axon. Because only the spaces in between the myelin, called **nodes of Ranvier**, are permeable to ions, action potentials do not propagate smoothly through myelinated axons but rather jump from node to node in "saltatory" fashion. This allows for much quicker signal transmission down the axon.

The **synapse** is the gap between the axon terminal of one neuron (called the presynaptic neuron because it is before the synapse) and the dendrites of another neuron (called the postsynaptic neuron). Neurons can also communicate directly with other postsynaptic cells, such as those on glands or muscles. The vast majority of synapses are chemical, whereby the electrical signal in the presynaptic neuron is converted into a chemical signal in the synapse that then incites a new electrical signal in the postsynaptic neuron. Electrical synapses can occur where gap junctions join the cytoplasms of neighboring cells, directly transmitting ionic signals.

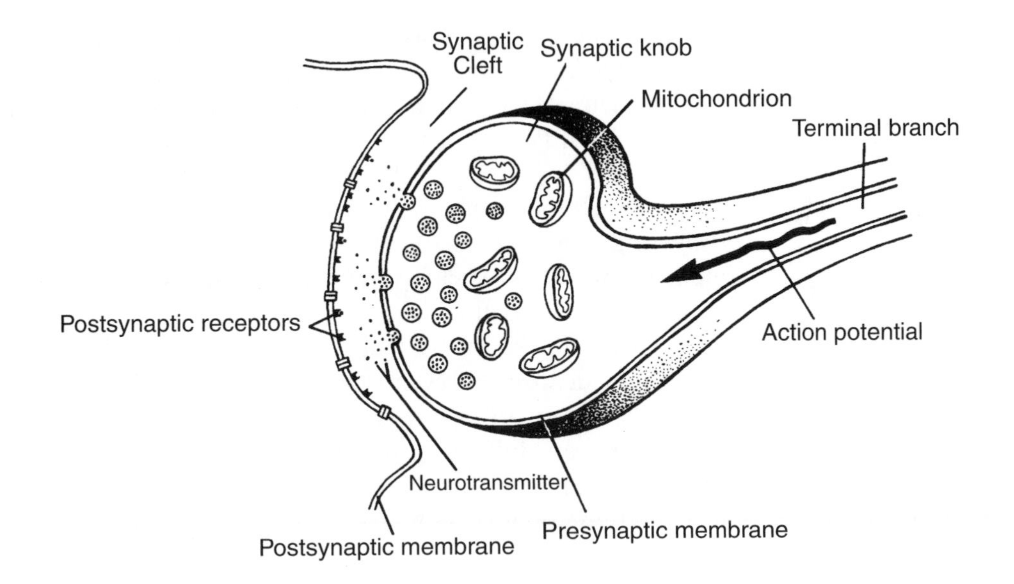

As the action potential reaches the synaptic knob (terminal), voltage-gated calcium (Ca^{2+}) channels at the terminal end open up, and the rapid influx of calcium ions causes membrane-bound vesicles filled with neurotransmitters to merge with the presynaptic membrane, releasing their contents into the synapse. The synapse is a very small space, comprising at most a few micrometers from the presynaptic neuron to the postsynaptic one. Excitatory neurotransmitters, such as **acetylcholine** (ACh) will bind to membrane receptors on the postsynaptic dendrites or cell body and cause the opening of sodium channels (ligand-gated channels) on the postsynaptic cell. This starts the process of an action potential all over again.

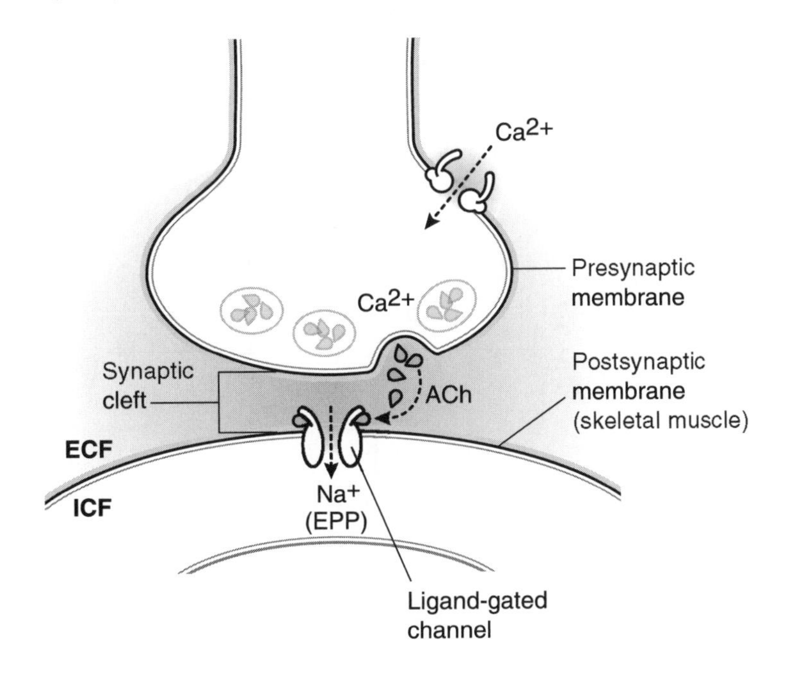

This excitatory impulse is known as an excitatory postsynaptic potential, or EPP for short, as can be seen in the diagram above. There are, however, inhibitory neurotransmitters that open up potassium channels or chlorine channels rather than sodium ones. This causes the outflow of positive K^+ ions from the cell or the inflow of negative Cl^- ions into the cell, making the inside even more negatively charged than it was at rest. This hyperpolarization effectively stops the action potential from traveling to the postsynaptic neuron. Inhibitory neurotransmitters, such as **GABA** and **glycine**, are used for example in pain pathways to moderate the amount of pain you feel when you are hurt. Notice in the diagram below how an inhibitory neuron can even synapse on an neuron that releases excitatory neurotransmitter, helping to control how that excitatory axon influences the postsynaptic cell. Excitatory and inhibitory signals are all integrated by the postsynaptic neuron to determine if that neuron will reach threshold and trigger a new action potential.

Motor neurons travel from the CNS to muscles and glands in the PNS (peripheral nervous system). They are *only* excitatory. **Sensory** neurons travel from the PNS to the CNS. They can be either excitatory or inhibitory.

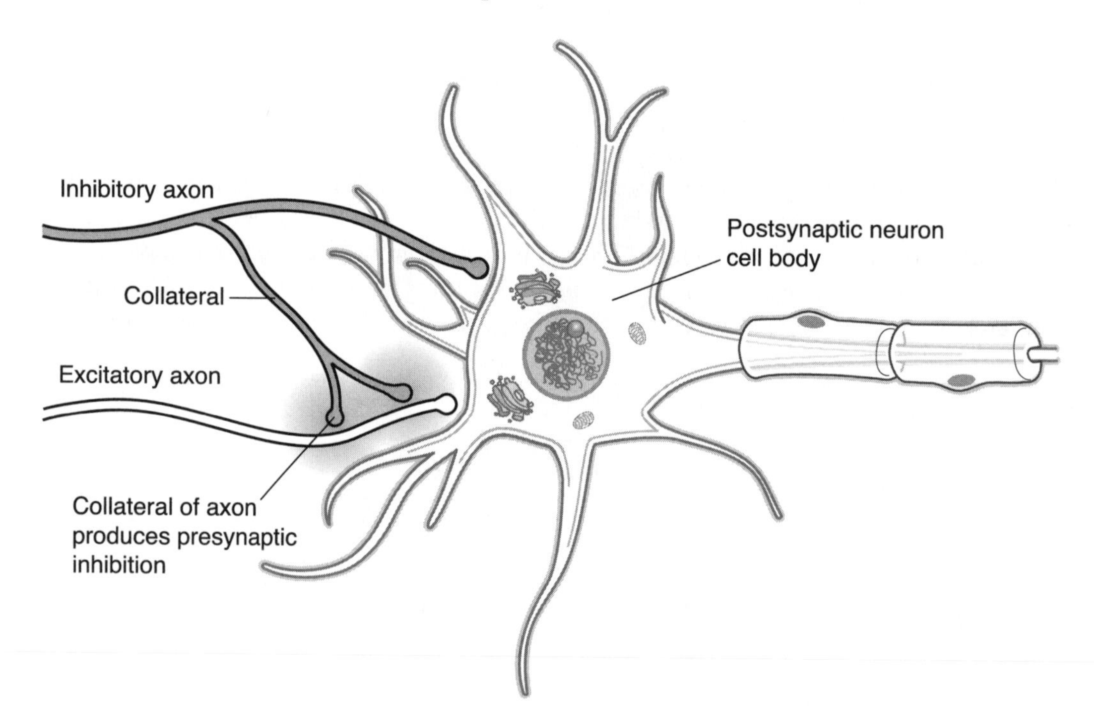

ORGANIZATION OF THE VERTEBRATE NERVOUS SYSTEM

There are many different kinds of neurons in the vertebrate nervous system. Neurons that carry information about the external or internal environment to the brain or spinal cord are called **afferent neurons.** Neurons that carry commands from the brain or spinal cord to various parts of the body (e.g., muscles or glands) are called **efferent neurons.** Some neurons (**interneurons**) participate only in local circuits; their cell bodies and their nerve terminals are in the same location.

Nerves are essentially bundles of axons covered with connective tissue. A nerve may carry only sensory fibers (**a sensory nerve**), only motor fibers (**a motor nerve**), or a mixture of the two (**a mixed nerve**). Neuronal cell bodies often cluster together; such clusters are called **ganglia** in the periphery; in the central nervous system they are called **nuclei.** The nervous system itself is divided into two major systems, the **central nervous system** and the **peripheral nervous system.**

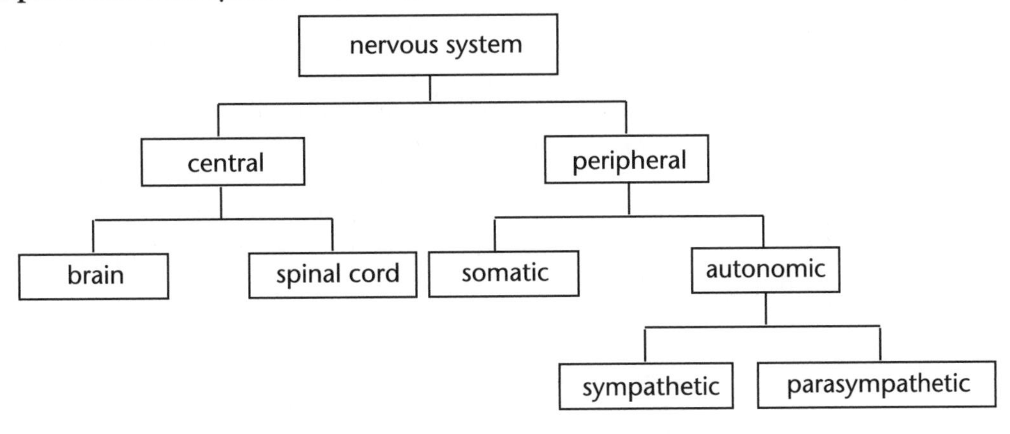

Central Nervous System

The central nervous system (CNS) consists of the **brain** and **spinal cord.**

Brain

The brain is a jellylike mass of neurons that resides in the skull. Its functions include interpreting sensory information, forming motor plans, and cognitive function (thinking). The brain consists of **gray matter** (cell bodies) and **white matter** (myelinated axons). The brain can be divided into the **forebrain**, **midbrain**, and **hindbrain.**

a. **Forebrain**

The forebrain consists of the **telencephalon** and the **diencephalon.** The telencephalon consists of right and left hemispheres; each hemisphere can be divided into four different lobes: **frontal, parietal, temporal,** and **occipital.** A major component of the telencephalon is the **cerebral cortex,** which is the highly convoluted gray matter that can be seen on the surface of the brain. The cortex processes and integrates sensory input and motor responses and is important for memory and creative thought. Right and left cerebral cortices communicate with each other through the **corpus callosum**.

The diencephalon contains the **thalamus** and **hypothalamus**. The thalamus is a relay and integration center for the spinal cord and cerebral cortex. The hypothalamus controls visceral functions such as hunger, thirst, sex drive, water balance, blood pressure, and temperature regulation. It also plays an important role in the control of the endocrine system.

b. Midbrain

The midbrain is a relay center for visual and auditory impulses. It also plays an important role in motor control.

c. Hindbrain

The hindbrain is the posterior part of the brain and consists of the **cerebellum**, the **pons** and the **medulla**. The cerebellum helps to modulate motor impulses initiated by the motor cortex, and is important in the maintenance of balance, hand-eye coordination, and the timing of rapid movements. One function of the pons is to act as a relay center to allow the cortex to communicate with the cerebellum. The medulla (also called the medulla oblongata) controls many vital functions such as breathing, heart rate, and gastrointestinal activity. Together, the midbrain, pons, and medulla constitute the **brainstem**.

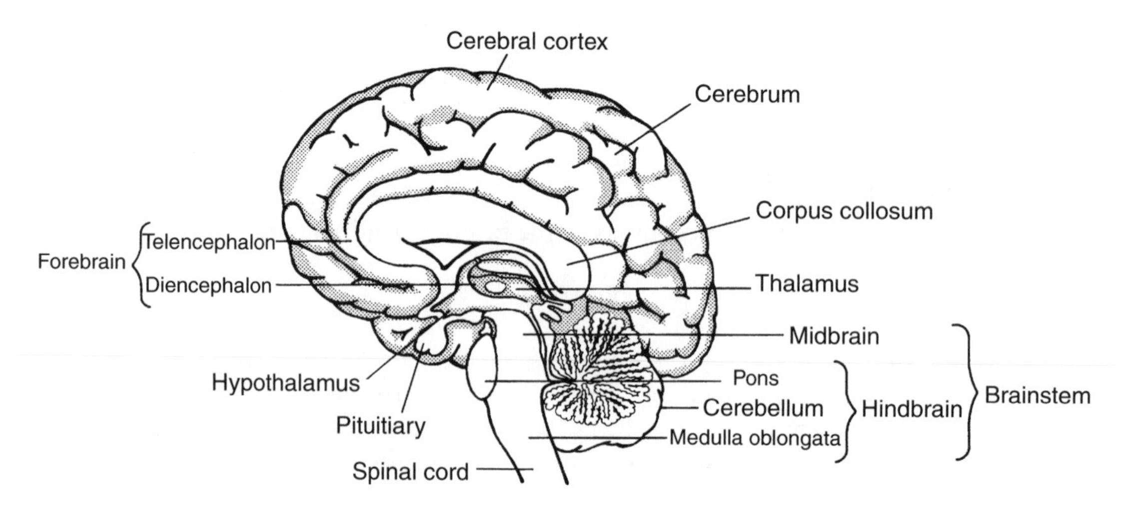

Spinal Cord

The spinal cord is an elongated structure continuous with the brainstem, that extends down the dorsal side of vertebrates. Nearly all nerves that innervate the viscera or muscles below the head pass through the spinal cord, and nearly all sensory information from below the head passes through the spinal cord on the way to the brain. The spinal cord can also integrate simple motor responses (e.g., reflexes) by itself. A cross-section of the spinal cord reveals an outer white matter area containing motor and sensory axons and an inner gray matter area containing nerve cell bodies. Sensory information enters the spinal cord dorsally; the cell bodies of these sensory neurons are located in the **dorsal root ganglia.** All motor information exits the spinal cord ventrally. Nerves branches entering and leaving the cord are called roots. The spinal cord is divided into four regions (going in order from the brainstem to the tail): **cervical, thoracic, lumbar,** and **sacral.**

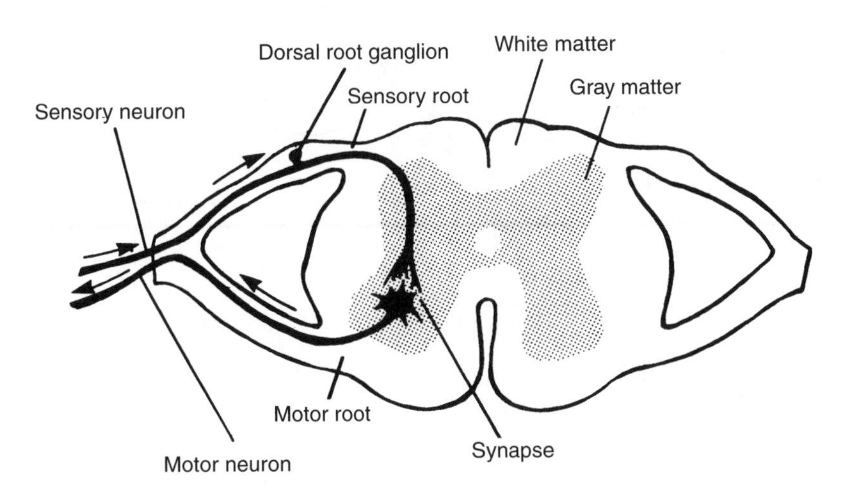

Peripheral Nervous System

The peripheral nervous system (**PNS**) consists of 12 pairs of cranial nerves, which primarily innervate the head and shoulders, and 31 pairs of spinal nerves, which innervate the rest of the body. Cranial nerves exit from the brainstem and spinal nerves exit from the spinal cord. The PNS has two primary divisions: the **somatic** and the **autonomic nervous systems,** each of which has both motor and sensory components.

Somatic Nervous System

The somatic nervous system (SNS) innervates skeletal muscles and is responsible for voluntary movement. Motor neurons release the neurotransmitter acetylcholine (ACh) onto ACh receptors located on skeletal muscle. This causes depolarization of the skeletal muscle, leading to muscle contraction. In addition to voluntary movement, the somatic nervous system is also important for reflex action. There are both **monosynaptic** and **polysynaptic** reflexes.

- Monosynaptic reflex pathways have only one synapse between the sensory neuron and the motor neuron. The classic example is the **knee-jerk reflex.** When the tendon covering the patella (kneecap) is hit, stretch receptors sense this and action potentials are sent up the sensory neuron and into the spinal cord. The sensory neuron synapses with a motor neuron in the spinal cord, which in turn, stimulates the quadriceps muscle to contract, causing the lower leg to kick forward.

- In polysynaptic reflexes, sensory neurons synapse with more than one neuron. A classic example of this is the **withdrawal reflex**. When a person steps on a nail, the injured leg withdraws in pain, while the other leg extends to retain balance.

Reflex Arc for Knee-Jerk

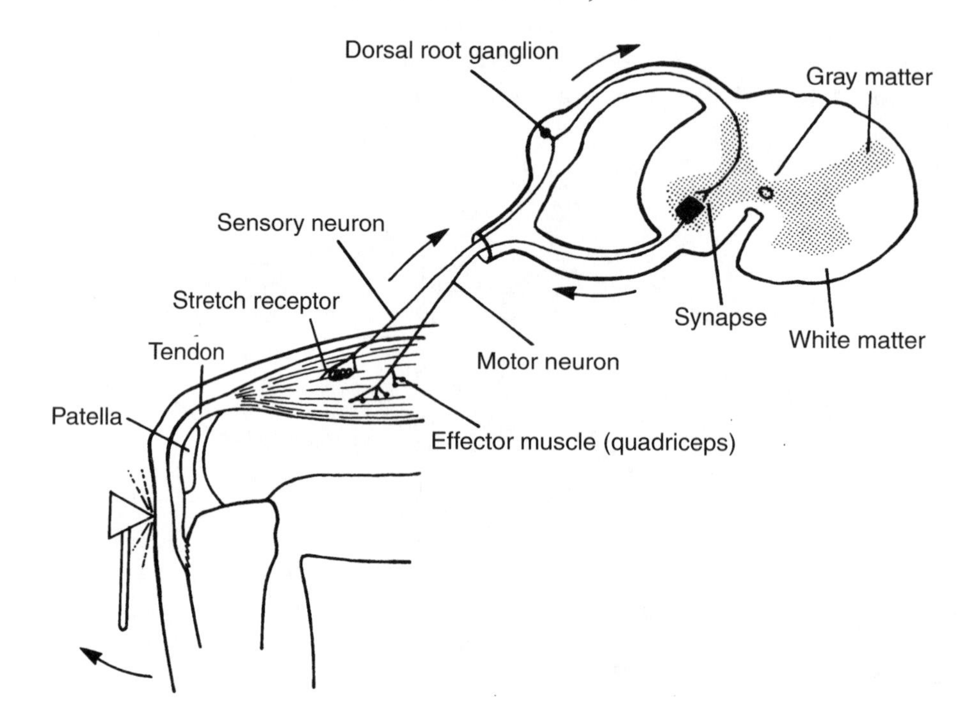

Autonomic Nervous System

The autonomic nervous system (**ANS**) is sometimes also called the involuntary nervous system because it regulates the body's internal environment without the aid of conscious control. Whereas the somatic nervous system innervates skeletal muscle, the ANS innervates cardiac and smooth muscle. Smooth muscle is located in areas such as blood vessels, the digestive tract, the bladder, and bronchi, so it isn't surprising that the ANS is important in blood pressure control, gastrointestinal motility, excretory processes, respiration, and reproductive processes. ANS pathways are characterized by a two-neuron system. The first neuron (preganglionic neuron) has a cell body located within the CNS and its axon synapses in peripheral ganglia. The second neuron (postganglionic neuron) has its cell body in the ganglia and then synapses on cardiac or smooth muscle. The ANS is comprised of two subdivisions, the **sympathetic** and the **parasympathetic nervous systems**, which generally act in opposition to one another.

a. Sympathetic nervous system

The sympathetic division is responsible for the "flight or fight" responses that ready the body for action. It basically does everything you would want it to do in an emergency situation. It increases blood pressure and heart rate, it increases blood flow to skeletal muscles and it decreases gut motility. The preganglionic neurons emerge from the thoracic and lumbar regions of the spinal cord and use acetylcholine as their neurotransmitter; the postganglionic neurons typically release norepinephrine. The action of preganglionic sympathetic neurons also causes the adrenal medulla to release adrenaline (epinephrine) into the bloodstream.

b. Parasympathetic nervous system

The parasympathetic division acts to conserve energy and restore the body to resting activity levels following exertion ("rest and digest"). It acts to lower heart rate and to increase gut motility. One very important parasympathetic nerve which innervates many of the thoracic and abdominal viscera is called the **vagus nerve**. Parasympathetic neurons originate in the brainstem (cranial nerves) and the sacral part of the spinal cord. Both the preganglionic and postganglionic neurons release acetycholine.

Special Senses

The body has three types of sensory receptors to monitor its internal and external environment: **interoceptors**, **proprioceptors**, and **exteroceptors**. Interoceptors monitor aspects of the internal environment such as blood pressure, the partial pressure of CO_2 in the blood, and blood pH. Proprioceptors transmit information regarding the position of the body in space. These receptors are located in muscles and tendons to tell the brain where the limbs are in space, and are also located in the inner ear to tell the brain where the head is in space. Exteroceptors sense things in the external environment such as light, sound, taste, pain, touch, and temperature.

The Eye

The eye detects light energy (as photons) and transmits information about intensity, color, and shape to the brain. The eyeball is covered by a thick, opaque layer known as the **sclera**, which is also known as the white of the eye. Beneath the sclera is the **choroid** layer, which helps to supply the retina with blood. The innermost layer of the eye is the **retina**, which contains the photoreceptors that sense the light.

The transparent **cornea** at the front of the eye bends and focuses light rays. The rays then travel through an opening called the **pupil**, whose diameter is controlled by the pigmented, muscular **iris**. The iris responds to the intensity of light in the surroundings (light makes the pupil constrict). The light continues through the lens, which is suspended behind the pupil. The lens, the shape of which is controlled by the **ciliary muscles**, focuses the image onto the retina. In the retina are photoreceptors that **transduce** light into action potentials. There are two main types of photoreceptors: **cones** and **rods**. Cones respond to high-intensity illumination and are sensitive to color, while rods detect low-intensity illumination and are important in night vision. The cones and rods contain various pigments that absorb specific wavelengths of light. The cones contain three different pigments that absorb red, green and blue wavelengths; the rod pigment, **rhodopsin**, absorbs one wavelength. The photoreceptor cells synapse onto **bipolar cells**, which in turn synapse onto **ganglion cells**. Axons of the ganglion cells bundle to form the right and left **optic nerves**, which conduct visual information to the brain. The point at which the optic nerve exits the eye is called the **blind spot** because photoreceptors are not present there. There is also a small area of the retina called the fovea, which is densely packed with cones, and is important for high acuity vision.

The eye also has its own circulation system. Near the base of the iris, the eye secretes aqueous humor, which travels to the anterior chamber of the eye from which it exits and eventually joins venous blood.

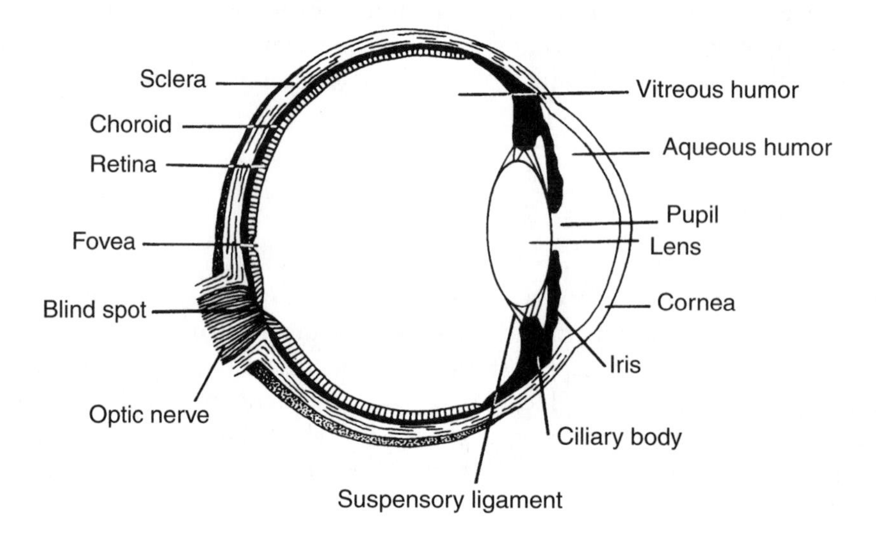

The Ear

The ear transduces sound energy (pressure waves) into impulses perceived by the brain as sound. The ear is also responsible for maintaining equilibrium (balance) in the body.

Sound waves pass through three regions as they enter the ear. First, they enter the **outer ear**, which consists of the **auricle** (pinna) and the **auditory canal**. At the end of the auditory canal is the **tympanic membrane** (**eardrum**) of the **middle ear**, which vibrates at the same frequency as the incoming sound. Next, the three bones, or ossicles (**malleus**, **incus**, and **stapes**), amplify the stimulus, and transmit it through the **oval window**, which leads to the fluid-filled **inner ear**. The inner ear consists of the **cochlea** and the **semicircular canals**. The cochlea contains the **organ of Corti**, which has specialized sensory cells called hair cells. Vibration of the ossicles exerts pressure on the fluid in the cochlea, stimulating the hair cells to transduce the pressure into action potentials, which travel via the **auditory (cochlear) nerve** to the brain for processing.

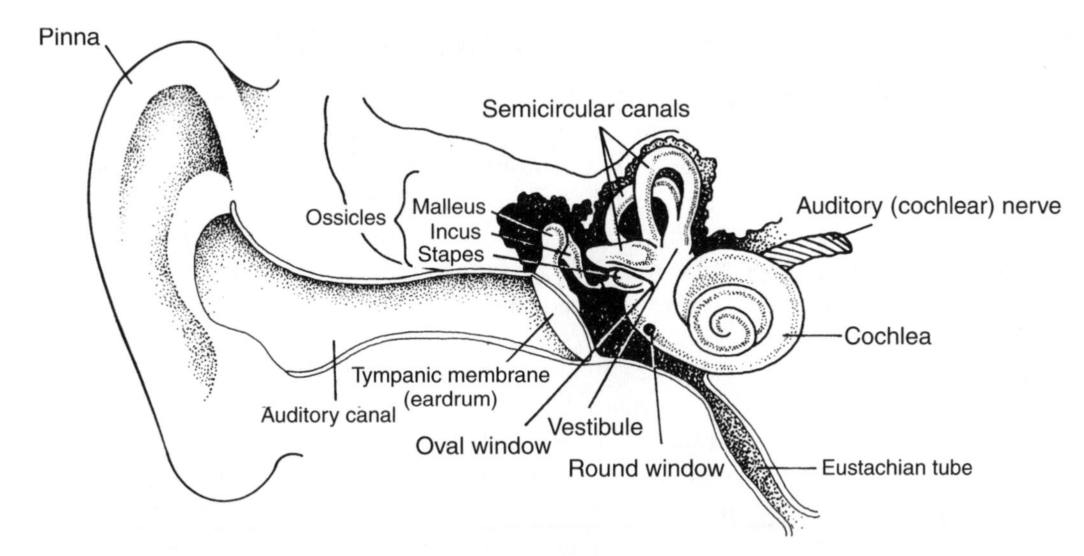

The three semicircular canals are each perpendicular to the other two and filled with a fluid called **endolymph**. At the base of each canal is a chamber with sensory hair cells; rotation of the head displaces endolymph in one of the canals, putting pressure on the hair cells in it. This changes the nature of impulses sent by the vestibular nerve to the brain. The brain interprets this information to determine the position of the head.

The Chemical Senses

The chemical senses are taste and smell. These senses transduce chemical changes in the environment, specifically in the mouth and nose, into **gustatory** and **olfactory** sensory impulses, which are interpreted by the nervous system.

a. Taste

Taste receptors, or **taste buds**, are located on the tongue, the soft palate, and the epiglottis. Taste buds are composed of approximately 40 epithelial cells. The outer surface of a taste bud contains a **taste pore**, from which microvilli, or **taste hairs**, protrude. The receptor surfaces for taste are on the taste hairs. Interwoven around the taste buds is a network of nerve fibers that are stimulated by the taste buds. These neurons transmit gustatory information to the brainstem via three cranial nerves. There are four kinds of taste sensations: sour, salty, sweet, and bitter. Although most taste buds will respond to all four stimuli, they respond preferentially; i.e., at a lower threshold, to one or two of them.

b. Smell

Olfactory receptors are found in the olfactory membrane, which lies in the upper part of the nostrils over a total area of about 5 cm^2. The receptors are specialized neurons from which **olfactory hairs**, or **cilia**, project. These cilia form a dense mat in the nasal mucosa. When odorous substances enter the nasal cavity, they bind to receptors in the cilia, depolarizing the olfactory receptors. Axons from the olfactory receptors join to form the **olfactory nerves**. The olfactory nerves project directly to the **olfactory bulbs** in the base of the brain.

THE ENDOCRINE SYSTEM AND ITS MESSENGERS

The endocrine system acts along with the nervous system as a means of internal communication, coordinating the activities of organ systems around the body. Endocrine glands synthesize and secrete chemical hormones that are dumped directly into the bloodstream and affect specific target organs or tissues. Compared to the nervous system, hormones take much longer to communicate with cells of the body. After all, they can travel only as fast as the blood flows. Yet hormonal signaling can cause behavioral effects that last for days, far longer than any nervous signal can last.

The GRE tends not to focus on whether you know every hormone that is secreted from every gland, though you should know the basics. Many hormones are named according to their function: Growth hormone, for example, causes cell growth and division and is commonly released in large amounts in juveniles, and follicle-stimulating hormone causes follicles in the ovary to mature and change into eggs. It's important to understand that built into the endocrine system are several controls based on **negative feedback**, so that the oversecretion of hormones can be avoided. In general, high levels of a particular hormone in the bloodstream inhibit further production of that hormone. In some cases, the hormone itself acts back on the cells that first produced it, and blocks the hormone biosynthesis pathways in those cells. In other cases, an antagonistic hormone will act to counter another's effects. We see this in the case of the pancreatic hormones **insulin** and **glucagon**, which together regulate the concentration of glucose in the bloodstream.

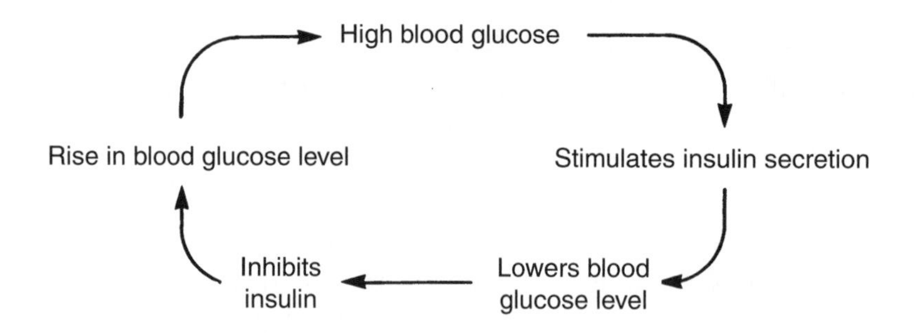

While insulin stimulates the uptake of glucose by muscle and adipose cells around the body, glucagon stimulates the breakdown of glycogen in the muscle and liver to increase blood glucose levels. In addition, glucagon stimulates **gluconeogenesis**, the process by which nonhexose substrates such as amino acids and fatty acids are converted into glucose. The combination of these two hormones working together in an antagonistic function serves to regulate blood glucose levels in a precise manner.

Central Nervous System Control

Two small, yet essential, areas of the brain tightly control the endocrine system: the hypothalamus and the pituitary. You should be aware of the following concerning these areas:

Hypothalamus

- Small part of the brain directly behind the eyes near the center of the head.
- Main job is to secrete "**releasing hormones**," which push the pituitary to release "**stimulating hormones.**"
- Ex: **GnRH**—gonadotropin releasing hormone causes the pituitary to release the gonadotropins (LH and FSH).

Pituitary

- Two sections: anterior (front), and posterior (rear), each release different hormones.
- Most pituitary hormones travel through the blood to cause other glands to release their hormones. Ex: **Gonadotropins** (Follicle Stimulating Hormone and Leutinizing Hormone) cause cells in the testes and ovaries to release their hormones (progesterone, testosterone, etc.).
- The pituitary gland is called the "**master gland**" for all the hormones it produces and the control it can exert over other glands.
- Notice in the following diagram how the blood supply goes directly through the pituitary for easy secretion of hormones. Also notice how the hypothalamus is situated right on top of the pituitary.

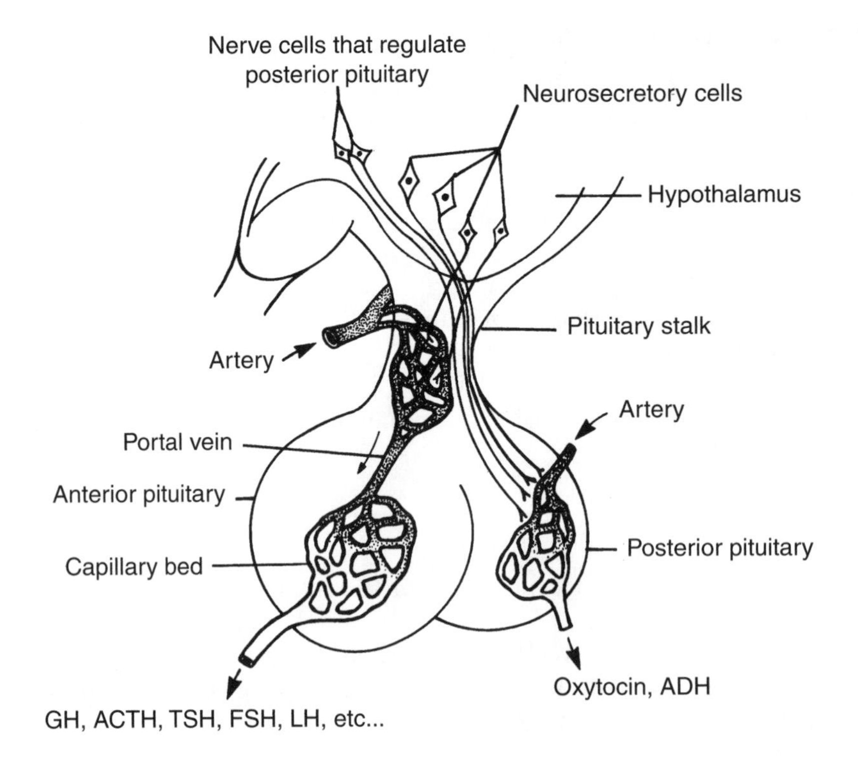
Nerve cells that regulate
posterior pituitary
Neurosecretory cells
Hypothalamus
Pituitary stalk
Artery
Artery
Portal vein
Anterior pituitary
Capillary bed
Posterior pituitary
Oxytocin, ADH
GH, ACTH, TSH, FSH, LH, etc...

What follows is a list of the principle hormones in humans that are not covered elsewhere in this book. Take a few moments to familiarize yourself with their names, the glands that produce them, and their functions.

Principal Hormones in Humans

Hormone	Source	Action
Growth hormone	Anterior pituitary	Stimulates bone and muscle growth
Prolactin	Anterior pituitary	Stimulates milk production and secretion
Adrenocorticotropic hormone (ACTH)	Anterior pituitary	Stimulates the adrenal cortex to synthesize and secrete glucocorticoids
Thyroid-stimulating hormone (TSH)	Anterior pituitary	Stimulates the thyroid to produce thyroid hormones
Luteinizing hormone (LH)	Anterior pituitary	Stimulates ovulation in females; testosterone synthesis in males
Follicle-stimulating hormone (FSH)	Anterior pituitary	Stimulates follicle maturation in females; spermatogenesis in males
Oxytocin	Hypothalamus; stored in posterior pituitary	Stimulates uterine contractions during labor, and milk secretion during lactation
Vasopressin (ADH)	Hypothalamus; stored in posterior pituitary	Stimulates water reabsorption in kidneys
Thyroid hormone	Thyroid	Stimulates metabolic activity
Calcitonin	Thyroid	Decreases the blood calcium level
Parathyroid hormone	Parathyroid	Increases the blood calcium level
Glucocorticoids	Adrenal cortex	Increases blood glucose level and decreases protein synthesis
Mineralocorticoids	Adrenal cortex	Increases water reabsorption in the kidneys
Epinephrine and Norepinephrine	Adrenal medulla	Increases blood glucose level and heart rate
Glucagon	Pancreas	Stimulates conversion of glycogen to glucose in the liver

ANIMAL BEHAVIOR AND LEARNING

Animal behavior is extremely tied into the evolutionary relationships between species, and into the phylogeny (evolutionary history) of a given species. We can ask questions about "why" a certain behavior exists, which are evolutionary questions in nature, or about "how" a certain behavior is carried out, which are mechanistic questions in nature and often much easier to answer. In order to answer these mechanistic questions, scientists look at the biological machinery that underlies the behavior and the stimuli that cause it. There are countless observations of animal behavior that make for fascinating stories, but we will focus here on the major types of behavior and learning that you may see on the GRE.

Fixed Action Patterns (FAPs)

Many animals exhibit highly stereotypical, innate behaviors known as fixed action patterns. These behaviors occur in response to a certain stimulus: They are not learned but rather neurologically hard-wired, and progress from start to finish in almost exactly the same manner every time exposure to that stimulus occurs. The triggering stimulus for FAPs is the **sign stimulus**. For instance, some moths receiving ultrasound impulses from bat sonar rapidly tuck in their wings and dive toward the ground as a survival impulse. Male stickleback fishes often viciously attack red-bellied intruders, since these, in nature, are often other stickleback males too close for comfort. In many birds, the gaping mouth of babies waiting in the nest for food occurs as a result of parents' landing on the nest's edge, and occurs later on in development at the mere sight of the parents nearby. The graylag goose is famous for its FAP that is used to push its eggs back into its nest when they have fallen out. In fact, to demonstrate how ritualized these FAPs are, if the egg is removed as the graylag goose pulls it back toward the nest with its beak, the goose will continue the movement for a certain length of time and then restart the movement after it sits down on the nest and realizes that an egg is still missing.

When animals are confronted with novel situations that test their FAPs, such as placing a doorknob outside the nest of the graylag goose, the doorknob will be retrieved as if it were an egg. The exaggeration of the sign stimulus, such as replacing a goose egg with a volleyball, causes an exaggeration of the FAP. The evolution of intelligence and learning is costly, because it involves not only the neural machinery to learn with but also a great deal of parental involvement and long juvenile development periods. Fixed action patterns and other innate behaviors help to minimize this cost and allow key survival behaviors to proceed without any parental investment in time and energy.

Imprinting

Imprinting is a process in which environmental patterns or objects presented to a developing organism during a brief critical period in early life become accepted permanently as an element of its behavioral environment. To put it another way, these patterns are "stamped in" and included in an animal's behavioral response. A duckling, for example, passes through a critical period in which it learns that the first large moving object it sees is its mother.

However, if a large, moving object other than its mother is the first thing it sees, the duckling will follow it, as Konrad Lorenz discovered when he was pursued by newborn ducklings that assumed that he was their mother.

Habituation and Sensitization

Habituation is one of the simplest learning patterns, involving the suppression of the normal responses to stimuli. When habituation occurs, repeated stimulation will result in decreased responsiveness to that stimulus. If the stimulus is no longer regularly applied, the response tends to recover over time. If you poke a snail once, it retracts into its shell, but if you poke it repeatedly, habituation occurs and the snail learns to ignore the stimulus. The receptors and sensory neurons involved become down-regulated over time and no longer signal, causing the habituation. Habituation can be due to reduced release of neurotransmitter by sensory neurons, by removal of neurotransmitter receptors from the postsynaptic membrane, or other mechanisms. Habituation also occurs in the CNS, allowing constant stimuli to be ignored. Habituation is reversible: for example, if you leave the snail alone for a while and then touch it again, its retraction response will return.

WHAT DOES HABITUATION LOOK LIKE IN OTHER ANIMALS?

When you first tap on the lab dish holding C. elegans, the small soil-dwelling worm, it will rapidly recoil from the tapping and swim rapidly backward. Gradually, it backs up less and less and finally stops backing up altogether. This habituation to the tapping response can last up to 24 hours or nearly 10 percent of the worm's lifespan!

Sensitization is related to habituation, but is the opposite response to a repeated stimulus. In sensitization, the repeated stimulation of a sensory neuron produces a more sensitive response rather than a reduced response. Like habituation, sensitization is a local response caused by a change in a neuron that is not lasting and does not alter the connections neurons form with each other. The neural circuits involved in habituation and sensitization can be quite simple, involving only a few cells.

HOW DOES TASTE-AVERSION LEARNING OCCUR?

A chemical compound essential to the brain's development, nerve growth factor (NGF) seems to play a crucial role in learning, specifically learning to avoid certain food sources. NGF causes the growth of new axons and dendrites from the neurons of developing brains. When NGF is blocked in rats, the animals are unable to associate certain tastes with feeling sick, which would affect the rats' survival in nature.

Associative Learning

A simple form of learning called associative learning occurs when an animal links in its mind two events or stimuli that appear to be related to produce the same response. The **conditioned reflex** is a type of associative learning that was studied by Pavlov and involves the association of a normally physical response with an environmental stimulus that is not normally related to the response. Pavlov studied the salivation reflex in dogs. In 1927, he discovered that if a dog was presented with an arbitrary stimulus (e.g., a bell) and then presented with food, it would eventually salivate on hearing the bell alone. He developed the following terminology:

- An **established (innate) reflex** is an unconditioned stimulus (e.g., food for salivation), and the response that it naturally elicits is the unconditioned response (e.g., salivation).
- A **neutral stimulus** is a stimulus that will not by itself elicit a response, prior to conditioning (e.g., a bell). During conditioning, the establishment of a new reflex, a neutral stimulus is presented with an unconditioned stimulus. When the neutral stimulus elicits a response in the absence of the unconditioned stimulus, it becomes the conditioned stimulus.
- The product of the conditioning experience is termed the **conditioned reflex** (e.g., salivation at the sound of a ringing bell).

Operant or Instrumental Conditioning

This form of conditioning involves conditioning responses to stimuli with the use of reward or reinforcement. When the organism exhibits a behavioral pattern that the experimenter would like to see repeated, the animal is rewarded, with the result that it exhibits this behavior more often. B. F. Skinner used the "Skinner Box" to show that animals in a cage could be conditioned to push down a lever to release food from a food dispenser. Instrumental conditioning can be performed through positive reinforcement (such as a food reward) or negative reinforcement (such as giving an animal a painful electric shock whenever it exhibits a certain behavior).

Organisms eventually "unlearn" conditioned responses, if they are not reinforced. Extinction is the gradual elimination of conditioned responses in the absence of reinforcement. The recovery of such a conditioned response after extinction is termed spontaneous recovery.

WHAT IS THE HIPPOCAMPUS?

Injury to this small, C-shaped part of the brain, found buried deep within the temporal lobe above the ears, causes forgetfulness and short-term memory loss so severe that someone might completely forget he has just met someone minutes before. Those with hippocampal lesions may retain the ability to reason logically, but lose the ability to remember recent experiences. The hippocampus seems to play a crucial role in other mammals and birds as well, allowing them to remember for weeks or even months where they have stashed food or built nests!

Intraspecific Interactions

Just as an organism communicates within itself via nervous and endocrine systems, it also requires methods to communicate with other members of its species. These methods include behavioral displays, pecking order, territoriality, and responses to chemicals.

Mechanisms of Communication

Many animal behaviors involve mechanisms of communication. Communication takes many different forms. Chemical signals appear to be a very ancient means of communication between organisms, just as hormones in the body allow cells to communicate with each other. In fact, some of the most primitive multicellular organisms, slime molds, use cyclic AMP as a chemical communication that causes the individual amoeboid cells to come together to form a multicellular fruiting body for sporulation. Insects sometimes use wind-born chemical pheromones for attraction of mates, and vertebrates also use chemical signals in urine to mark territory. Auditory signals can travel long distances, tactile communication is useful at close range in crowded conditions, and visual communication is useful in species with highly evolved visual systems like birds and mammals.

A **display** may be defined as an innate behavior that has evolved as a signal for communication between members of the same species. According to this definition, a song, a call, or an intentional change in an animal's physical characteristics is considered a display.

Pecking Order and Territoriality

Frequently, the relationships among members of the same species living as a contained social group become stable for a period of time. When food, mates, or territory are disputed, a dominant member of the species will prevail over a subordinate one. This social hierarchy is often referred to as the **pecking order**. This established hierarchy minimizes violent intraspecific aggressions by defining the stable relationships among members of the group; subordinate members only rarely challenge dominant individuals.

Members of many terrestrial vertebrate species defend a limited area or territory from intrusion by other members of the same species. These territories are typically occupied by a male or a male-female pair. The territory is frequently used for mating, nesting, and feeding. **Territoriality** serves the adaptive function of distributing members of the species so that the environmental resources are not depleted in a small region. Furthermore, intraspecific competition is reduced. Although there is frequently a minimum size for a species' territory, that size varies with population size and density. The larger the population, and the scarcer the resources (e.g., food) available to it, the smaller the territories are likely to be.

CHAPTER SIXTEEN

Support and Movement

While some organisms are **sessile**, remaining anchored in place, most depend upon locomotion, or movement from place to place, for food, shelter, and mating. Sessile organisms use cilia to sweep food and other floating debris into their mouths from water passing by, as a hydra might. Some, such as plants, move their leaves or stems to catch unsuspecting insects or to point themselves toward a light source. Organisms that do move around need to fight against both gravity, which acts to pull them down toward the ground, and friction between air or water. Muscles, bones, and specialized structures such as wings aid in their movement, and smaller organisms use cilia, flagella, or pseudopods to get from place to place. This chapter looks at movement in organisms and the ways in which they support their bodies in order to get around.

SKELETONS AND MUSCLE CONTRACTION

Vertebrate skeletal muscles are bundles of parallel fibers. Each fiber is a multinucleated cell created by the fusion of mononucleate embryonic cells. The nuclei are usually found at the periphery of a muscle cell. Embedded in the fibers are filaments called **myofibrils**, which are further divided into contractile units known as **sarcomeres**. The myofibrils are enveloped by a modified ER (endoplasmic reticulum) that holds concentrated stores of calcium (Ca^{2+}) ions and is called the **sarcoplasmic reticulum**. The membrane surrounding a muscle fiber, the **sarcolemma**, is folded into a series of transverse tubules (**T-tubules**) that can propagate an action potential almost instantly through the entire fiber.

Skeletal Muscle

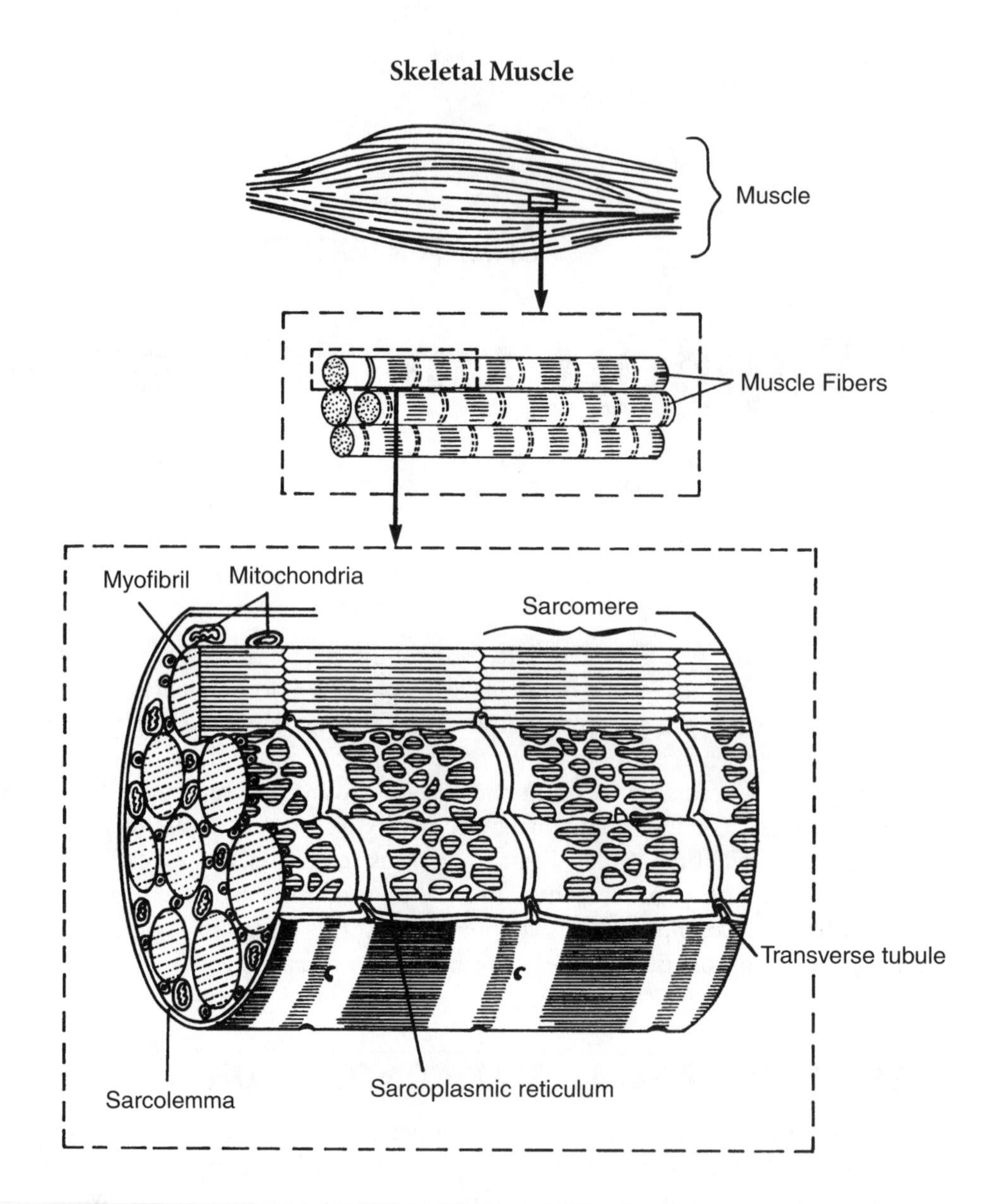

The sarcomere is composed of thin and thick filaments. The thin filaments are chains of globular actin molecules associated with two other proteins, troponin and tropomyosin. The thick filaments are composed of organized bundles of myosin molecules; each myosin molecule has a head region and a tail region.

Electron microscopy reveals that the sarcomere is organized as follows: Z lines define the boundaries of a single sarcomere and anchor the thin filaments. The M line runs down the center of the sarcomere. The I band is the region containing thin filaments only. The H zone is the region containing thick filaments only. The A band spans the entire length of the thick filaments and any overlapping portions of the thin filaments. Note that during contraction, the A band is not reduced in size, while the H zone and I band are.

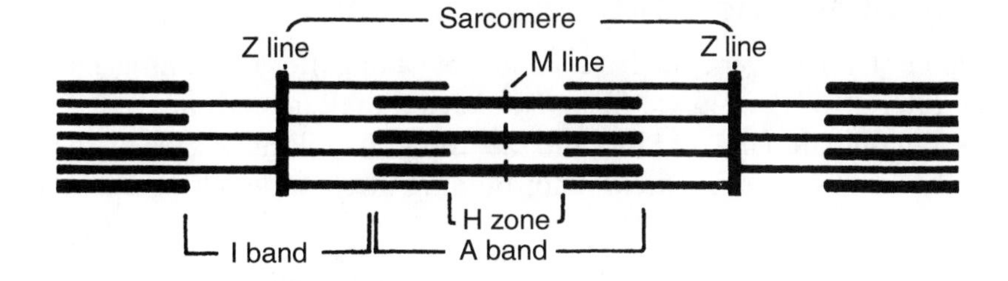

Acetylcholine at the Neuromuscular Junction

A single muscle fiber is usually innervated by only one motor nerve axon. The neurotransmitter that is released by this motor axon at the synapse with skeletal muscle is ACh, acetylcholine, and the receptors on the surface of the muscle cells are ACh receptors. The special region where the motor nerve synapses on the muscle is called the **motor end plate.** As the motor axon approaches the end plate, it loses its myelin sheath and splits into branches. At the ends of these branches are presynaptic terminals that contain vesicles filled with ACh. Calcium influx causes the ACh vesicles to release their contents into the synapse between the neuron and muscle fiber. The binding of the ACh to receptors on the muscle fibers causes depolarization of the muscle fiber as Na^+ ions flood across into the muscle cell. This depolarization triggers muscle contraction.

Curare, a mixture of plant toxins that some native tribes use on their arrowheads to paralyze animals they hunt, is an irreversible inhibitor of the receptor site for ACh at the motor end plate. Curare poisoning prevents muscle cells from depolarizing because they do not bind to ACh when it is released by nearby nerves. The paralysis of respiratory muscles leads to rapid death.

The synthesis of ACh in motor neurons requires choline, acetyl-CoA, and the enzyme choline acetyl-transferase (CAT). The availability of choline is the rate-limiting factor for the building of the ACh. Removal and breakdown of ACh, once it is released into the synapse, uses the enzyme **acetylcholinesterase** (AchE), which breaks the ACh into choline and acetate. The removal of ACh is required to terminate the nervous signal that leads to muscle contraction. The choline can then be taken back into the presynaptic cell via high-affinity active transport pumps:

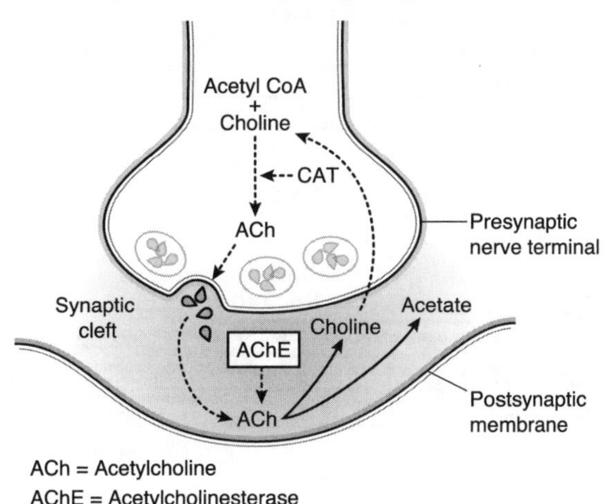

ACh = Acetylcholine
AChE = Acetylcholinesterase
CAT = Choline acetyltransferase

Mechanism of Muscle Contraction and Muscle Response

Once an action potential is generated, it is conducted along the sarcolemma and the T system, and into the interior of the muscle fiber. This causes the sarcoplasmic reticulum to release Ca^{2+} into the sarcoplasm. The Ca^{2+} binds to the troponin molecules, causing the tropomyosin strands to shift, thereby exposing the myosin-binding sites on the actin filaments.

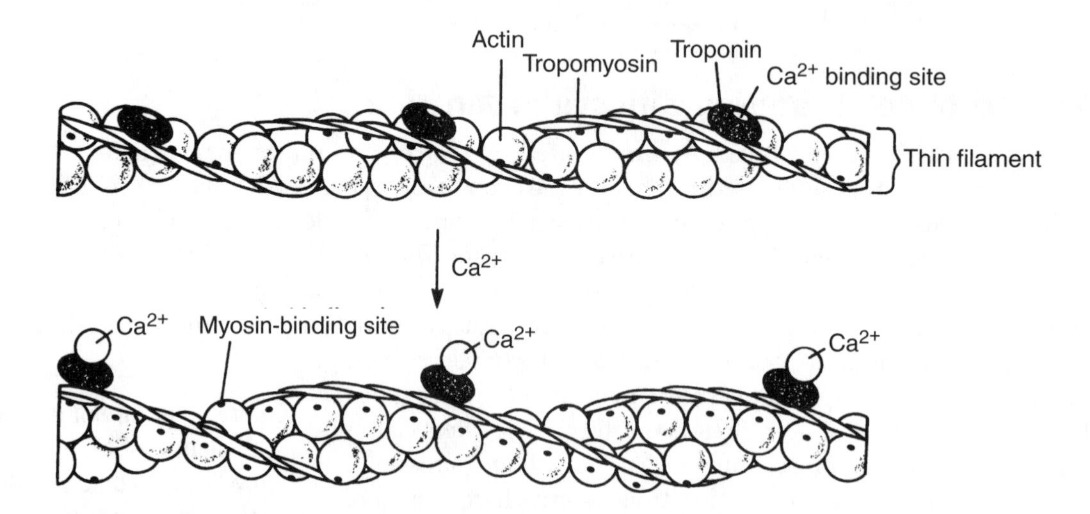

The free globular heads of the myosin molecules move toward and then bind to the exposed binding sites on the actin molecules, forming actin-myosin cross-bridges. In creating these cross-bridges, the myosin pulls on the actin molecules, drawing the thin filaments toward the center of the H zone and shortening the sarcomere (see contraction figure below.). ATPase activity in the myosin head provides the energy for the powerstroke that results in the dissociation of the myosin head from the actin. (An ATPase is an enzyme that hydrolyzes ATP.) The myosin returns to its original position and is now free to bind to another actin molecule and repeat the process, thus further pulling the thin filaments towards the center of the H zone.

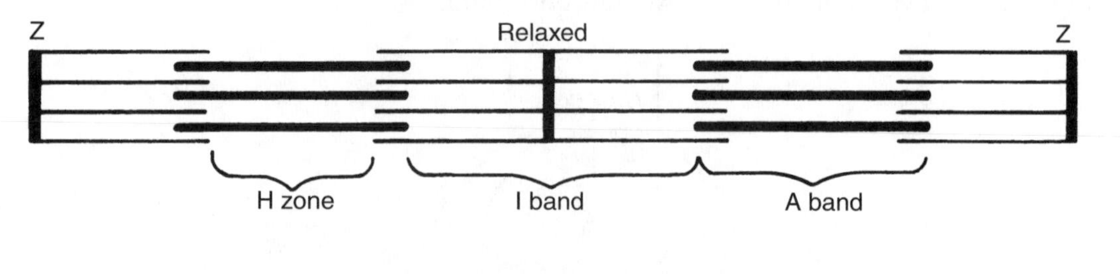

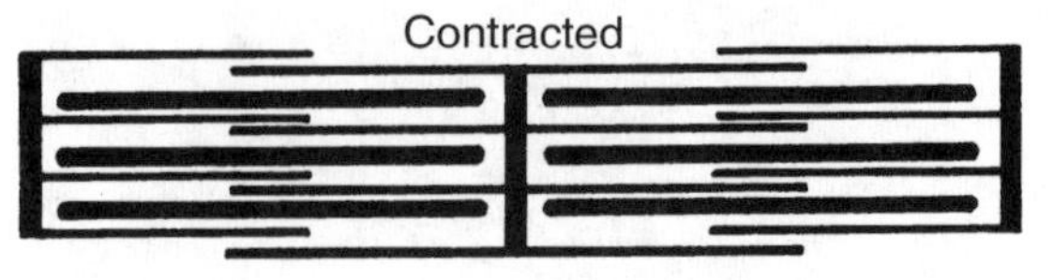

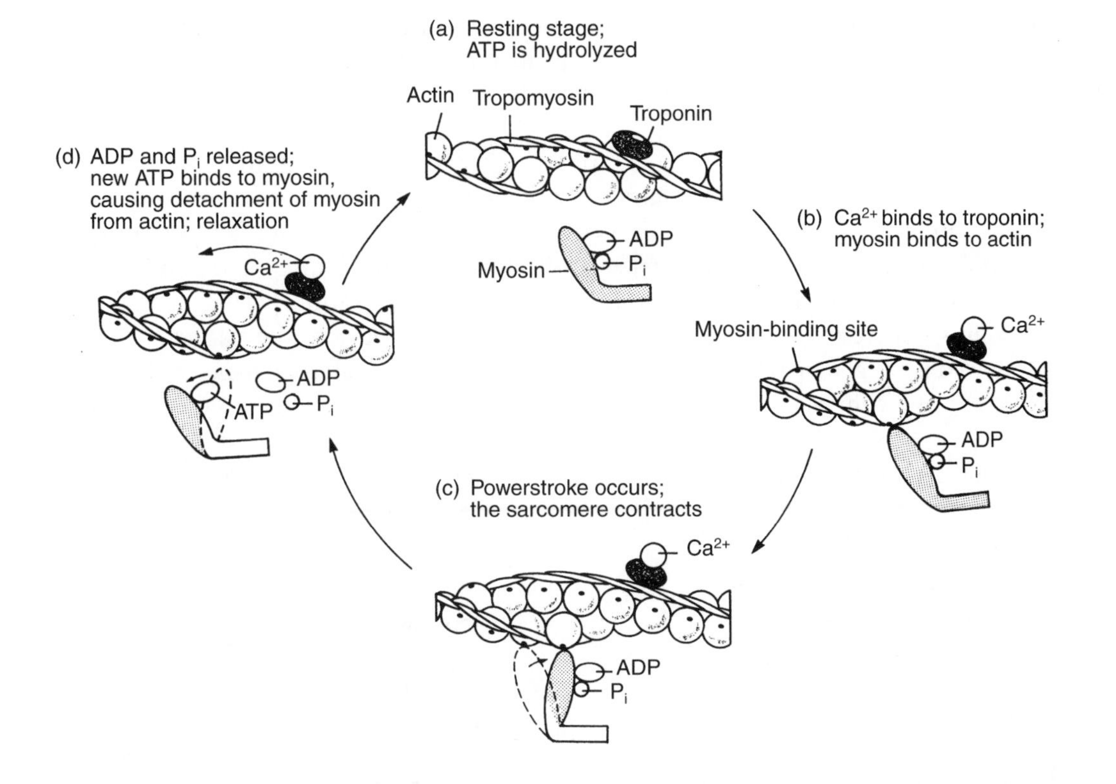

When the sarcolemmic receptors are no longer stimulated, the Ca^{2+} is pumped back into the sarcoplasmic reticulum. The products of ATP hydrolysis are released from the myosin head, a new ATP binds to the head, resulting in the dissociation of the myosin from the thin filament, and the sarcomere returns to its original width. In the absence of Ca^{2+}, the myosin-binding sites on the actin are again covered by tropomyosin molecules, thereby preventing further contraction. After death, ATP is no longer produced, and the myosin heads can't detach from actin, and therefore the muscle cannot relax. This condition is known as rigor mortis.

Stimulus and Muscle Response

Individual muscle fibers generally exhibit an all-or-none response; only a stimulus above a minimal value called the threshold value can elicit contraction. The strength of the contraction of a single muscle fiber cannot be increased, regardless of the strength of the stimulus. Whole muscle, on the other hand, does not exhibit an all-or-none response. Although there is a minimal threshold value needed to elicit a muscle contraction, the strength of the contraction can increase as stimulus strength is increased by involving more fibers. A maximal response is reached when all of the fibers have reached the threshold value and the muscle contracts as a whole.

Think of tropomyosin as the chaperone responsible for protecting actin's binding sites from the advances of the myosin head. In the presence of Ca2+, troponin changes its conformation and moves tropomyosin away from its guard position, thereby allowing myosin to bind to actin and the action to begin.

Tonus refers to the continual low-grade contractions of muscle, which are essential for both voluntary and involuntary muscle contraction. Even at rest, muscles are in a continuous state of tonus.

A simple twitch is the response of a single muscle fiber to a brief stimulus at or above the threshold stimulus, and consists of a latent period, a contraction period, and a relaxation period. The latent period is the time between stimulation and the onset of contraction. During this time lag, the action potential spreads along the sarcolemma and Ca^{2+} ions are released. Following the contraction period, there is a brief relaxation period in which the muscle is unresponsive to a stimulus; this period is known as the absolute refractory period. This is followed by a relative refractory period, during which a greater-than-normal stimulus is needed to elicit a contraction.

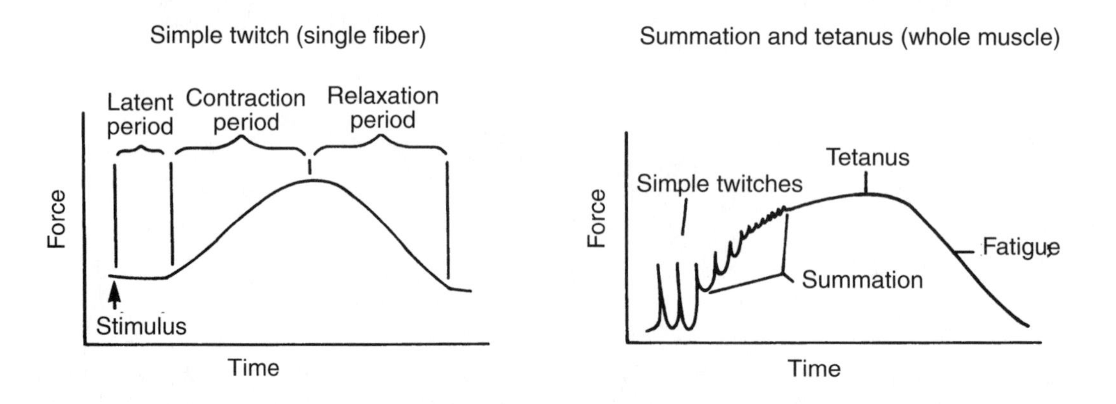

When the fibers of a muscle are exposed to very frequent stimuli, the muscle cannot fully relax. The contractions begin to combine, becoming stronger and more prolonged. This is known as frequency summation. The contractions become continuous when the stimuli are so frequent that the muscle cannot relax. This type of contraction is known as tetanus and is stronger than a simple twitch of a single fiber. If tetanization is prolonged, the muscle will begin to fatigue.

Smooth Muscle

Smooth muscle is responsible for involuntary actions and is innervated by the autonomic nervous system. Smooth muscle is found in the digestive tract, bladder, uterus, and blood vessel walls, among other places. Smooth muscle cells possess one centrally located nucleus unlike skeletal muscles. While smooth muscle cells also contain actin and myosin filaments, these filaments lack the organization of skeletal sarcomeres; consequently, smooth muscles lack the striations of skeletal muscle.

As in skeletal muscle, smooth muscle contractions result from the sliding of actin and myosin over one another, and are regulated by an influx of calcium ions. However, smooth muscle contractions are slower, and are capable of being sustained longer than skeletal muscle contractions. Smooth muscle typically has both inhibitory and excitatory synapses that regulate contraction via the nervous system. Smooth muscle also has the property of reflexively contracting without nervous stimulation; this is called myogenic activity.

Cardiac Muscle

The muscle tissue of the heart is composed of cardiac muscle fibers. These fibers possess characteristics of both skeletal and smooth muscle fibers. As in skeletal muscle, actin and myosin filaments are arranged in sarcomeres, giving cardiac muscle a striated appearance. However, cardiac muscle cells generally have only one or two centrally located nuclei. Cardiac muscle is innervated by the autonomic nervous system, which serves only to modulate its inherent beat, since cardiac muscle, like smooth muscle, is myogenic. Cardiac muscle cells are linked by gap junctions like smooth muscle, allowing action potentials to spread directly from cell to cell throughout the heart.

Skeletal Systems

There are three main functions of any skeleton: support, aid in movement, and protection. Moreover, we see three main types of skeletons across the animal kingdom: a **hydrostatic**, or fluid, skeleton, in which fluid is held under pressure within a closed body cavity (e.g., in earthworms); an **exoskeleton**, defined as a hard shell or casing deposited on the surface on an organism (e.g., in insects); and, an **endoskeleton** or internal system of bones and cartilage that support surrounding soft tissues (e.g., in humans and other vertebrates). **Cartilage** is a type of connective tissue that is softer and more flexible than bone. It is composed of an elastic form matrix called chondrin that is secreted by specialized cells known as **chondrocytes**. Embryonic skeletons in mammals are mostly made of cartilage before turning into bone due to calcification and hardening. Most cartilage is avascular (lacking blood vessels) and devoid of nerves, but receives nourishment from nearby capillaries.

Bone is a mineralized connective tissue that comes in two basic types: compact and spongy. It is in the spongy bone that we find bone marrow, where the production of red and white blood cells takes place. The bones of the appendages, the long bones, are characterized by a cylindrical shaft called a **diaphysis** and dilated ends called epiphyses. The diaphysis is composed primarily of compact bone surrounding a cavity containing bone marrow. The epiphyses are made of spongy bone surrounding a layer of compact bone. The **epiphyseal plate**, or growth plate, is a disk of cartilaginous cells separating the diaphysis from the epiphysis, and it is the site of bone growth longitudinally (in length). The epiphyseal plate remains cartilaginous as long as growth continues. When the plate seals and calcifies, growth ceases. A fibrous sheath known as the **periosteum** protects the bone.

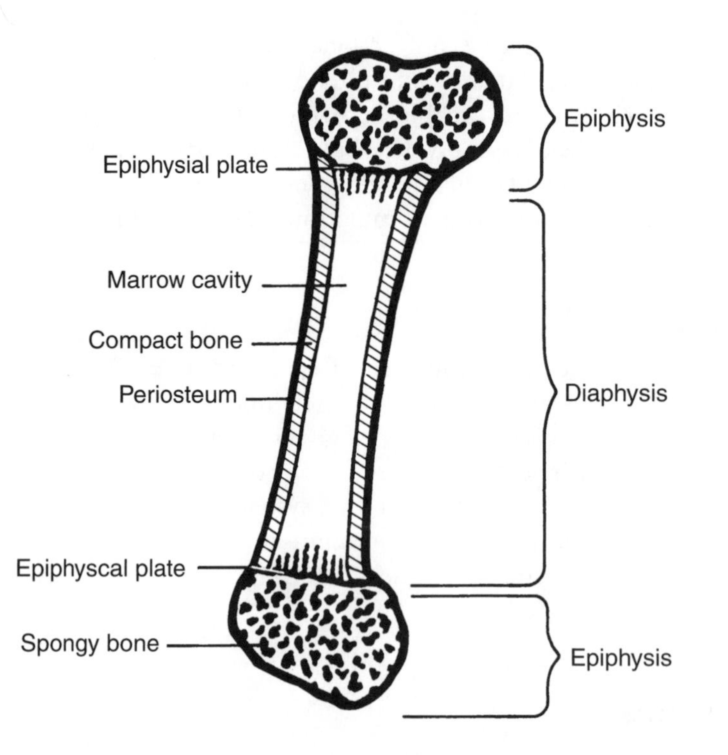

Bone is much more dynamic than you might think! Bones exist in a dynamic equilibrium between being broken down by osteoclasts and being built up by osteoblasts. Osteoclasts eat away at the bony matrix, adding calcium to the blood, and osteoblasts take calcium out of the blood to deposit it onto the bones. This constant bone remodeling takes place throughout one's entire life.

The bony matrix that builds bone is deposited in structural units called osteons, each of which consists of a tiny channel known as a **Haversian canal** that is surrounded by concentric circles called **lamellae.** Blood vessels, nerves, and lymph vessels travel through these Haversian canals to supply the bone tissue with nutrients, gases, and innervation. Also within the bone are mature bone cells, osteocytes, which are involved in the maintenance of the bone's structure. **Canaliculi** that radiate from gaps (lacunae) in the bony matrix serve to connect indirectly all the Haversian canals for exchange of nutrients and waste.

Microscopic Bone Structure

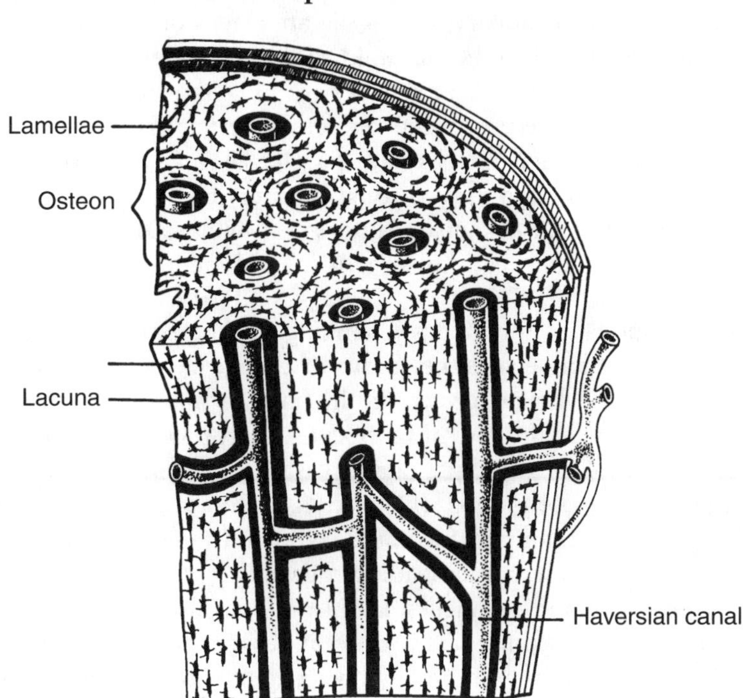

CILIA, FLAGELLA, AND PSEUDOPODIA

We have already discussed the basic structure of both cilia and flagella in chapter 4 of the last section. However, the ways in which cilia, flagella, and their counterparts in ameboid organisms—**pseudopods**—aid in movement are important to understand for the GRE. You will recall that cilia and flagella are both structured internally in the same way: a circle of nine doublet microtubules, one complete and one incomplete; **dynein** arms joining neighboring doublets; and, **nexin** links to hold the doublets together.

The dynein arms have ATPase activity, meaning that in the presence of ATP, they can move from one tubulin monomer to another. The dynein arms enable the microtubules to slide past each other in a highly regulated manner so that the cilium bends. As the dynein releases, the cilium snaps back to its original upright position. In such a way, cilia can move back and forth in an oar-like motion many times per second to cause the movement of water past the organism. This motion can sweep water and particles into the mouths of organisms, such as in the paramecium, can move particles and mucous out of the lungs as in mammals, or can permit locomotion from place to place as we see in many single-celled protists that are covered in thousands of tiny cilia.

Flagella are like extra-long cilia. While bacterial flagella are attached into the cell membrane and are unrelated in structure, eukaryotic flagella are extensions of the cell cytoplasm from basal bodies. Because of their length, the rapid sliding of dynein in between microtubules causes the flagellum to move back and forth in a whiplike motion. Most organisms that use flagella to move have somewhere between one and five flagella along the length of their bodies. Euglena and some other protists have flagella. Flagella are also used to propel the sperm cells of lots of different organisms through a watery medium toward the egg. If flagella are sheared off the body of an organism, they will usually rapidly regrow from the basal bodies that are found just underneath the organism's cell membrane.

Actin filaments are responsible for the fast-changing surface extensions called pseudopodia seen in organisms like the amoeba as they move or change shape. Many vertebrate cells can also use pseudopods to crawl over surfaces in the body or move through capillary walls (as in the case of monocyte-macrophages). Actin molecules are able to rapidly polymerize at the "leading edge" of a growing pseudopod (literally, false foot). As longer and longer actin filaments form, the cell membrane of the organism pushes out toward this leading edge taking with it the organism's cytoplasm and organelles. In this way, organisms such as amoebae can use pseudopodia to move in response to environmental stimuli or in search of food. Food is engulfed within one of these pseudopods as phagocytosis takes place.

There are some evolutionary differences across species in terms of cilia and flagella structure: Although most species have a 9+2 arrangement of microtubules, some flatworms have a 9+1 array while some arthropods use a 9+0 array.

CHEMOTAXIS AND OTHER MOVEMENT

Movement must be controlled carefully, and because of the role of actin filaments in controlling a great deal of cellular movement, actin can be considered part of any cell's signal-transduction machinery. **Chemotaxis** is defined as movement in a direction based on a gradient of a diffusible chemical that is sensed by a cell. White blood cells called neutrophils can "sense" when bacteria are nearby by using receptor proteins on their cell surfaces. Bacteria produce proteins that are formed using the amino acid methionine that has N-formyl groups added to it. These N-formylated methionines, found at the start of bacterial proteins, are easily picked up by the neutrophil receptors, and other organisms do not add formyl groups to their starting methionines; thus, they are a sure sign of bacteria nearby.

At times of low food supply, the slime mold *Dictyostelium* switches from life as single-celled, amoeboid organisms to a multicellular sluglike form created when these amoebae group together. The signal for aggregating into a large creature from single-celled individuals is the release of cyclic AMP (cAMP) by the starving amoebae, and nearby cells move using pseudopodia toward the source of cAMP. No one is sure exactly how the increase in local cAMP concentration can cause single cells to move toward the cAMP release site; however, it is likely that receptors for cAMP on the cell surfaces of the amoebae cause the activation of

G-proteins in the cell's cytoplasm, allowing the formation of ATP within the cell, causing polymerization of actin monomers. Certain G proteins have also been shown to affect **chemotaxis** in other types of cells.

Plants are able to respond to a variety of environmental stimuli, including light, touch, and gravity. While we will cover plant **tropisms** in chapter 18, one form of movement bears mentioning here: the rapid movement of traps used by insectivorous plants such as the Venus' flytrap or bladderworts are known as **nastic movements**. Insectivorous plants, also known as carnivorous plants, capture insects, spiders, protozoans, and other organisms as a source of nitrogen. While some of these plants, such as the pitcher plant, have passive traps whereby hapless insects fall into water-filled pits within bucket-shaped leaves to be digested, the plants mentioned above use active traps. In the Venus' flytrap, its leaves are bivalved and convex (like oyster shells but backward, with the bulge toward the middle). There are six sensitive hairs per leaf. The action of an insect or other organism touching these hairs causes a rapid closure of the leaf around the insect as the bulge reverses itself suddenly. An insect caught in the trap will be digested within 3–5 days. Bladderworts use underwater, air-filled "bladders" that have a trap door at one end. As hairs on the trap door are triggered, the door opens and water is pulled rapidly into the bladder, sweeping inside any small organisms nearby. These organisms will be digested with enzymes secreted into the bladder.

Both the leaf motion in flytraps and the trap door activation in bladderworts are accomplished through action potentials conveyed through the plant tissue that result in rapid loss of turgor pressure in upper epidermal cells in the affected leaves. In other words, cells in these leaves rapidly lose water within them and shrivel, causing the leaves to curl upward or inward. The loss of turgor pressure is accomplished by expulsion of water from cells, using large amounts of ATP. Therefore, these traps can be fatigued if stimulated more than a few times in succession (as sufficient ATP runs out). It is likely that the means to expel the water comes from the active transport of ions such as potassium out of certain cells and into the extracellular space or into other cells nearby.

CHAPTER SEVENTEEN

Animal Reproduction and Development

Reproduction is the process by which an organism perpetuates itself and its species through ensuring the precise duplication of genetic material and its representation in successive generations. There are three main areas within reproduction that you should be familiar with for the GRE: meiotic cell division and the formation of gametes, asexual reproduction, and sexual reproduction. This chapter takes a look at reproduction across the animal kingdom and specifically in humans.

REPRODUCTIVE STRATEGIES AND COMPARATIVE ANATOMY

The strategies that organisms use to pass on genes to offspring are quite diverse. The two main methods consist of single parent, **asexual** reproduction and two-parent, **sexual** reproduction. Some organisms can do both of these at different times in their life, and complexities exist within both asexual and sexual strategies.

Asexual: Binary Fission

As with all asexual reproduction, binary fission results in offspring with genes from one parent only. The simple definition of fission is that it is the splitting of one parent into at least two organisms that are of equal size—that means, for example, that a parent splitting into two offspring will create two equally sized cells that begin as half the size of the original cell. Remember that the parents must copy all DNA and organelles before fission, so that each "daughter" cell produced by the binary fission is identical.

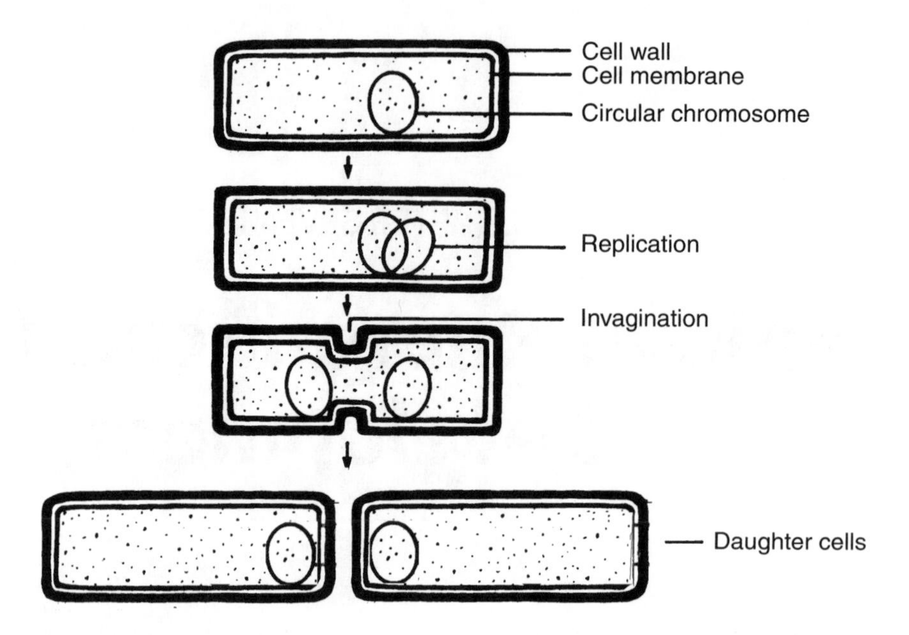

Asexual: Budding and Gemmules

Similar to binary fission, the process of budding involves offspring arising from the cell/cells of a single parent, and it is used by many marine invertebrates (hydra and other cnidarians are most commonly cited) to literally "bud" or grow an offspring from part of the parent. These buds can develop into fully grown adults either attached to the parent in a **colonial arrangement**, or they can separate as individuals. This is as if you could reproduce by having your child slowly morph out of your shoulder. The offspring is a smaller, but identical replica of the parent. This process is different from binary fission in that the body of the parent does not actually divide itself into two parts, but buds as plants might do from one of their stems. In some cases, these buds form internally and can be released en masse or individually—the internal buds are referred to as gemmules.

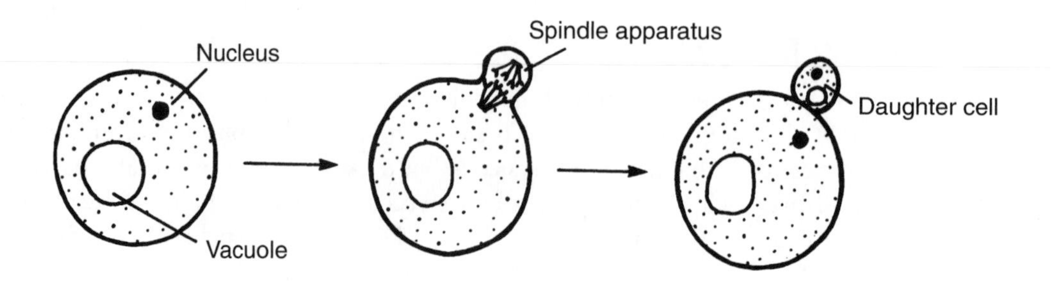

Asexual: Fragmentation and Regeneration

Many organisms, such as planaria (flatworms), reproduce by fragmentation, a process in which their body breaks apart into many fragments, each of which can develop into another organism. In some animals, only certain fragments have the ability to develop into offspring, and other fragments disintegrate. Planarians and some other animals (e.g., echinoderms like sea stars) also have the ability to regenerate—either to grow back a part of their body that has been lost or torn off, or to grow an entirely new organism from a piece of their body. Even certain amphibians (e.g., species of salamander) have a limited ability to regenerate. It is thought that the ability to do this has to do with developmental **homeotic genes** that remain capable of activation long after embryologic development ends.

Asexual: Parthenogenesis

Production of a new individual from a virgin female without intervention from a male is the process of parthenogenesis, whereby unfertilized eggs develop fully into adults without fertilization. Rotifers, small crustaceans like *Daphnia*, and certain small insects commonly use this process for reproduction. Parthenogenesis has also been reported in certain lizards and snakes. Because the eggs are not fertilized, the resulting offspring are haploid and must produce their own eggs via mitosis, not meiosis. In almost all cases, offspring are genetically identical females.

Many parthenogenetic organisms can reproduce both asexually (with unfertilized eggs) or sexually, and they can willingly switch between these two reproductive modes at different points in their lives according to the environment around them. More research must still be done to determine what causes this change between sexual and asexual modes. Climate change seems to play a crucial role, as more *stable and unchanging environments tend to favor asexual reproduction and the birth of large numbers of identical offspring; seasonal or rapidly fluctuating environments favor sexual reproduction and the recombination of alleles*, allowing the increased likelihood of favorable variants surviving best in the changing climate.

Asexual: Sporulation

In addition, some plants and fungi produce and release **spores**, reproductive cells held within a hardened coating, that can survive harsh environments and leave their casings only when the surrounding conditions are right for growth and development. This process, known as **sporulation**, often results in the growth of new organisms from one parent, yet in many cases spores from two different parents will combine in a sexual fashion. There is a great variety of asexual and sexual reproduction within the plant and fungi kingdoms, the specifics of which you do not need to worry about for the GRE. Be aware, however, that sporulation does form an integral part of the life cycle of many plants and fungi. Bacteria can also form spores, hardened casings around themselves, which enable them to survive in adverse conditions and become infectious years after they become dormant and spore-encased.

There are clearly advantages to asexual reproduction over sexual reproduction. In stable climates, where there is not a great deal of natural selection for successful genotypes taking place, asexual reproduction allows for rapid proliferation of successful genotypes that do exist. Yet in less stable environments, where variation of genotypes within the population is essential for evolutionary success (see section IV), sexual reproduction provides the necessary shuffling of alleles through meiosis that allows this variation to build up. Sexual reproduction involves the genetic contribution of two parents in the form of **gametes**, specialized sex cells that are produced by each parent and combine during fertilization. *Meiosis is the process that creates these gametes*, resulting in the creation of four haploid cells from every diploid cells that starts the process. The two gametes, sperm and egg, will combine at fertilization to restore the diploid number of chromosomes in the zygote, or fertilized egg.

Sexual Reproduction

You should be aware of the basics of meiosis for the GRE. Remember that this process takes place only in areas where gametes are produced (the testes and ovaries in mammals, for example). The first round of cell division (meiosis I) produces two intermediate daughter cells, while the second division (meiosis II) involves the separation of sister chromatids (copied chromosomes), similar to what happens in mitosis but with haploid gametes as the result. Each meiotic division has the same stages as mitosis.

Meiosis I

a. Prophase I

The chromatin condenses into chromosomes, the spindle apparatus forms, and the nucleoli and nuclear membrane disappear. Homologous chromosomes (chromosomes that code for the same traits, one inherited from each parent), come together and intertwine in a process called synapsis. Since at this stage each chromosome consists of two sister chromatids, each synaptic pair of homologous chromosomes contains four chromatids, and is therefore often called a tetrad. Sometimes chromatids of homologous chromosomes break at corresponding points and exchange equivalent pieces of DNA; this process is called crossing over. Note that crossing over occurs between homologous chromosomes, and not between sister chromatids of the same chromosome. (The latter are identical, so crossing over would not produce any change.) Those chromatids involved are left with an altered but structurally complete set of genes. The chromosomes remain joined at points called chiasmata where the crossing over occurred. Such genetic recombination can "unlink" linked genes, thereby increasing the variety of genetic combinations that can be produced via gametogenesis. Recombination among chromosomes results in increased genetic diversity within a species. Note that sister chromatids are no longer identical after recombination has occurred.

Synapsis

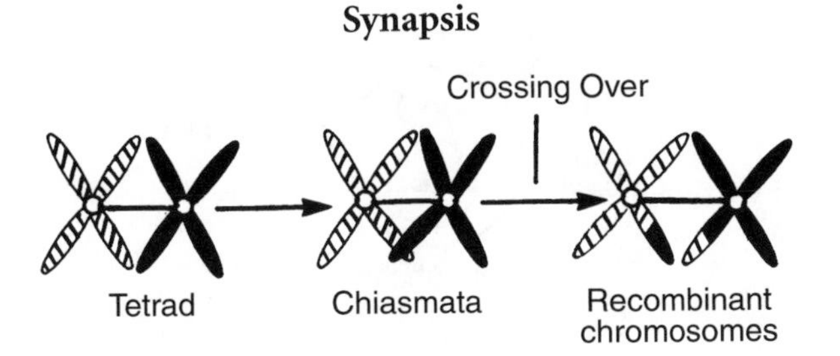

b. Metaphase I

Homologous pairs (tetrads) align at the equatorial plane, and each pair attaches to a separate spindle fiber by its kinetochore.

c. Anaphase I

The homologous pairs separate and are pulled to opposite poles of the cell. This process is called disjunction, and it accounts for a fundamental Mendelian law: independent assortment. During disjunction, each chromosome of paternal origin separates (or disjoins) from its homologue of maternal origin, and either chromosome can end up in either daughter cell. Thus, the distribution of homologous chromosomes to the two intermediate daughter cells is random with respect to parental origin. Each daughter cell will have a unique pool of alleles (genes coding for alternative forms of a given trait; e.g., yellow flowers or purple flowers), from a random mixture of maternal and paternal origin.

d. Telophase I

A nuclear membrane forms around each new nucleus. At this point each chromosome still consists of sister chromatids joined at the centromere. The cell divides (by cytokinesis) into two daughter cells, each of which receives a nucleus containing the haploid number of chromosomes. Between cell divisions there may be a short rest period, or interkinesis, during which the chromosomes partially uncoil.

Meiosis II

This second division is very similar to mitosis, except that meiosis II is not preceded by chromosomal replication.

a. Prophase II

The centrioles migrate to opposite poles and the spindle apparatus forms.

b. Metaphase II

The chromosomes line up along the equatorial plane. The centromeres divide, separating the chromosomes into pairs of sister chromatids.

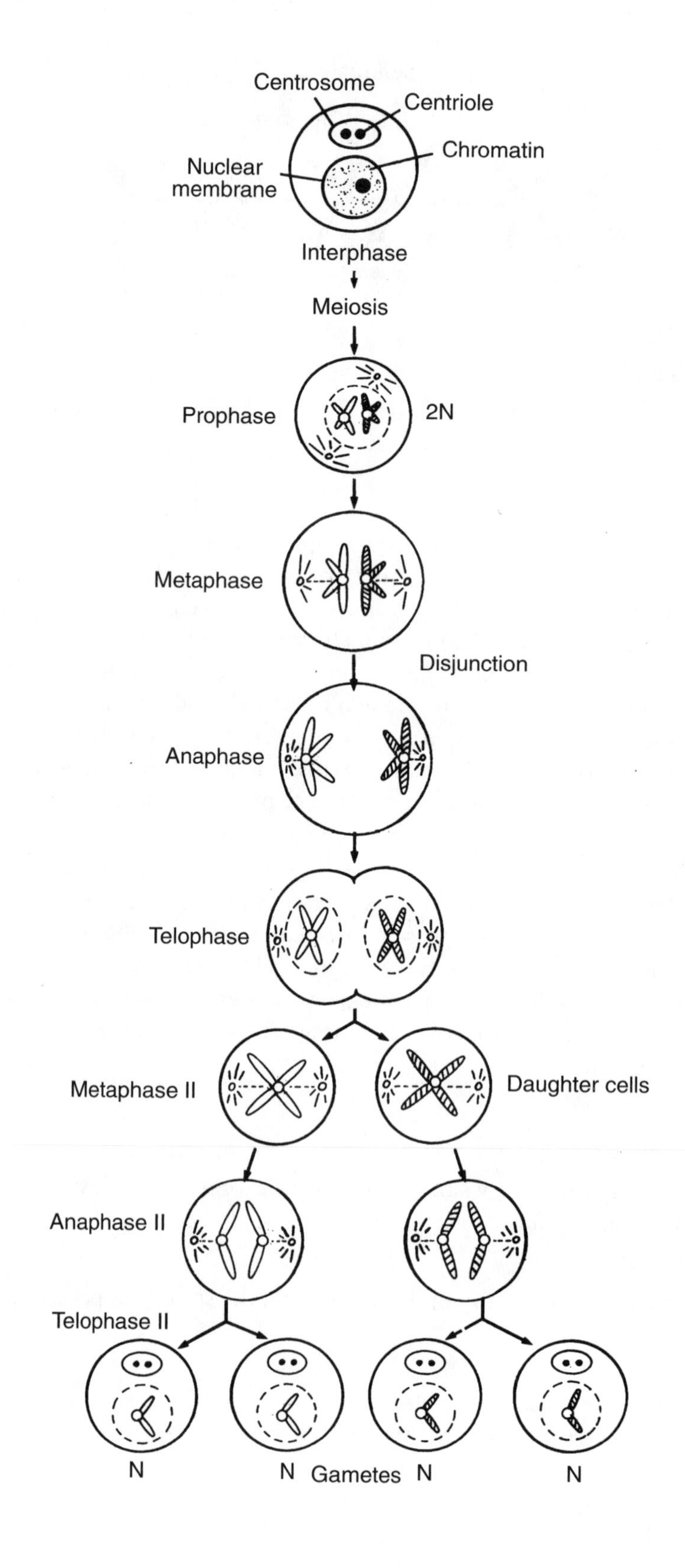
Centrosome
Centriole
Chromatin
Nuclear membrane
Interphase
Meiosis
Prophase
2N
Metaphase
Disjunction
Anaphase
Telophase
Metaphase II
Daughter cells
Anaphase II
Telophase II
N
N
Gametes
N
N

c. **Anaphase II**

The sister chromatids are pulled to opposite poles by the spindle fibers.

d. **Telophase II**

A nuclear membrane forms around each new (haploid) nucleus. Cytokinesis follows and two daughter cells are formed. Thus, by the completion of meiosis II, four haploid daughter cells are produced per gametocyte. (In women, only one of these becomes a functional gamete.)

HOW DO THE PROCESSES OF MEIOSIS AND MITOSIS DIFFER?

Difference 1: The chromosomes in meiosis line up in homologous pairs during metaphase, while they line up in single file for mitosis.

Difference 2: Crossing over occurs in meiosis but not in mitosis.

The combination of these two differences means not only that cells undergoing meiosis have to divide twice to split their sister chromatids, but also that genetic variation is vastly increased.

The random distribution of chromosomes in meiosis, coupled with crossing over in prophase I, enables an individual to produce gametes with many different genetic combinations. Thus, as opposed to asexual reproduction, which produces identical offspring, sexual reproduction produces genetic variability in offspring. The possibility of so many different genetic combinations is believed to increase the capability of a species to evolve and adapt to a changing environment.

Reproductive cycles are often related to seasons. Many organisms produce sperm cells and egg cells only in the spring, for example. This allows organisms to conserve resources and permits reproduction to take place only when the most energy or the most favorable conditions are available.

Many animals use a variety of sexual strategies to maximize both the number of offspring they produce and the fitness of those offspring. Barnacles, earthworms, tapeworms, and many other burrowing or relatively sessile animals are often **hermaphroditic**, possessing both male and female copulatory organs. Most need to mate with another member of the species in order to use these organs (i.e., they cannot mate with themselves), but when they do mate, they can effectively produce twice the number of offspring, since two independent mating events will take place. This is extremely advantageous for organisms that tend to remain fairly isolated (such as tapeworms alone in a human's digestive tract) or cannot move around easily to find a mate (such as barnacles), because when they do find a mate, the matings are far more productive than they would be if each parent contributed only one set of sperm or eggs rather than both. Some hermaphroditic organisms can self-mate as well, which precludes the need to find mates. Keep in mind that this is sexual reproduction, since the production of male sex cells at the male organs is an independent meiotic event from the production of egg cells at the female organ.

Sequential hermaphroditism is a related phenomenon whereby organisms have the ability to alter their gender according to the needs of the population they are in. The Caribbean blue-headed wrasse is an example. All fish are born female, and only the largest female will turn into a male, produce sperm cells, and fertilize the rest of the "**harem**." If this male is killed off or dies, the next largest female changes gender!

Some organisms fertilize their eggs externally, outside of the body, while others fertilize internally. Both males and females in the animal world can carry eggs as they develop, although most commonly females are the egg-carriers. For external fertilization to take place, the environment must be moist, and the production of a large number of egg cells is necessary to ensure that just a few offspring survive. Parents that use **external fertilization** rarely invest much parental care into the growth and development of their offspring, and the offspring are often on their own from fertilization onward. Eggs laid externally in a watery environment are usually coated in a gelatinous coating for protection, through which the sperm cells penetrate for fertilization. Animals that fertilize **internally** have fewer zygotes (fertilized eggs), which are often cared for by the parents for some period of time during development. The organisms will lay their fertilized eggs outside of their bodies to develop (e.g. reptiles or birds) or will carry the offspring to term within their bodies (e.g. most mammals). Fertilized eggs laid outside of the body often have the added protection of parents watching over them and relatively hard, waterproof shells. The **amniote egg** was a crucial evolutionary development allowing birds, reptiles, and some mammals to take over terrestrial (land-based) habitats, away from the protection of water, since these often hard-shelled eggs cushion and protect the developing embryo and nourish it with a self-contained supply of water and food.

Sexual reproduction is far more costly in energy expenditure than asexual reproduction, yet so many species reproduce sexually because of the enormous advantages it provides through increasing diversity within populations.

Human Sexual Reproduction

Human reproduction is a highly complex process involving not only sexual intercourse between male and female, but interactions between the reproductive and endocrine systems within the body. Children are the product of fertilization—the fusion of sperm and egg (the gametes) in the female reproductive tract. The gametes are produced in the primary reproductive organs, or gonads.

Male Reproductive Anatomy

The male gonads, called the testes, contain two functional components: the seminiferous tubules and the interstitial cells (cells of Leydig). Sperm are produced in the highly coiled seminiferous tubules, where they are nourished by Sertoli cells. The interstitial cells, located between the seminiferous tubules, secrete testosterone and other androgens (male sex hormones). The testes are located in an external pouch called the scrotum, which maintains testes temperature 2–4°C lower than body temperature, a condition essential for sperm survival. Sperm pass from the seminiferous tubules into the coiled tubules of the epididymis. Here they acquire motility, mature, and are stored until ejaculation. During ejaculation they travel through the vas deferens to the ejaculatory duct and then to the urethra. The urethra passes through the penis and opens to the outside at its tip. In males, the urethra is a common passageway for both the reproductive and excretory systems.

Sperm is mixed with seminal fluid as it moves along the reproductive tract; seminal fluid is produced by three glands: the seminal vesicles, the prostate gland, and the bulbourethral glands. The paired seminal vesicles secrete a fructose-rich fluid that serves as an energy source for the highly active sperm. The prostate gland releases an alkaline milky fluid that protects the sperm from the acidic environment of the female reproductive tract. Finally, the bulbourethral glands secrete a small amount of viscous fluid prior to ejaculation; the function of this secretion is not known. Seminal fluid aids in sperm transport by lubricating the passageways through which the sperm will travel. Sperm plus seminal fluid is known as semen.

Male Reproductive Tract

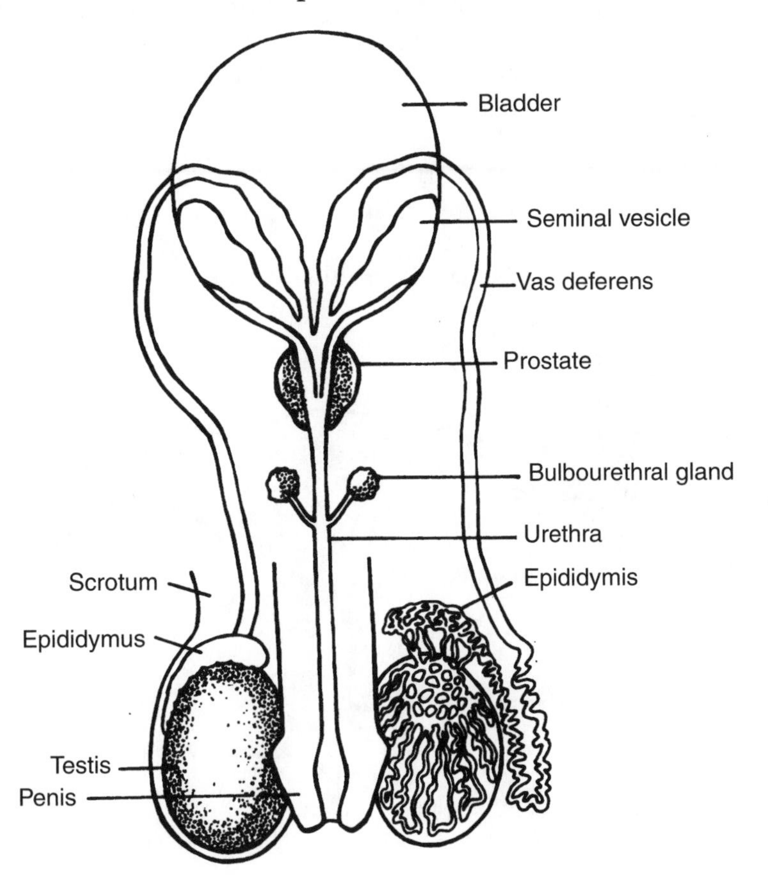

Female Reproductive Anatomy

The female gonads, called the ovaries, produce eggs (ova), and secrete the hormones estrogen and progesterone. The ovaries are found in the abdominal cavity, below the digestive system. The ovaries consist of thousands of follicles; a follicle is a multilayered sac of cells that contains, nourishes, and protects an immature ovum. It is actually the follicle cells that produce estrogen. Once a month, an immature ovum is released from the ovary into the abdominal cavity and drawn into the nearby fallopian tube. The inner surface of the fallopian tube is lined with cilia that create currents that move the ovum into and along the tube. Each fallopian tube opens into the upper end of a muscular chamber called the uterus, which is the site of fetal development. The lower, narrow end of the uterus is called the cervix. The cervix connects with the vaginal canal, which is the site of sperm deposition during intercourse and is also the passageway through which a baby is expelled during childbirth. The external female genitalia is referred to as the vulva.

Note that in the mammalian (placental) female, the reproductive and excretory systems are distinct from one another; i.e., the urethra and the vagina are not connected.

Female Reproductive Tract

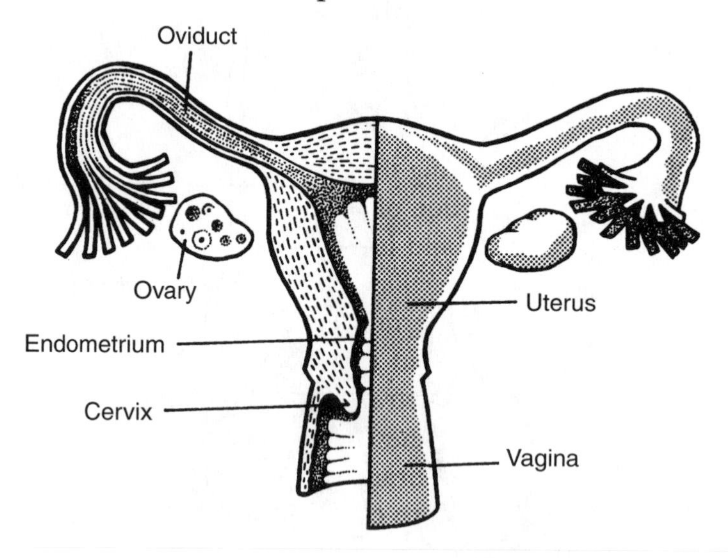

GAMETE PRODUCTION, MENSTRUATION, AND KEY HORMONES

Spermatogenesis

Spermatogenesis, or sperm production, occurs in the seminiferous tubules. Diploid cells called spermatogonia differentiate into diploid cells called primary spermatocytes, which undergo the first meiotic division to yield two haploid secondary spermatocytes of equal size; the second meiotic division produces four haploid spermatids of equal size. Following meiosis, the spermatids undergo a series of changes leading to the production of mature sperm, or spermatozoa, which are specialized for transporting the sperm nucleus to the egg, or ovum. The mature sperm is an elongated cell with a head, neck, body, and tail. The head consists almost entirely of the nucleus. The tail (flagellum) propels the sperm, while mitochondria in the neck and body provide energy for locomotion. A caplike structure called the acrosome, derived from the Golgi apparatus, develops over the anterior half of the head. The acrosome contains enzymes needed to penetrate the tough outer covering of the ovum. After a male has reached sexual maturity, approximately 3 million primary spermatocytes begin to undergo spermatogenesis per day, the maturation process taking a total of 65–75 days.

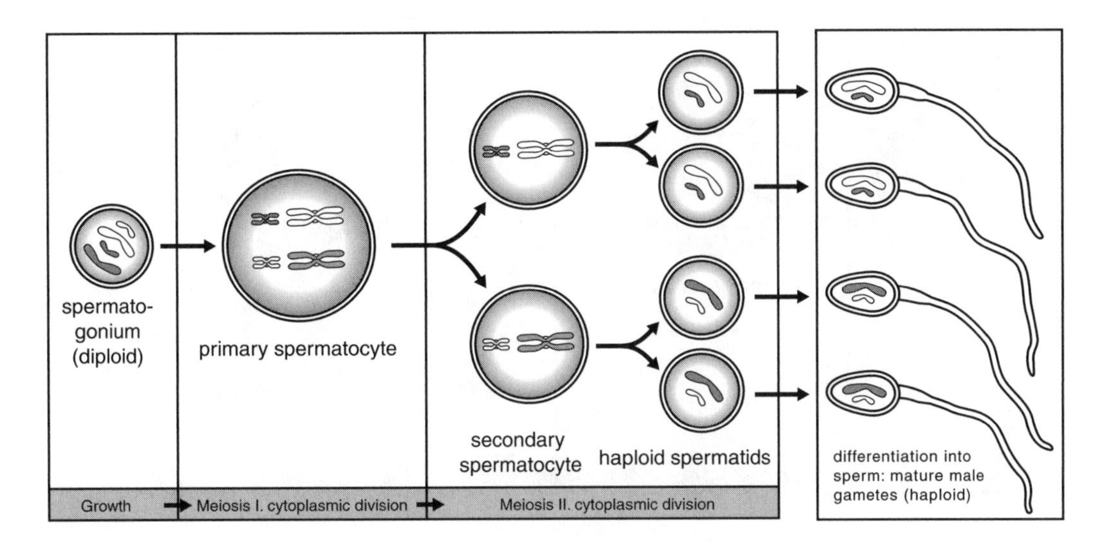

Oogenesis

Oogenesis, which is the production of female gametes, occurs in the ovarian follicles. At birth, all of the immature ova, known as primary oocytes, that a female will produce during her lifetime are already in her ovaries. Primary oocytes are diploid cells that form by mitosis in the ovary. After menarche (the first time a female gets her period), one primary oocyte per month completes meiosis I, yielding two daughter cells of unequal size—a secondary oocyte and a small cell known as a polar body. The secondary oocyte is expelled from the follicle during ovulation. Meiosis II does not occur until fertilization. The oocyte cell membrane is surrounded by two layers of cells; the inner layer is the zona pellucida, the outer layer is the corona radiata. Meiosis II is triggered when these layers are penetrated by a sperm cell, yielding two haploid cells—a mature ovum and another polar body. (The first polar body may also undergo meiosis II; eventually, the polar bodies die.) The mature ovum is a large cell containing a lot of cytoplasm, RNA, organelles, and nutrients needed by a developing embryo.

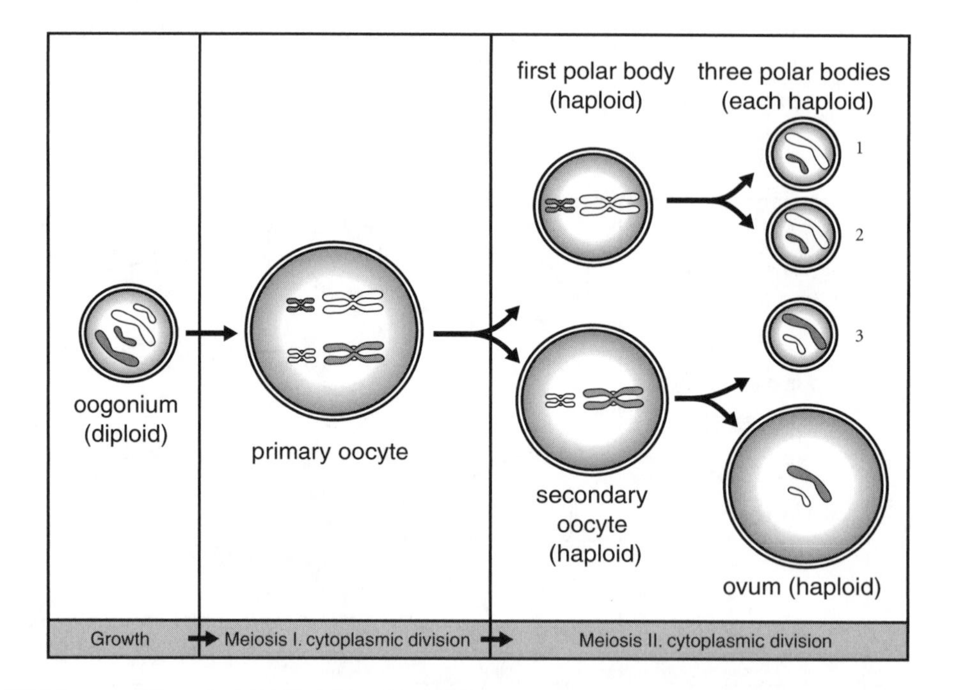

Women ovulate approximately once every four weeks (except during pregnancy and, usually, lactation) until menopause, which typically occurs between the ages of 45 and 50. During menopause, the ovaries become less sensitive to the hormones that stimulate follicle development (FSH and LH), and eventually they atrophy. The remaining follicles disappear, estrogen and progesterone levels greatly decline, and ovulation stops. The profound changes in hormone levels are often accompanied by physiological and psychological changes that persist until a new balance is reached.

Fertilization

An egg can be fertilized during the 12–24 hours following ovulation. Fertilization occurs in the lateral, widest portion of the fallopian tube. Sperm must travel through the vaginal canal, cervix, uterus, and into the fallopian tubes to reach the ovum. Sperm remain viable and capable of fertilization for one or two days following intercourse.

The first barrier that the sperm must penetrate is the corona radiata. Enzymes secreted by the sperm aid in penetration of the corona radiata. The acrosome is responsible for penetrating the zona pellucida; it releases enzymes that digest this layer, thereby allowing the sperm to come into direct contact with the ovum cell membrane. Once in contact with the membrane, the sperm forms a tubelike structure called the acrosomal process, which extends to the cell membrane and penetrates it, fusing the sperm cell membrane with that of the ovum. The sperm nucleus now enters the ovum's cytoplasm. It is at this stage of fertilization that the ovum completes meiosis II.

The acrosomal reaction triggers a cortical reaction in the ovum, causing calcium ions to be released into the cytoplasm; this, in turn, initiates a series of reactions that result in the formation of the fertilization membrane. The fertilization membrane is a hard layer that surrounds the ovum cell membrane and prevents multiple fertilizations. The release of Ca^{2+} also stimulates metabolic changes within the ovum, greatly increasing its metabolic rate. This is followed by the fusion of the sperm nucleus with the ovum nucleus to form a diploid zygote. The first mitotic division of the zygote soon follows.

Multiple Births

Monozygotic (identical) twins

Monozygotic twins result when a single zygote splits into two embryos. If the splitting occurs at the two-cell stage of development, the embryos will have separate chorions and separate placentas; if it occurs at the blastula stage, then the embryos will have only one chorionic sac and will therefore share a placenta and possibly an amnion. Occasionally the division is incomplete, resulting in the birth of "Siamese" twins, which are attached at some point on the body, often sharing limbs and/or organs. Monozygotic twins are genetically identical, since they develop from the same zygote. Monozygotic twins are therefore of the same sex, blood type, and so on.

Dizygotic (fraternal) twins

Dizygotic twins result when two ova are released in one ovarian cycle and are fertilized by two different sperm. The two embryos implant in the uterine wall individually, and each develops its own placenta, amnion, and chorion (although the placentas may fuse if the embryos implant very close to each other). Fraternal twins share no more characteristics than any other siblings, since they develop from two distinct zygotes.

Menstruation and Key Hormones

In females, two types of reproductive cycles can occur: the **menstrual cycle** and the **estrous cycle**. The purpose of each is to prepare the uterus to receive a fertilized ovum and to support an implanted embryo. Levels of different hormones in the bloodstream carefully control the cycles.

An estrous cycle occurs in most mammals other than primates and is often known as being "in heat" because of the periodic increase in the body temperature of ovulating females. If pregnancy does not occur, the uterine lining, which has thickened due to hormonal signals, will be reabsorbed into the body and there is no shedding of the lining (no bleeding). Animals in this period of heightened sexual activity, called "estrus," have very pronounced behavioral changes. Note that the spelling of the estrous cycle has an "o" but the spelling of the time period estrus does not. The seasons have a powerful effect on estrous cycles, and most mammals will mate only during while in estrus, at the time of ovulation. The length and frequency of estrous cycles varies tremendously throughout the animal kingdom.

In many primates, including humans, a menstrual cycle takes place, during which fertilization can occur. Because a menstrual cycle is not seasonal, sexual activity and fertilization can occur nearly year-round. The human menstrual cycle lasts approximately 28 days, and most females vary within their cycles from 20–40 days from start to finish. Hormones carefully regulate the menstrual and ovarian cycles in such a way that the growth of the follicle (the new egg and cells that surround it) and ovulation (the release of the egg from the ovary) coincide with the preparation of the uterine lining for embryo implantation. There are five major hormones that participate in this process and you should be aware of them for the GRE.

Hormone	*Secreted by the*	*Purpose*
GnRH (gonadotropin-releasing hormone)	Hypothalamus	Stimulates pituitary to release LH and FSH
FSH (follicle-stimulating hormone)	Pituitary	Stimulates growth and development of follicle
LH (luteinizing hormone)	Pituitary	Stimulates ovulation and corpus luteum formation
Estrogen	Ovaries; Adrenal gland	Signals uterine lining to thicken; female sex characteristics; regulation of brain hormones
Progesterone	Ovaries	Endometrium (uterine lining) maintenance; regulation of brain hormones

Below is a diagram of the hormonal and other changes that take place during the human menstrual cycle. You will notice that the cycle is really four separate cycles in one. Although you will not have to know small details from these diagrams for the GRE, it is best to understand how to read a series of complex graphs such as these.

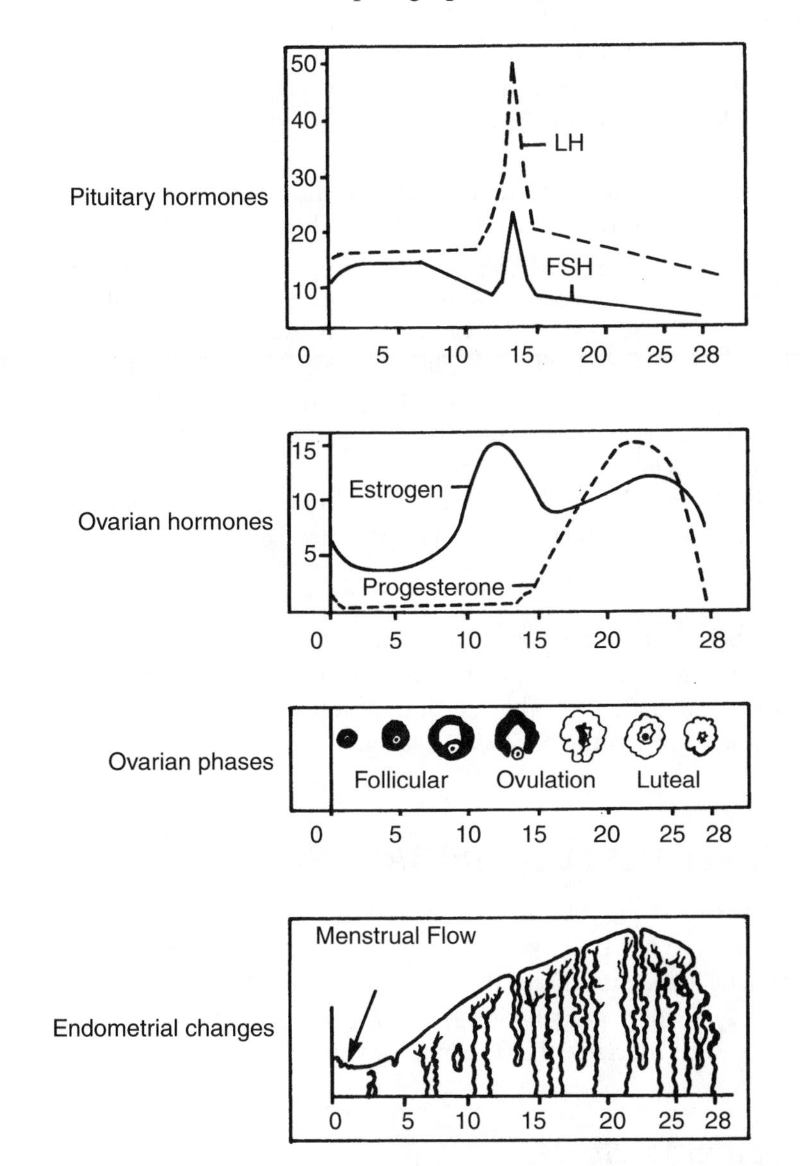

You have already learned that the hypothalamus secretes "releasing hormones," substances that control the pituitary's release of its own hormones. Gonadotropic releasing hormone (GnRH) from the hypothalamus causes the pituitary to release follicle-stimulating hormone (FSH) and luteinizing hormone (LH). These chemicals stimulate the maturation of sperm and the production of testosterone in the testes. In the female, levels of these hormones carefully control the cycles pictured above. At day 0 of the menstrual cycle, levels of FSH and

Birth control pills are made of certain concentrations of estrogen and progesterone, which are high enough to block the brain's secretion of GnRH, FSH, and LH, but not high enough to allow ovulation. Therefore, the uterine lining thickens up and the female stops taking the pills for several days each month to allow the lining to be expelled (to have her period), though ovulation almost never takes place as long as the pills are taken regularly.

LH in the female's blood begin to rise due to increased secretion of GnRH by the hypothalamus. As a result, follicles in the ovary begin to mature and an egg cell within the follicle "ripens." As follicles grow, their cells release estrogen to build up the uterine lining. Notice how the uterine lining continues to thicken throughout the cycle due to high levels of estrogen and, later, progesterone. Higher levels of estrogen actually block the secretion of GnRH, LH, and FSH.

However, as estrogen continues to be released, it reaches a threshold concentration within the bloodstream, after which the blocking of GnRH in the brain stops and both LH and FSH spike very high in concentration. This spiking causes ovulation, the release of the egg cell from the mature follicle. The follicle left over after release becomes known as the **corpus luteum** ("yellow body"), which secretes progesterone and some estrogen for approximately 12 days to keep the uterine lining thick. If the egg is not fertilized by day 25 or so, the corpus luteum disintegrates, levels of estrogen and progesterone drop rapidly, and menstruation or the sloughing off of the uterine lining takes place. If the egg is fertilized, secretion of human chorionic gonadotropin (hCG) will maintain the corpus luteum and, thus, the uterine lining. Pregnancy tests look for levels of hCG in the urine. Only fertilized eggs release hCG.

The "big picture" here is that there are two major cycles taking place: an ovarian cycle (maturation of the egg and formation of the corpus luteum); and, the uterine cycle (maintenance and dissolution of the uterine lining). Both cycles are regulated and coordinated primarily through the actions of estrogen.

EARLY AND LATE DEVELOPMENTAL PROCESSES

Implantation of a fertilized egg typically occurs about seven or eight days after ovulation. At implantation, the embryo has usually reached a stage where it is known as a blastula, a ball of hundreds of genetically identical cells that has a hollow, fluid-filled interior space. Just after implantation, a process called gastrulation occurs, when genetic differentiation of the cells will lead to the growth of specialized tissues and organs.

Early Developmental Stages

Embryology is the study of the development of a unicellular zygote into a complete multicellular organism. In the course of nine months, a unicellular human zygote undergoes cell division, cellular differentiation, and morphogenesis in preparation for life outside the uterus. Much of what is known about mammalian development stems from the study of less complex organisms such as sea urchins and frogs.

Cleavage

Early embryonic development is characterized by a series of rapid mitotic divisions known as **cleavage.** These divisions lead to an increase in cell number without a corresponding growth in cell protoplasm, i.e., the total volume of cytoplasm remains constant. Thus, cleavage results in progressively smaller cells, with an increasing ratio of nuclear-to-cytoplasmic material. Cleavage also increases the surface-to-volume ratio of each cell, thereby improving gas and nutrient exchange. An **indeterminate cleavage** is one that results in cells that maintain the ability to develop into a complete organism. Identical twins are the result of an indeterminate cleavage. A **determinate cleavage** results in cells whose future **differentiation** pathways are determined at an early developmental stage. Differentiation is the specialization of cells that occurs during development.

Clinical Correlate:

Sometimes the blastula implants itself outside of the uterus, a situation referred to as an ectopic pregnancy. Ectopic pregnancies occur most often in the fallopian tube (over 95%). The embryo usually aborts spontaneously. The tube may rupture, and a considerable amount of hemorrhaging may occur.

The first complete cleavage of the zygote occurs approximately 32 hours after fertilization. The second cleavage occurs after 60 hours, and the third cleavage after approximately 72 hours, at which point the 8-celled embryo reaches the uterus. As cell division continues, a solid ball of embryonic cells, known as the **morula**, is formed. **Blastulation** begins when the morula develops a fluid-filled cavity called the **blastocoel**, which by the fourth day becomes a hollow sphere of cells called the **blastula**. The mammalian blastula is called a **blastocyst** and consists of two cell groups: the **inner cell mass**, which protrudes into the blastocoel, and the **trophoblast**, which surrounds the blastocoel and later gives rise to the chorion and placenta. The inner cell mass becomes the embryo.

Mammalian Blastocyst

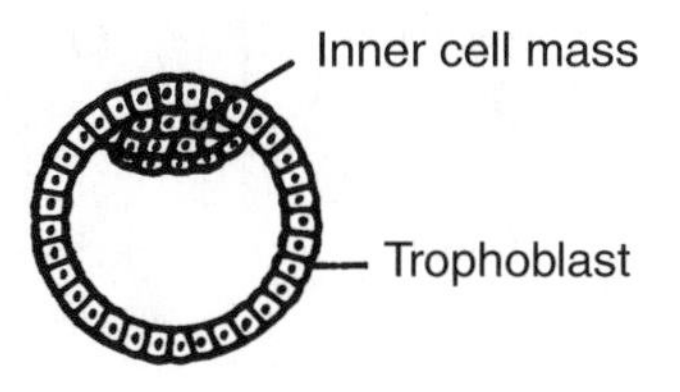

Implantation

The embryo implants in the uterine wall during blastulation, approximately 5–8 days after fertilization. The uterus is prepared for implantation by the hormone progesterone which causes glandular proliferation in the **endometrium**—the mucosal lining of the uterus. The embryonic cells secrete proteolytic enzymes that enable the embryo to digest tissue and implant itself in the endometrium. Eventually, maternal and fetal blood exchange materials at this site, later to be the location of the placenta.

Gastrulation

Once implanted, cell migrations transform the single cell layer of the blastula into a three-layered structure called a **gastrula.** In the amphibian, gastrulation begins with the appearance of a small invagination on the surface of the blastula. An inpocketing forms as cells continue to move toward the invagination, eventually eliminating the blastocoel. The result is a two-layered cup, with a differentiation between an outer cellular layer—the **ectoderm**, and an inner cellular layer—the **endoderm.** The newly formed cavity of the two-layered gastrula is called the **archenteron**, and later develops into the gut. The opening of the archenteron is called the **blastopore**. (In organisms classified as **deuterostomes**, such as humans, the blastopore is the site of the future anus; whereas in organisms classified as **protostomes**, the blastopore is the site of the future mouth). Proliferation and migration of cells into the space between the ectoderm and the endoderm gives rise to a third cell layer, called the **mesoderm.** These three **primary germ layers** are responsible for the differential development of the tissues, organs, and systems of the body at later stages of growth (see list below).

- **Ectoderm**—integument (including the epidermis, hair, nails, and epithelium of the nose, mouth, and anal canal), the lens of the eye, and the nervous system.
- **Endoderm**—epithelial linings of the digestive and respiratory tracts (including the lungs), and parts of the liver, pancreas, thyroid, and bladder.
- **Mesoderm**—musculoskeletal system, circulatory system, excretory system, gonads, connective tissue throughout the body, and portions of digestive and respiratory organs.

Amphibian Cleavage and Gastrulation

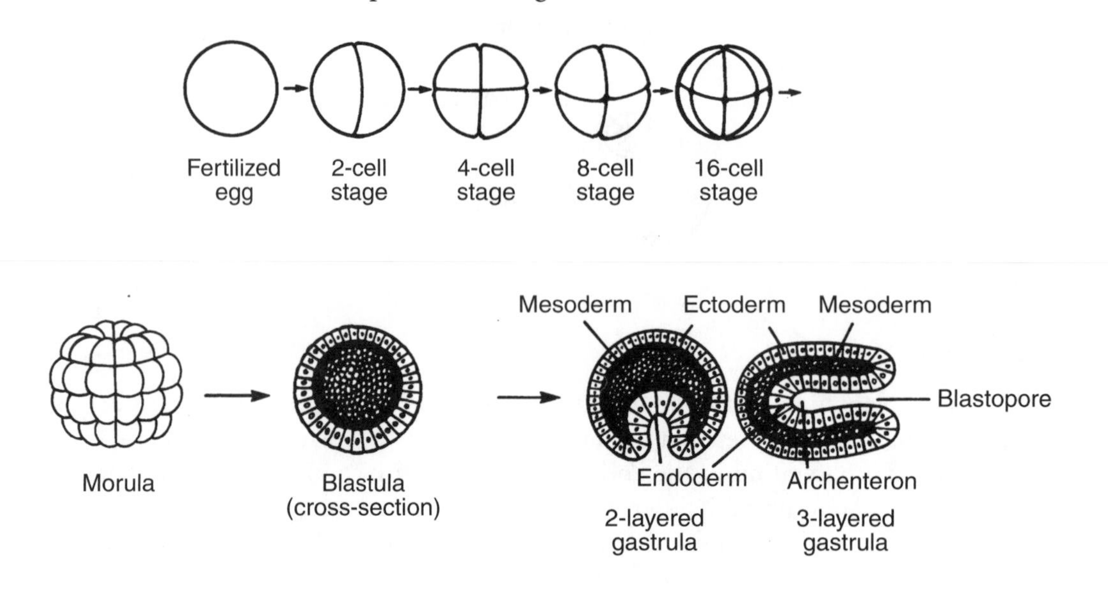

Despite the fact that all embryonic cells are derived from a single zygote and therefore have the same DNA, cells and tissues differentiate to perform unique and specialized functions. Most of this differentiation is accomplished through selective transcription of the genome. As the embryo develops, different tissue types express different genes. Most of the genetic information within any given cell is never expressed.

Induction is the influence of a specific group of cells (sometimes known as the **organizer**) on the differentiation of another group of cells. Induction is most often mediated by chemical substances (inducers) passed from the organizer to adjacent cells. In development of the eyes, lateral outpocketings from the brain (optic vesicles) grow out and touch the overlying ectoderm. The optic vesicle induces the ectoderm to thicken and form the lens placode. The lens placode then induces the optic vesicle to flatten and invaginate inward, forming the optic cup. The optic cup then induces the lens placode to invaginate and form the cornea and lens. Experiments with frog embryos show that if this ectoderm is transplanted to the trunk (after the optic vesicles have grown out), a lens will develop in the trunk. If, however, the ectoderm is transplanted before the outgrowth of the optic vesicles, it will not form a lens. Once induction has occurred (in this example), the lens cells will follow their developmental program to completion regardless of changes in their surroundings.

Late Developmental Stages

Neurulation

By the end of gastrulation, regions of the germ layers begin to develop into a rudimentary nervous system; this process is known as **neurulation**. A rod of mesodermal cells, called the **notochord**, develops along the longitudinal axis just under the dorsal layer of ectoderm. The notochord has an inductive effect on the overlying ectoderm, causing it to bend inward and form a groove along the dorsal surface of the embryo. The dorsal ectoderm folds on either side of the groove; these **neural folds** grow upward and finally fuse, forming a closed tube. This is the **neural tube**, which gives rise to the brain and spinal cord (**central nervous system**). Once the neural tube is formed, it detaches from the surface ectoderm. The cells at the tip of each neural fold are called the **neural crest** cells. These cells migrate laterally and give rise to many components of the **peripheral nervous system**, including the sensory ganglia, autonomic ganglia, adrenal medulla, and Schwann cells.

Amphibian Neurulation

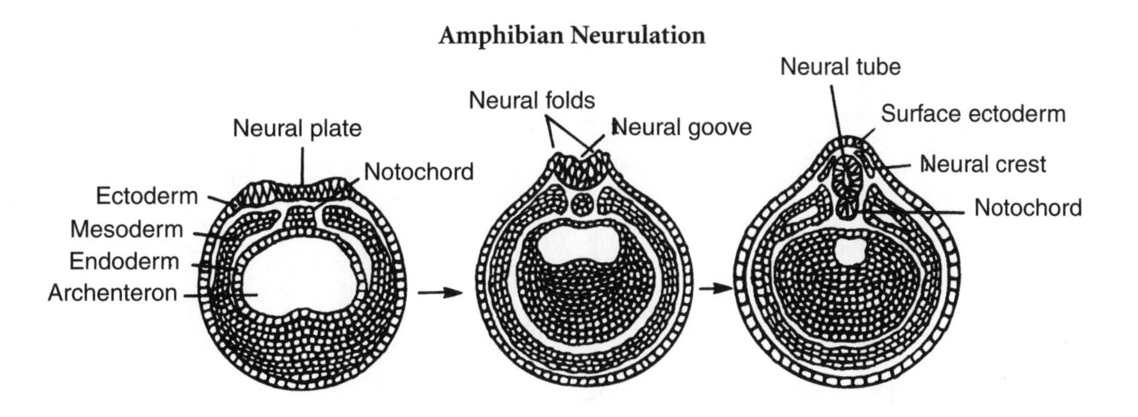

Fetal Respiration

The growing fetus (the embryo is referred to as a fetus after eight weeks of gestation) receives oxygen directly from its mother through a specialized circulatory system. This system not only supplies oxygen and nutrients to the fetus, but removes carbon dioxide and metabolic wastes as well. The two components of this system are the placenta and the umbilical cord, which both develop in the first few weeks following fertilization.

The placenta and the umbilical cord are outgrowths of the four extra-embryonic membranes formed during development: the amnion, chorion, allantois, and yolk sac. The amnion is a thin, tough membrane containing a watery fluid called amniotic fluid. Amniotic fluid acts as a shock absorber of external and localized pressure from uterine contractions during labor. Placenta formation begins with the chorion, a membrane that completely surrounds the amnion. About two weeks after fertilization the chorion extends villi into the uterine wall. These chorionic villi become closely associated with endometrial cells, developing into the spongy tissue of the placenta. A third membrane, the allantois, develops as an outpocketing of the gut. The blood vessels of the allantoic wall enlarge and become the umbilical vessels, which will connect the fetus to the developing placenta. The yolk sac, the site of early development of blood vessels, becomes associated with the umbilical vessels. At some point the allantois and yolk sac are enveloped by the amnion, forming the primitive umbilical cord, which is the initial connection between the fetus and the placenta. The mature umbilical cord consists of the umbilical vessels, which developed from the allantoic vessels, surrounded by a jellylike matrix. As the embryo grows, it remains attached to the placenta by the umbilical cord, which permits it to float freely in the amniotic fluid.

Clinical Correlate

Amniocentesis is the process of aspirating some amniotic fluid by inserting a needle into the amniotic sac. The amniotic fluid contains fetal cells that can be examined for chromosomal abnormalities as well as gender determination. Amniocentesis is recommended for pregnant women over 35, as women in this age group have a higher rate of meiotic nondisjunction which can result in genetic aberrations.

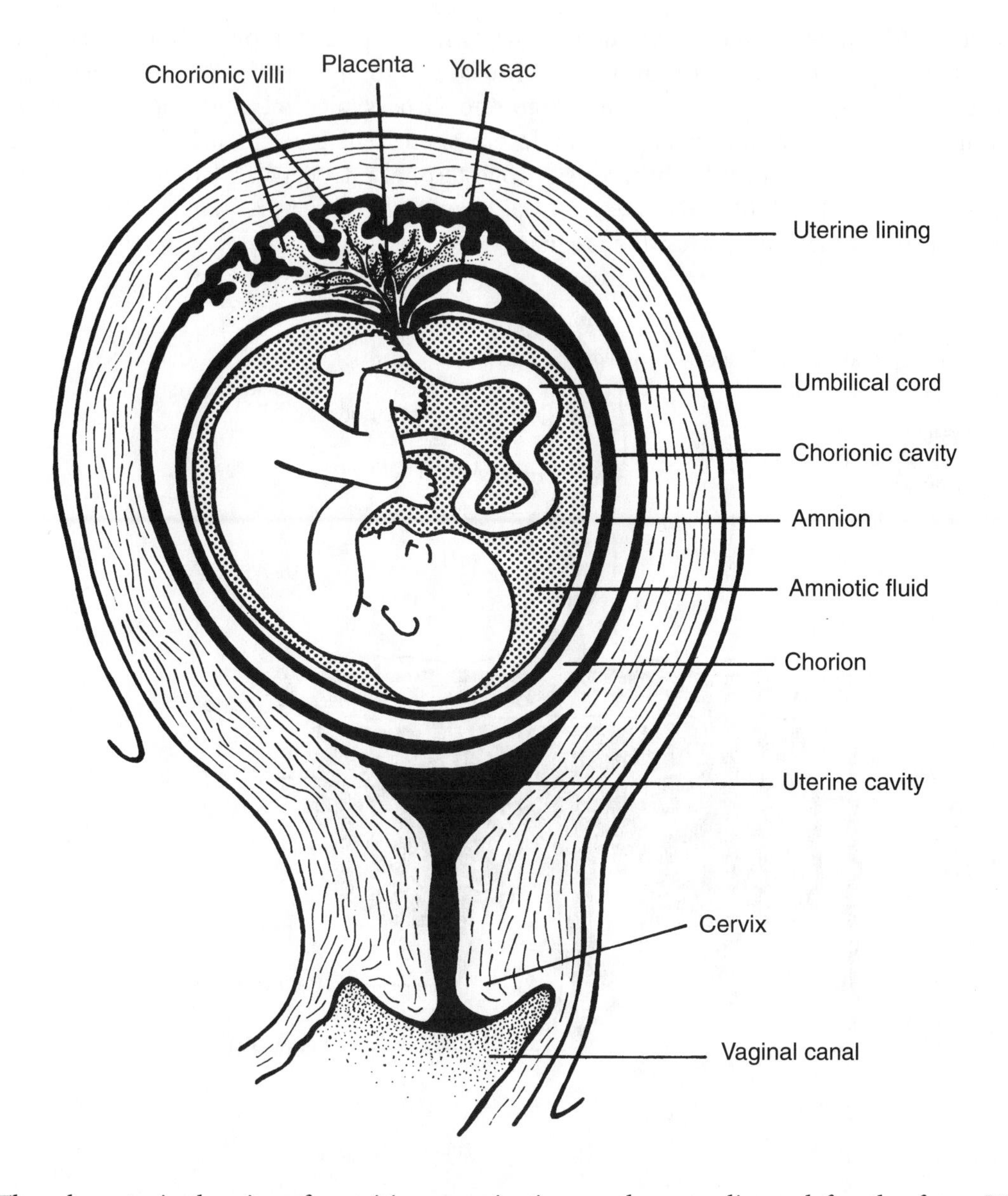

The placenta is the site of nutrition, respiration, and waste disposal for the fetus. Water, glucose, amino acids, vitamins, and inorganic salts diffuse across maternal capillaries into fetal blood. Fetal hemoglobin (Hb-F) has a greater affinity for oxygen than does adult hemoglobin (Hb-A); consequently, oxygen preferentially diffuses into fetal blood. Concurrently, metabolic wastes and carbon dioxide diffuse in the opposite direction—from fetal blood into maternal blood. Note that the circulatory systems of the mother and the fetus are not directly connected, so maternal and fetal blood do not mix. As can be seen in the following figure, all exchange of material between maternal and fetal blood vessels occurs in the placenta via diffusion.

In addition to nutritive and respiratory functions, the placenta offers the fetus some immunological protection by preventing the diffusion of foreign matter (e.g., bacteria) into fetal blood. However, the placenta is permeable to viruses, alcohol, and many drugs and toxins, all of which can adversely affect fetal development. The placenta also functions as an endocrine gland, producing the hormones progesterone, estrogen, and human chorionic gonadotropin (hCG)—all of which are essential for maintaining a pregnancy. The presence of hCG in urine is the simplest test for pregnancy.

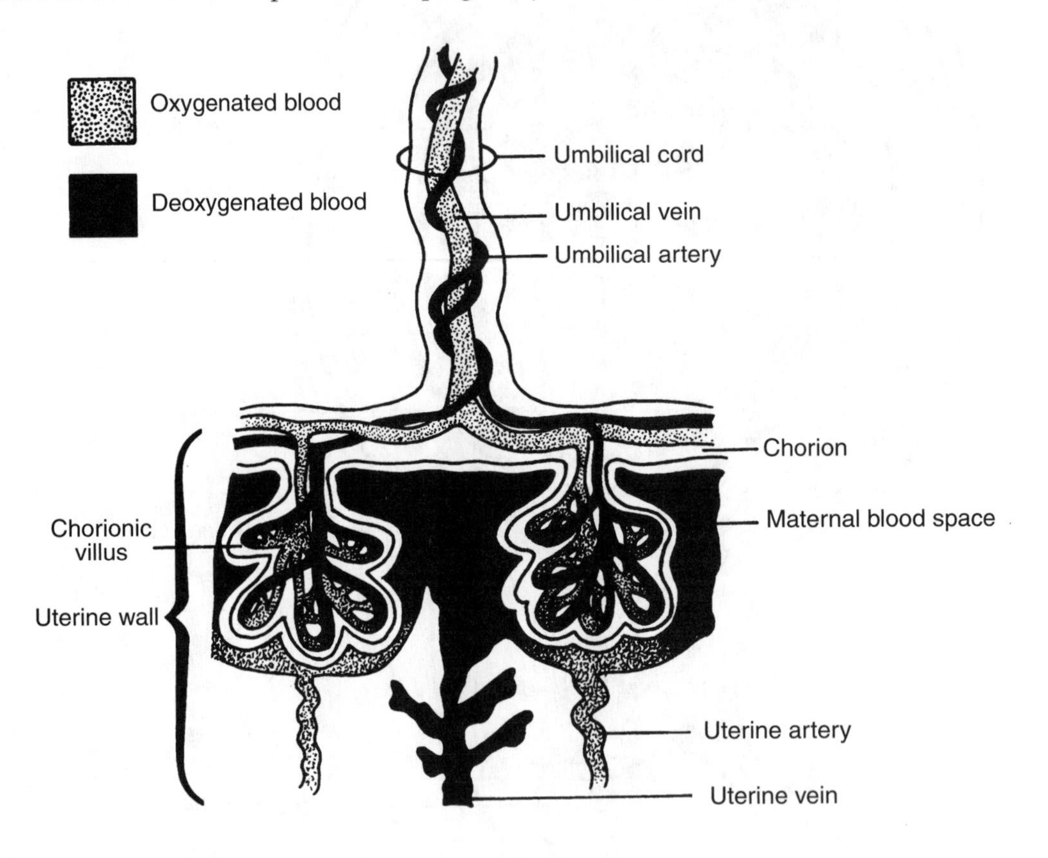

Fetal Circulation

Fetal circulation differs from adult circulation in several important ways. The major difference is that in fetal circulation, blood is oxygenated in the placenta (because fetal lungs are nonfunctional prior to birth), while in adult circulation, blood is oxygenated in the lungs. In addition, the fetal circulatory route contains three shunts that divert blood flow away from the developing fetal liver and lungs. The umbilical vein carries oxygenated blood from the placenta to the fetus. The blood bypasses the fetal liver by way of a shunt called the ductus venosus, before converging with the inferior vena cava. The inferior and superior venae cavae return deoxygenated blood to the right atrium. Since the oxygenated blood from the umbilical vein mixes with the deoxygenated blood of the venae cavae, the blood entering the right atrium is only partially oxygenated. Most of this blood bypasses the pulmonary circulation and enters the left atrium directly from the right atrium by way of the foramen ovale, a shunt that diverts blood away from the pulmonary arteries. The remaining blood in the right atrium empties into the right ventricle and is pumped into the pulmonary artery. Most of this blood is shunted directly from the pulmonary artery to the aorta via the ductus arteriosus, diverting even more blood away from the lungs. This means that in the fetus, the pulmonary arteries carry partially oxygenated blood to the lungs. The blood that does reach the lungs is further deoxygenated as the blood unloads its oxygen to the developing lungs. Remember, gas exchange does not occur in the fetal lungs—it occurs in the placenta. The deoxygenated blood then returns to the left atrium via the pulmonary veins. Despite the fact that this blood mixes with the partially oxygenated blood that crossed over from the right atrium (via the foramen ovale) before being pumped into the systemic circulation by the left ventricle, the blood delivered via the aorta has an even lower partial pressure of oxygen than the blood that was delivered to the lungs. This deoxygenated blood is returned to the placenta via the umbilical arteries.

In contrast, in adult circulation, deoxygenated blood enters the right atrium, the right ventricle pumps this blood to the lungs via the pulmonary arteries (those same arteries that carried partially oxygenated blood in the fetus), and gas exchange occurs in the lungs. Oxygenated blood is returned to the left atrium via the pulmonary veins (those same veins that carried deoxygenated blood in the fetus), and the left ventricle pumps the blood into circulation via the aorta.

The fetal umbilical artery and the adult pulmonary arteries are the only arteries that carry deoxygenated blood. The fetal umbilical vein and adult pulmonary veins are the only veins that transport oxygenated blood.

Clinical Correlate

A patent ductus arteriosus (PDA) occurs if the ductus arteriosus does not close after birth. The direction of blood flow through the PDA will be determined by the relative resistance of the pulmonary and systemic circulations. Normally, the pulmonary resistance is lower than systemic, so the pressure generated on the left side of the heart is greater than that on the right side and the blood will flow down its pressure gradient in a left-to-right direction from the aorta, through the patent ductus, back to the pulmonary artery. If pulmonary pressure rises over time and exceeds the systemic pressure, the flow will reverse and go in a right-to-left direction from the pulmonary artery, through the PDA, to the aorta. With such a right-to-left "shunt" the child will turn blue, since the deoxygenated venous blood bypasses the lungs and mixes with the oxygenated blood being pumped to the body through the aorta. A patent ductus arteriosus can be closed with medication or surgery.

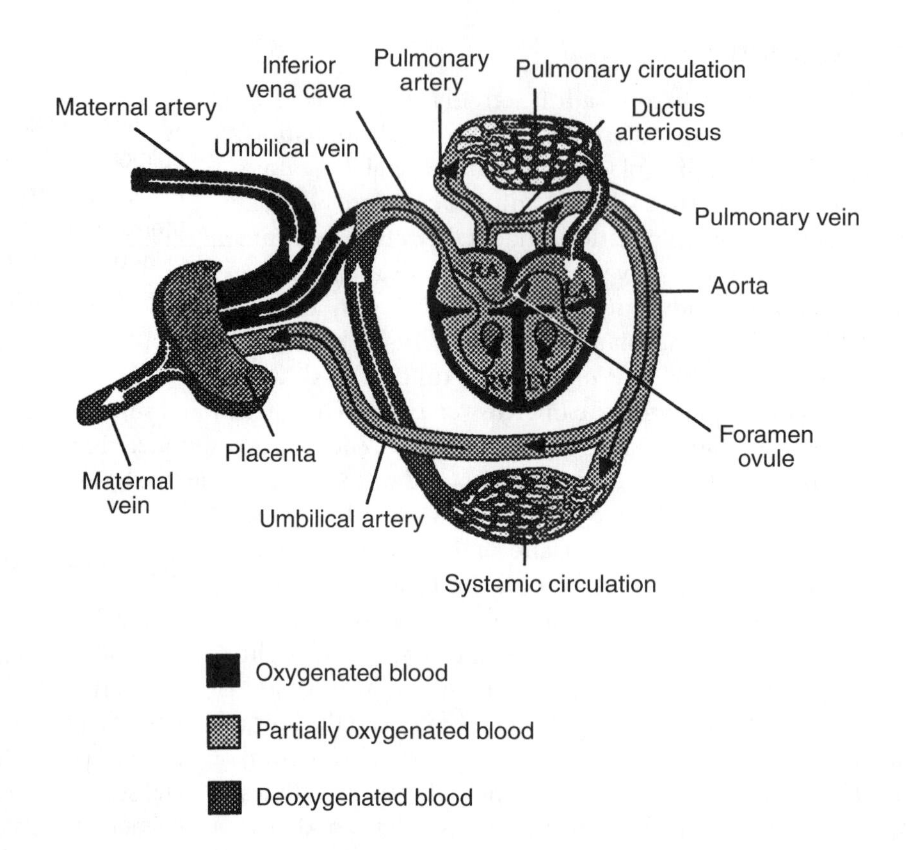

After birth, a number of changes occur in the circulatory system as the fetus adjusts to breathing on its own. The lungs expand with air and rhythmic breathing begins. Resistance in the pulmonary blood vessels decreases, causing an increase in blood flow through the lungs. When umbilical blood flow stops, blood pressure in the inferior vena cava decreases, causing a decrease in pressure in the right atrium. In contrast, left atrial pressure increases due to increased blood flow from the lungs. Increased left atrial pressure coupled with decreased right atrial pressure causes the foramen ovale to close. In addition, the ductus arteriosus constricts and later closes permanently. The ductus venosus degenerates over a period of time, completely closing in most infants three months after birth. The infant begins to produce adult hemoglobin, and by the end of the first year of life little fetal hemoglobin can be detected in the blood.

CHAPTER EIGHTEEN

Plant Structure, Function, and Organization

Members of the plant kingdom are photoautotrophs that utilize the energy of the sun, carbon dioxide, water, and minerals to manufacture chemical energy used in respiration and stored in carbohydrates. They are multicellular, eukaryotic, and autotrophic organisms. The direct ancestor of modern-day plants was likely a green algae. Many of the green algae species possess the same types of chlorophyll and pigments (chlorophyll b and beta-carotene) that plants possess. The first land plants were the **bryophytes**, in the division Bryophyta. Small and living close to a source of water, this group includes the mosses, liverworts, and hornworts. These nonvascular plants have two key adaptations to help them live on land: they are covered by a protective waxy cuticle that prevents desiccation (drying out), and they produce flagellated sperm cells within their male gametangium (reproductive structure) that can swim some distance through droplets of water to fertilize eggs produced by the female gametangium. As plants evolved, vascular tissues such as xylem and phloem allowed them to grow taller and live further away from abundant water sources. Other adaptations besides a waxy cuticle to protect stems and leaves from water loss included stomata, tiny holes in the undersides of leaves for gas exchange. In order to continue the move onto land, it was necessary to protect embryos from desiccation and to be able to spread gametes in a non-aqueous environment.

All plants exhibit alternation of generations in their life cycle between diploid and haploid forms. Within the plant kingdom are several phyla, with one of the key distinguishing characteristics between phyla being the presence or absence of vascular tissue. Within the tracheophytes, which have vascular tissue, two of the important modern phyla are the **gymnosperms** and the **angiosperms**. The gymnosperms such as the conifers have "naked seeds" that do not have endosperm and are not located in true fruits. The angiosperms are the flowering plants. They have flowers, true fruits, and a double fertilization to create endosperm to nourish the plant embryo.

Monocots vs. Dicots: Two Angiosperm Classes

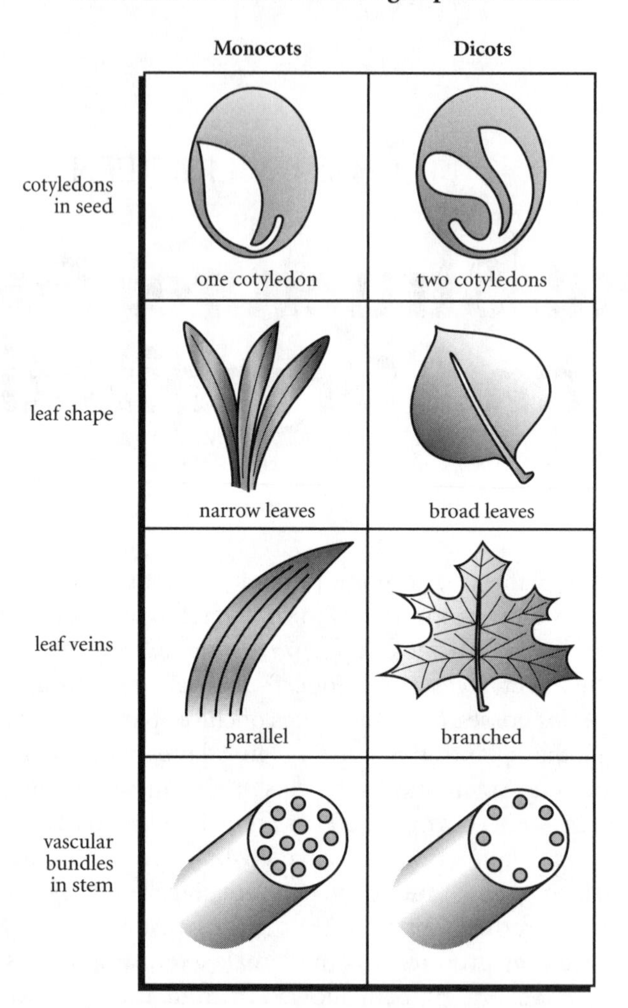

Angiosperms are the most successful form of plant life on the planet, and the diagram above distinguishes between two main classes of angiosperm: **dicots** and **monocots**. You should be familiar with the basic differences between the two types of flowering plants for the GRE.

ROOTS, SHOOTS, AND TISSUES

Structure of Plants

Plants with vascular tissues usually have three types of structures, or organs. These are **leaves**, **roots** and **branches**. The leaves provide most of the photosynthesis of the plant, the roots provide support in the soil, water and minerals, and the branches hold the leaves up to light and convey nutrients and water between the leaves and the roots. Each of these structures can be specialized in many ways.

Plants with taproots have long roots with a single extension deep into the soil while other plants have highly branched roots. Cells on the surface of roots often have long extensions called **root hairs** that increase the surface area of roots. Some plants without root hairs have a symbiotic relationship with fungi that increase the surface area of the root to absorb water and minerals. In legumes, nitrogen-fixing *Rhizobium* species of bacteria infect roots and form root nodules in symbiosis with the plant. The roots of plants often play an important role in preventing erosion. Tropical rain forest that is cleared is highly vulnerable to erosion of the thin soil if the plants and their root systems are absent.

Leaves can have a variety of shapes. Monocot leaves are usually very narrow with veins that run parallel to the length of the leaf, while dicot leaves are broad and with veins that are arranged in a net in the leaf. Modified leaves form thorns in cacti, tendrils in pea plants and petals in flowers. The leaves produce all of the energy of the plant and are specialized to gather sunlight. The broad shape helps to gather sunlight for themselves and in some cases to block sunlight for competing plants. The shape and arrangement of leaves is one of the key features used to distinguish plant species.

Terrestrial plants have broad leaves that maximize the absorption of sunlight but also tend to increase loss of water by evaporation. Leaves of terrestrial plants have a waxy cuticle on top to conserve water. The lower epidermis of the leaf is punctuated by **stomata:** openings that allow diffusion of carbon dioxide, water vapor, and oxygen between the leaf interior and the atmosphere. A loosely packed **spongy layer** of cells inside the leaf contains chloroplasts with air spaces around cells. Another photosynthetic layer in the leaf, the **palisade layer**, consists of more densely packed elongated cells spread over a large surface area. A moist surface that lines the photosynthetic cells in the spongy layer is necessary for diffusion of gases into and out of cells. Air spaces in leaves increase the surface area available for gas diffusion by the cells. The size of stomata is controlled by **guard cells** around that can open and close the opening. These cells open during the day to admit CO_2 for photosynthesis and close at night to limit loss of water vapor through **transpiration**, the evaporation of water from leaves that draws water up through the plant vascular tissues from the soil. The upper surface layer of cells in leaves has no openings, an adaptation that reduces water loss from the leaf.

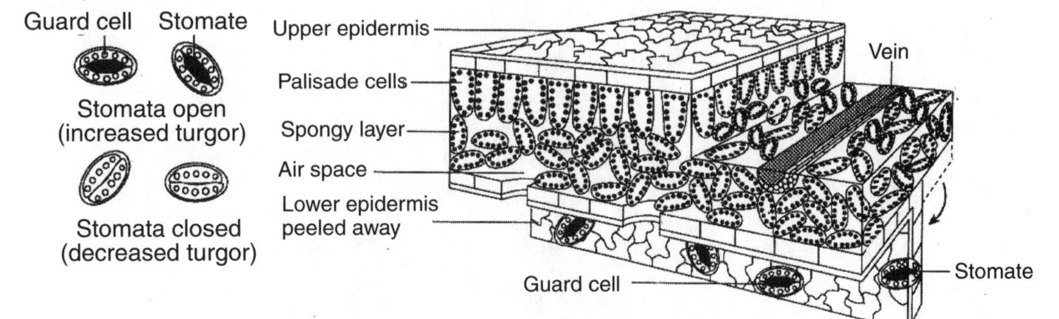

As mentioned in chapter 16, rapid removal of potassium ions (K^+) from cells in the Venus' flytrap allows the closing of a leaf trap on unsuspecting insects. The *shuttling of potassium ions* also seems to be the mechanism for the opening and closing of the guard cells surrounding the stomata. The guard cells are kidney-shaped cells in dicots and dumbbell-shaped cells in monocots that change their shape according to the amount of water that exists within them. This water exerts a pressure called **turgor pressure**. When water is abundant, the guard cells swell and when water is sparse, they clamp down and shut the stomata. This makes perfect sense, as the guard cells allow transpiration when there is plenty of water in the leaves and water conservation when there is not. Guard cells absorb water in response to potassium ions that are driven into the cells. Water follows these ions due to the osmotic gradient that is created by the presence of extra particles inside the cell as compared to outside the cell. The presence of *blue-light receptors* on the guard cell membranes is what drives the movement of K^+ into the guard cell. Sunlight, particularly blue wavelengths, causes K^+ ion channels to open and potassium to flood in. This explains why guard cells are usually open during the day and closed at night, helping photosynthesis to take place as carbon dioxide gas can then enter through the open stomata.

Growth in higher plants is restricted to areas of perpetually embryonic and undifferentiated tissue known as **meristems**. Meristems are self-renewing populations of cells that divide and cause plant growth either in height or width. **Apical meristems** exist at the tips of roots and stems, whereas **lateral meristems** (also known as **cambium**) are found within the stem between layers of xylem and phloem on the sides of the plant. The trunk of the plant thickens each year because the embryonic cells of the cambium produce more and more xylem and phloem, supporting the growth of a larger tree with more leaves. Growth upward that occurs as a result of cell division within apical meristems is called **primary growth**, while growth outward is called **secondary growth**. It is secondary growth that causes the well-known concentric tree rings that one sees when a large, woody tree is cut down.

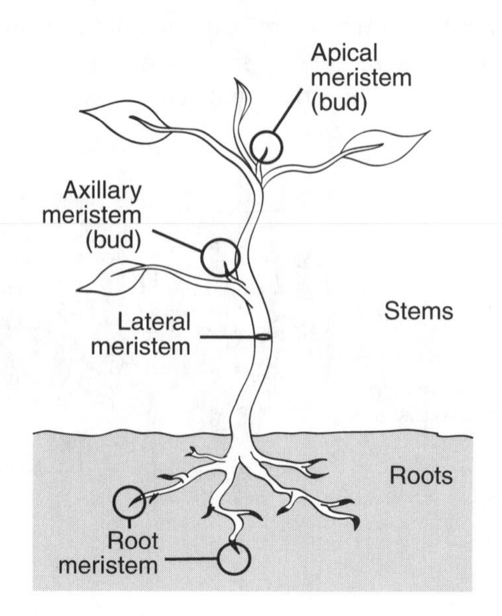

Control of Growth in Plants

The regulation of growth patterns is largely accomplished by plant hormones, which are almost exclusively devoted to this function. These hormones are produced by actively growing parts of the plant, such as the meristematic tissues in the apical region (apical meristem) of shoots and roots. They are also produced in young, growing leaves and developing seeds. Some of these hormones and their specific functions are discussed below, including auxins, gibberelins, cytokinins, ethylene and abscisic acid.

Auxins

This is an important class of plant hormone associated with several growth patterns, including the following types.

Auxins are responsible for **phototropism**, the tendency of the shoots of plants to bend toward light sources, particularly the sun. When light strikes the tip of a plant from one side, the auxin supply on that side is reduced and the illuminated side of the plant grows more slowly than the shaded side. This asymmetrical growth in the cells of the stem causes the plant to bend toward the slower growing light side; thus the plant turns toward the light.

Geotropism is the term given to the growth of portions of plants towards or away from gravity. With negative geotropism, shoots tend to grow upward, away from the force of gravity. If the plant is turned on its side (horizontally), the shoots will eventually turn upward again. Gravity increases the concentration of auxin on the lower side of the horizontally placed plant, while the concentration on the upper side decreases. This unequal distribution of auxins stimulates cells on the lower side to elongate faster than cells on the upper side. Thus the shoots turn upward until they grow vertically once again.

With positive geotropism, roots, unlike shoots, grow toward the pull of gravity. In a horizontally placed stem, however, the effect on the root cells is the opposite. Those exposed to a higher concentration of auxin (the lower side) are inhibited from growing, while the cells on the upper side continue to grow. In consequence, the root turns downward. Auxins produced in the terminal bud of a plant's growing tip move downward in the shoot and inhibit development of lateral buds. Auxins also initiate the formation of lateral roots, while they inhibit root elongation.

Gibberellins

Gibberellins are plant hormones that stimulate rapid stem elongation, particularly in plants that normally do not grow tall. Gibberellins also inhibit the formation of new roots, and stimulate the production of new phloem cells by the cambium (where the auxins stimulate the production of new xylem cells). Finally, these hormones terminate the dormancy of seeds and buds, and induce some biennial plants to flower during their first year of growth.

Cytokinins

Cytokinins are a class of hormones that promote cell division. Kinetin is an important type of cytokinin, and is involved in general plant growth, breaking seed dormancy, and expanding leaves. The ratio of cytokinin to auxin is of particular importance in the differentiation of buds and roots. If auxins are dominant then roots form but if cytokinins are more abundant then buds form from a stem.

Ethylene

Ethylene stimulates the ripening of fruit and the loss of leaves during seasonal changes. Fruit is often harvested before it is ripe and then sprayed with ethylene when it reaches the market to stimulate ripening. Ethylene inhibitors are important to the maintenance of dormancy in the lateral buds and seeds of plants during autumn and winter. They break down gradually with time and, in some cases, are destroyed by the cold, so that buds and seeds can become active in the next growing season.

Abscisic acid

Abscisic acid is a hormone that blocks the growth of stems and is commonly produced by the plant to promote dormancy and reduce damage that would be caused otherwise by harsh weather during cold periods in temperate climates.

Plant Tissues

There are five major types of plant cells that you should be familiar with for the GRE. They each help to build various tissues throughout the plant and serve a variety of functions.

Parenchyma Cells

These are the least specialized of all plant cells and can perform most of the metabolic functions of the plant. If you look at the diagram of leaf structure above, you can see where the mesophyll layer of cells exists. Here, among other processes, photosynthesis takes place, and this mesophyll layer is made almost entirely of parenchymal cells.

Collenchyma Cells

These cells help growing, young plants to remain upright, as these cells are usually formed into long vertical strands and elongate with the growing stems of the young plants.

Sclerenchyma Cells

Far more rigid than the juvenile collenchymal cells, these cells form hardened fibers throughout the plant and often lose their cytoplasm as they mature, forming thick, hard cell walls for support. The coats of seeds and the hard outsides of nuts are formed from these cells.

Tracheids and Vessel Elements

These are specialized cells that make up the xylem, which is discussed in detail in the next part of this chapter. Xylem tissues transport water throughout the plant, and these cells form long, often spiral tubes that can stretch as the plant elongates. At maturity, tracheids and vessel elements are dead, but water continues to flow through them and they retain their capability to stretch.

Sieve-Tube Members

Phloem, also discussed in the next part of the chapter, is made from these specialized cells. Usually losing most of their organelles (including their nuclei) as they mature, these cells carry glucose-rich fluid from cell to cell and from leaves down to the roots.

You should also make the distinction between **dermal tissue** and **vascular tissue.** Dermal tissue (think epidermal) forms the outer coverings of the plant, such as the waxy cuticle of leaves that protects them from water loss. Root hairs, discussed later in more detail, are extensions of dermal cells in the roots of plants. Xylem and phloem, however, are vascular tissues (think cardiovascular system—veins and arteries), since these tissues are used for transport of material around the plant.

You might see the words *symplast* and *tonoblast* on the GRE.

symplast: the cytoplasmic compartment made continuous by the presence of plasmodesmata in plant cell walls

tonoplast: the membrane of the central vacuole, inside of which water and starch are stored.

XYLEM, PHLOEM, AND TRANSPORT

Xylem

The vascular tissue of the xylem contains a continuous column of water from the roots to the leaves, extending into the veins of the leaves. The leaves regulate the amount of water that is lost through transpiration. Water is transported from the roots to the shoots, in other words from roots to stems, through two different mechanisms: **root pressure** and **cohesion-tension**.

Osmotic pressure in the roots tends to build up due to water absorption (see the next section on cation exchange for more details). This pressure pushes water up from the roots into the stem of the plant. Root pressure functions best in extremely humid conditions when lots of water is in the ground or at night. The drops of water, known as "dew," that appear on blades of grass or other small plants in the morning is due to this root pressure. During the day, transpiration occurs at a high enough rate that dew is generally not seen, since the water that leaves the tips of the grass is evaporating before it can build up. Root pressure is not sufficient to move water up tall plants, however.

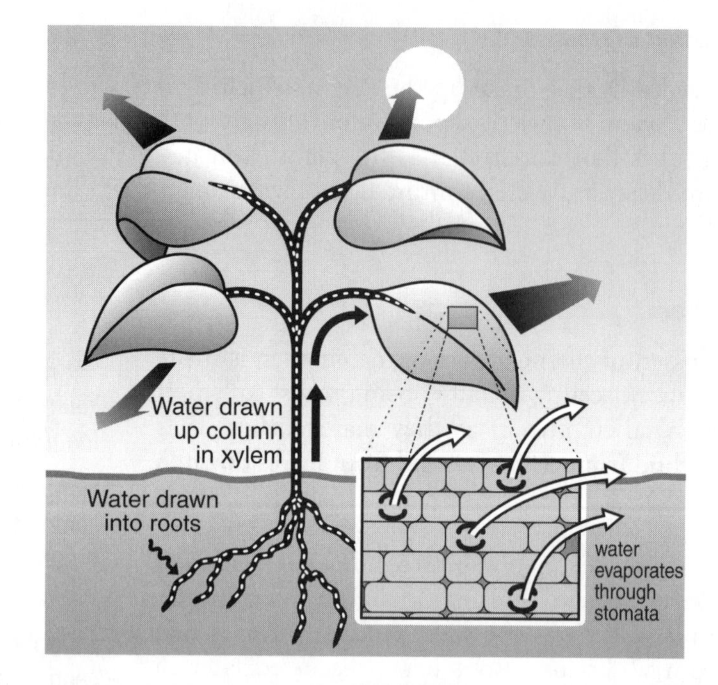

For taller plants, rather than being pushed from the bottom as occurs from root pressure, water in the xylem must be pulled from above by water evaporating from the leaves. The pull of water leaving the leaves by transpiration provides a negative pressure (tension), while the tendency for water molecules to stick together (cohesion) due to their polarity transmits water up the plant towards the leaves. Transpiration occurs because there is less water in the air surrounding the leaves than there is within the leaves, yet on humid days, transpiration is limited or nonexistent. Water molecules are also attracted to the walls of the xylem tubes, and the thin tubes coupled with water's cohesion help to fight gravity and lift water to the tops of trees.

Phloem

Phloem tubes are much thinner-walled than xylem and are found toward the outside edges of stems. They begin up in the leaves, where sugars are made by photosynthesis and stored as starch within mesophyll cells. Much of this sugar and starch, though, is transported down phloem tubes from shoots to roots (from leaves and stems to the roots). Active transport between the mesophyll cells, where sugar is made, and the nearby phloem tubes is what allows sugar to move into the phloem's sieve tubes. The active transport pumps are symport pumps that move one H^+ ion along with every sucrose molecule. The H^+ that is used to push the sucrose across into the phloem is rapidly transported back out to the mesophyll cells by an H^+ ion ATPase pump, so that more sucrose can be moved. As sugar is loaded into the phloem, the **water potential**, or the amount of water pressure, in the phloem is effectively reduced. The influx of sugar into the phloem creates an osmotic potential that pulls more water into the phloem, generating a water pressure that forces sap (essentially sugar-rich water) down the phloem toward the roots. The xylem will recycle this water back from the roots.

As you can see in the following figure, vascular cambium (C) divides laterally to create new phloem (P) on the outer edges of the tree and new Xylem (X) near the inner core. This results in the formation of annual tree rings.

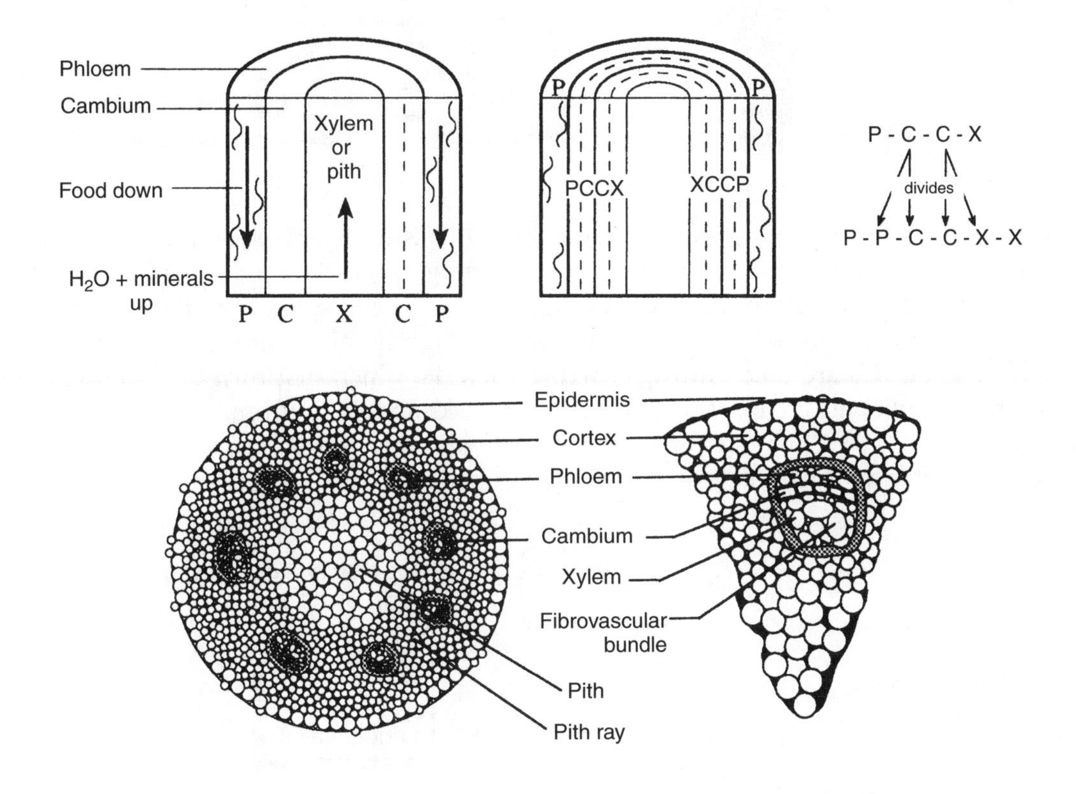

CATION EXCHANGE AND NUTRITION

The essential nutrients that plants need to live and grow include macronutrients in relatively large amounts, and micronutrients in trace amounts. **Macronutrients** include the biomolecules that make up the majority of lipids, carbohydrates, proteins, and nucleic acids: carbon, hydrogen, nitrogen, oxygen, phosphorous, and sulfur (CHNOPS) as well as calcium, potassium, and magnesium (needed to build chlorophyll). **Micronutrients** include iron, chlorine, copper, manganese, and zinc as well as boron, nickel, and miniscule amounts of molybdenum. Many of these metals are used as cofactors for enzymes, particularly those involved in the electron transport chains within mitochondria and chloroplasts.

Hydrogen ions (H^+) that float freely in the water surrounding root hairs help plants absorb some of the key nutrients described above. Soil clumps around the roots tend to attract certain positively charged metal ions, such as potassium, magnesium, and copper. The H^+ ions are able to knock these cations free of the soil particles so that the ions can be absorbed into the root hairs, a process called **cation exchange**. The H^+ ions are secreted by the root hairs in the form of carbonic acid (H_2CO_3), which is made from CO_2 mixing with water in the soil. The CO_2 is the product of local cellular respiration.

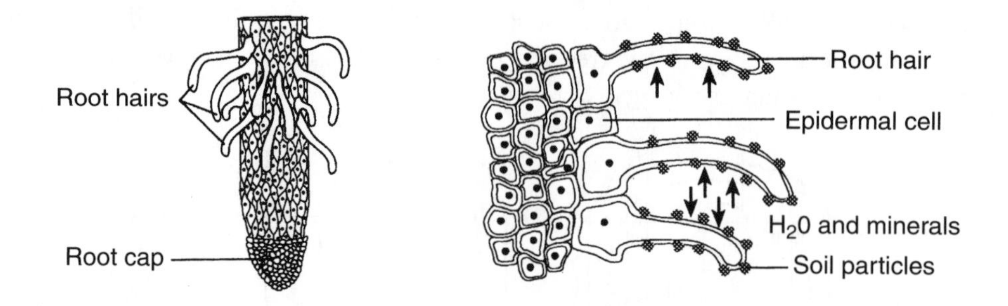

One of the most important molecules needed by plants is nitrogen, which they cannot use without outside help from symbiotic bacteria living on their roots. Needed for the production of proteins and nucleic acids, nitrogen must be converted from its atmospheric form, N_2, into a more plant-friendly form, either NH_4^+ (ammonium) or NO_3^- (nitrate). The bacteria that do this are called **nitrogen-fixing bacteria**, which first turn the atmospheric nitrogen into ammonia (NH_3). The ammonia then picks up free H^+ from the soil to become ammonium ion, NH_4^+. Nitrate ions are produced by other soil bacteria that oxidize the ammonium after it forms, and most plants use nitrate preferentially over ammonium.

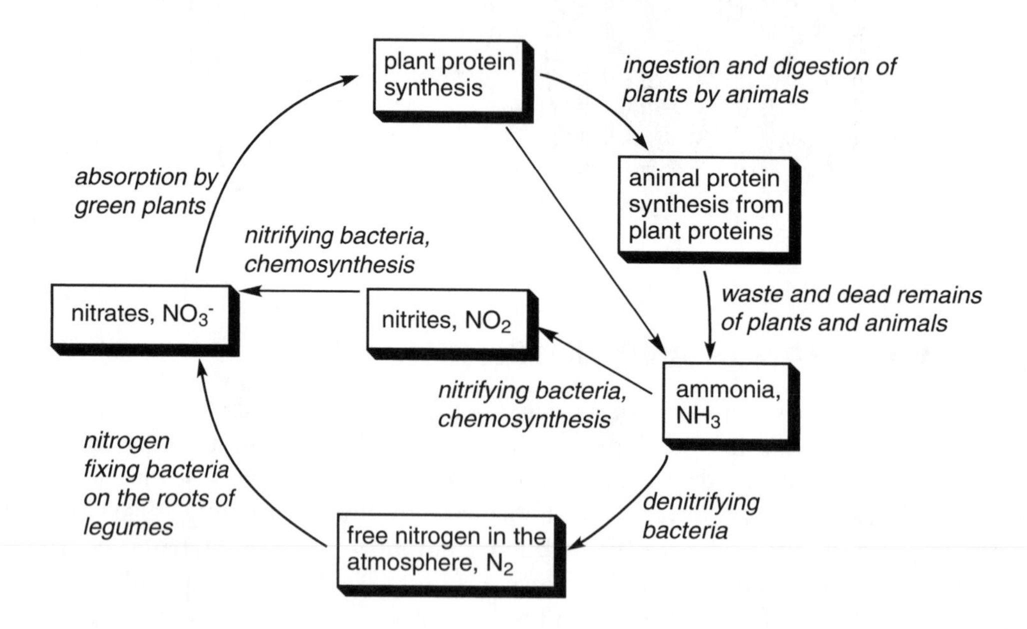

Most animal life on Earth depends upon the interaction between plants and nitrogen-fixing bacteria. Animals eat the plants and synthesize their own proteins from plant proteins. Nitrogen is returned back into the atmosphere by **denitrifying bacteria** that convert nitrogenous wastes back into N_2. The main source of these nitrogenous compounds is the decomposition of dead organisms by bacteria and fungi.

CHAPTER NINETEEN

Plant Reproduction and Development

The life cycles of plants are characterized by an alternation of the diploid **sporophyte** generation and the haploid **gametophyte** generation. The relative lengths of the two stages vary according to the type of plant, but the evolutionary trend has been toward the increased dominance of the sporophyte generation in making up the majority of the plant life cycle. Key evolutionary adaptations among land plants include the seed and the fruit. In addition, there are external controls over plant growth that vary according to the environment the plant is in. This chapter investigates aspects of plant reproduction and development that you should be familiar with for the GRE.

LIFE CYCLE OVERVIEW—GAMETOPHYTES VERSUS SPOROPHYTES

Algae, the ancestors of today's plants, produce both haploid cells and diploid cells. These cells arise because of an alternation in generations that arises from the algae's production of haploid generations followed by diploid generations. We see this in modern-day plants as well.

The major advantage of producing a haploid generation in between diploid ones is the possible weeding out of deleterious (harmful) recessive alleles. Haploid plants may have genotype R or r, but *not* RR, Rr, or rr. Therefore, if r represents a harmful recessive allele, those haploid plants with it will be less likely to reproduce and pass it onto the next, diploid, generation. In nonvascular plants such as mosses, the plant you *see* on the ground, the predominant form, is the gametophyte stage, while in vascular plants (ferns, conifers such as pine trees, and all flowering plants such as maples and oaks) the plant you see is the sporophyte generation. The gametophyte produces gametes (sperm and egg), and these gametes will produce fertilized eggs that grow within seeds in two types of plants only: the **gymnosperms** (pine trees and related conifers); and, the **angiosperms** (flowering plants).

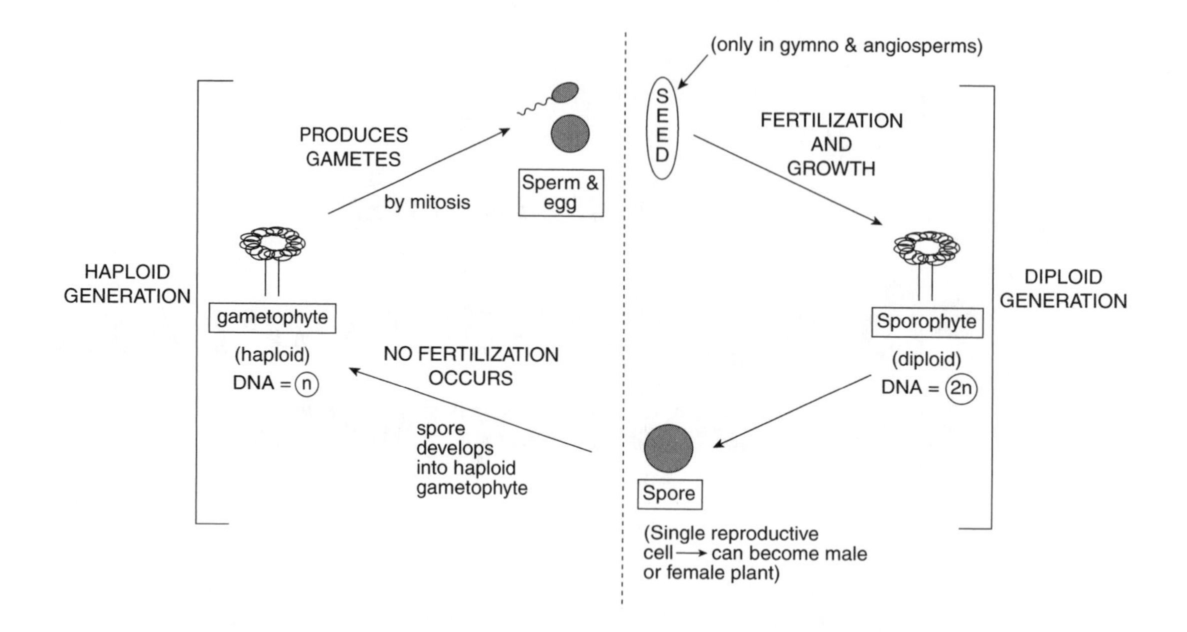

FRUITS AND SEED DEVELOPMENT

In conifers, most species have both male (pollen) cones and female (ovulate) cones. The cones you are used to picking up off the ground are the larger, female cones ("pinecones"). Pollen cones contain hundreds of cells that can undergo meiosis and become sperm cells contained within encapsulated pollen grains. Ovulate cones, covered in large, thick scales, contain two **ovules** on the inside edge of each scale. Each ovule can contain many eggs, yet generally only one zygote will develop from one ovule. When pollen grains fall on an ovule (usually blown by the wind), fertilization can take up to a year to occur as the pollen grains slowly dig a tube into the ovule to fertilize eggs within. The new embryo (sporophyte) that begins its growth within the fertilized ovule will, in fact, grow a very basic root system and some embryonic leaves before the seed even reaches the soil.

The Flower

The flower is the organ for sexual reproduction of angiosperms and consists of male and female organs. The flower's male organ is known as the stamen. It consists of a thin, stalklike filament with a sac at the top. This structure is called the anther, and produces haploid spores. The haploid spores develop into pollen grains. The haploid nuclei within the spores will become the sperm nuclei, which fertilize the ovum. The flower's female organ is termed the pistil. It consists of three parts: the stigma, the style, and the ovary. The stigma is the sticky top part of the flower, protruding beyond the flower, which catches the pollen. The tubelike structure connecting the stigma to the ovary at the base of the pistil is known as the style; this organ permits the sperm to reach the ovules. The ovary, the enlarged base of the pistil, contains one or more ovules. Each ovule contains the haploid egg nucleus.

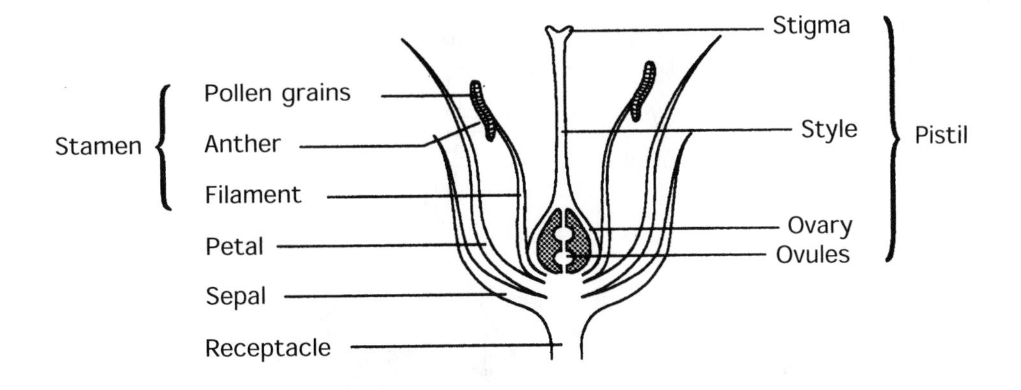

Petals are specialized leaves that surround and protect the pistil. They attract insects with their characteristic colors and odors. This attraction is essential for cross-pollination—the transfer of pollen from the anther of one flower to the stigma of another. Note that some species of plants have flowers that contain only stamens ("male plants") and other flowers that contain only pistils ("female plants").

The pollen grain develops from the spores made by the sporophyte. Pollen grains are transferred from the anther to the stigma. Agents of cross-pollination include insects, wind, and water. The flower's reproductive organ is brightly colored and fragrant in order to attract insects and birds, which help to spread these male gametophytes. Carrying pollen directly from plant to plant is more efficient than relying on wind-borne pollen and helps to prevent self-pollination, which does not create diversity. When the pollen grain reaches the stigma (pollination), it releases enzymes that enable it to absorb and utilize food and water from the stigma and to generate a pollen tube. The pollen tube is the remains of the evolutionary gametophyte. The pollen's enzymes proceed to digest a path down the pistil to the ovary. Contained within the pollen tube are the tube nucleus and two sperm nuclei; all are haploid.

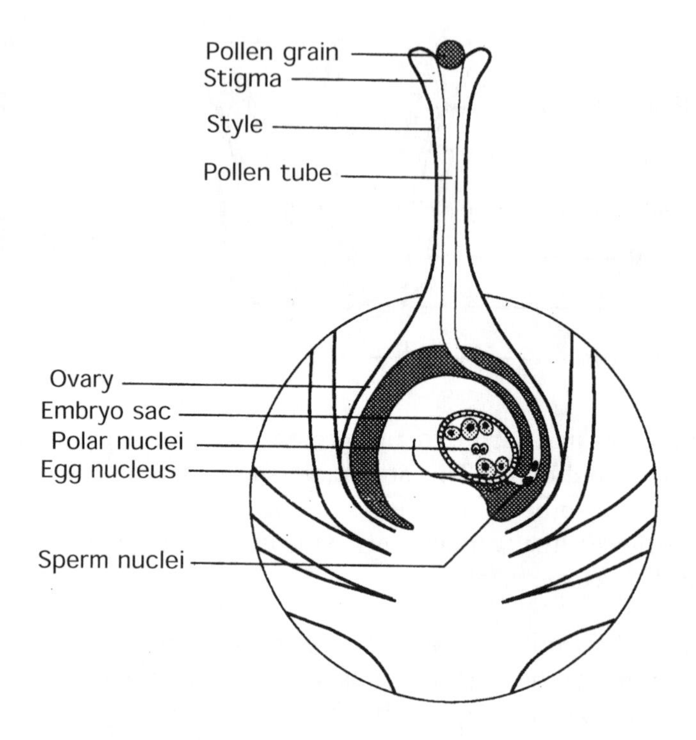

The female gametophyte in angiosperms develops in the ovule from one of four spores. This embryo sac contains nuclei, including the two polar (endosperm) nuclei and an egg nucleus. During fertilization (see figure) the sperm nuclei of the male gametophyte (pollen tube) enters the female gametophyte (embryo sac), and a double fertilization occurs. One sperm nucleus fuses with the egg nucleus to form the diploid zygote, which develops into the embryo. The other sperm nucleus fuses with the two polar bodies to form the endosperm (triploid or 3n). The endosperm provides food for the embryonic plant. In dicot plants, the endosperm is absorbed by the seed leaves (cotyledons).

1 sperm nucleus (n) + 1 egg nucleus (n) = zygote (2n) = embryo (2n)
1 sperm nucleus (n) + 2 polar nuclei (n) = endosperm (3n)

The zygote produced in the sequence above divides mitotically to form the cells of the embryo. This embryo consists of the following parts, each with its own function:

- The **epicotyl** develops into leaves and the upper part of the stem.
- The **cotyledons** or seed leaves store food for the developing embryo.
- The **hypocotyl** develops into the lower stem and root.
- The **endosperm** grows and feeds the embryo. In dicots, the cotyledon absorbs the endosperm.
- The **seed coat** develops from the outer covering of the ovule. The embryo and its seed coat together make up the seed. Thus, the seed is a ripened ovule.

The fruit, in which most seeds develop, is formed from the ovary walls, the base of the flower, and other consolidated flower pistil components. Thus, the fruit represents the ripened ovary. The fruit may be fleshy (as in the tomato) or dry (as in a nut). It serves as a means of seed dispersal; it enables the seed to be carried more frequently or effectively by air, water, or animals, through ingestion and subsequent elimination. Eventually, the seed is released from the ovary, and will germinate under proper conditions of temperature, moisture, and oxygen.

For seeds to germinate, they need water! Water allows key enzymes to become active and causes cell respiration to start up within the seed. Temperature and pH, as you learned in Section II, are also important for these enzymes to activate.

EXTERNAL CONTROL MECHANISMS

Environmental stimuli and seasonal changes help regulate plant life cycles and survival. These influences include light, water, temperature, and predators. Perhaps the single most important factor guiding plants in when to reproduce and flower (in the case of angiosperms) is the number of daylight hours that the plant receives. Because daylight hours shorten and lengthen with the change of season in most parts of the world, most plants are able to sense the season by the amount of light they are receiving during a given period. As such, some plants are known as **short-day plants**, and will flower when exposed to daylight hours shorter than a certain threshold period (i.e., shorter than 17 hours). Others are known as **long-day plants**, since they will flower with days longer than a certain threshold.

Experiments have shown, however, that it is really the amount of *darkness* a plant receives that matters, not the amount of daylight. Short-day plants really need a night that lasts a certain number of hours to flower, and if the night is longer than that, it doesn't matter the number of daylight hours. You may see the terms "short-day" and "long-day" used on the GRE when *short-day plants are really long-night plants and vice versa.* In many plants, exposure of just one leaf on the plant to the proper amount of nighttime hours is sufficient to induce flowering, yet without their leaves, plants are effectively blind and cannot detect day length.

CHAPTER TWENTY

Monera

The Monera, or bacteria, are simple single-celled organisms that have colonized every inhabitable spot on this planet. They demonstrate a huge variety of shapes and sizes, and their metabolic needs are met by a diversity of mechanisms that can serve their needs in almost every climate and habitat. Bacteria tend to be classified based on their shape, from rodlike structures to spherical cells that cluster together in colonies. This chapter gives a brief look at the material you'll need to know about bacteria for the GRE.

CELL STRUCTURE AND CLASSIFICATION

The prokaryotes include two distinct kingdoms of organisms that are recognized today: bacteria and archaebacteria. Only recently, in part through analysis of the genomes of these two groups, have the differences between the bacteria and archaebacteria been realized. In fact, analysis of the ribosomal RNA of these organisms suggests that archaebacteria are more closely related to eukaryotes than they are to bacteria. All prokaryotes are single-celled organisms that lack membrane-bound internal structures.

Although some prokaryotes live in groups of cells as filaments or colonies, each cell is a separate independent organism. Bacteria have several distinct external forms that are often used to distinguish them. Rod-shaped bacteria are known as bacilli and spherical bacteria are known as cocci. Prokaryotes have an outer lipid bilayer cell membrane, but lack membrane-bound organelles and also lack a cytoskeleton. They have no nucleus, but instead, a region of the cell called the **nucleoid** region in which the genome, consisting of a single circular molecule of DNA, is concentrated. The lack of organelles means that the interior of prokaryotes is one continuous compartment, the **cytosol**, in which all of the activities of life occur. Prokaryotes may also contain

Don't forget: Prokaryote = Monera = Bacteria. These words can be used interchangeably.

plasmids, small circular extrachromosomal DNAs containing few genes. Plasmids replicate independently and often incorporate genes that allow the prokaryotes to survive adverse conditions, making them very useful in modern molecular biology to carry genes in and out of bacteria. Prokaryotes have lipid bilayer cell membranes as do eukaryotes, with transmembrane proteins involved in the communication between the interior and the exterior of the cell. The cytoplasm of prokaryotes contains many ribosomes for protein production as does the cytoplasm of eukaryotes, although prokaryotic ribosomes are smaller.

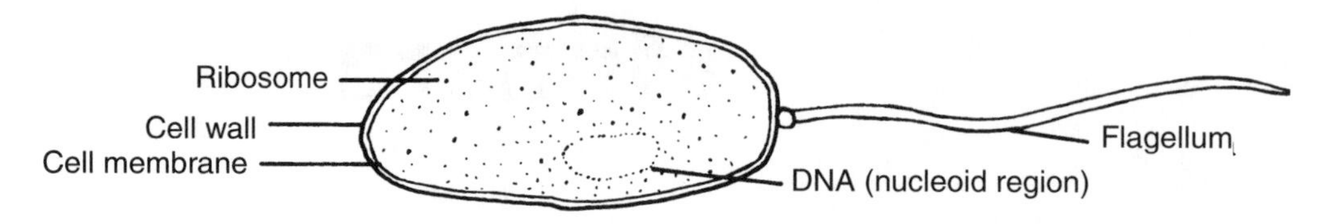

Most prokaryotes have a strong porous cell wall for support of the cell. In bacteria the cell wall is made of peptidoglycan polymers woven into a single large molecule that surrounds the cell. Peptidoglycans are polysaccharides cross-linked by peptides that contain amino acids with the D stereochemistry, while proteins found in organisms normally have L-amino acids. Archaebacteria lack peptidoglycan in their cell wall, one indication of their uniqueness. Some bacteria have an additional outer membrane beyond the cell wall that often contains lipid toxins in pathogenic bacteria. Gram-staining is commonly used to separate bacteria into two groups: gram-negative and gram-positive. Gram-positive bacteria have a thick peptidoglycan cell wall that holds the stain but no outer membrane, while gram-negative bacteria have a thin cell wall and an outer membrane.

A layer of sticky polysaccharides called the capsule is often found around the cell membrane and the cell wall. The polysaccharides may help in the attachment of bacterial cells to outside surfaces and may help pathogenic bacteria to avoid the immune system since the cell wall can be hidden underneath the polysaccharides. Projections from the bacterial surface include flagella involved in movement (although unrelated in structure to the flagella found in some eukaryotic cells) and pili that are involved in attaching cells to other cells, in some cases for the exchange of genetic information. Bacterial flagella have a motor in the base anchored to the cell that causes the flagella to whip around; the flagella itself is composed of a protein called flagellin. Eukaryotic flagella are formed from microtubules with motor proteins to drive movement and are surrounded by the plasma membrane.

Prokaryotes reproduce by a mechanism distinct from eukaryotes. Eukaryotic cells divide by either mitosis or meiosis, and many eukaryotes reproduce sexually, with the union of genetic information from two parents. Prokaryotes reproduce asexually by a mechanism called binary fission, in which a cell replicates its DNA and divides in two. Binary fission does not involve any of the structures or processes of mitosis, which eukaryotes use to perform asexual reproduction. Bacteria do exchange genetic information through conjugation and other methods, and they also can perform recombination, but these occur apart from cell division.

MODES OF NUTRITION

Although prokaryotic cells are simple in structure, they are diverse in their metabolic activities and habitats. One means of distinguishing the life styles of organisms is according to their use of oxygen. **Obligate anaerobes** cannot survive in the presence of oxygen, **facultative aerobes** can survive with or without oxygen, and **obligate aerobes** require oxygen to survive. Another means of classifying organisms is according to their mode of nutrition. **Photoautotrophs** are photosynthetic, using light to generate the energy to produce their own nutrient molecules. Photosynthetic bacteria use the plasma membrane as the site of photosynthesis, although some have infoldings of the membrane called mesosomes. Another type of nutrition is used by **chemoautotrophs.** These organism produce their own nutrient molecules but use energy derived from inorganic molecules to drive nutrient production rather than the power of the sun. Nitrogen fixation, essential to all life, is carried out by chemoautotrophic prokaryotes. Chemoautotrophs are responsible for deep sea thermal vent ecosystems that are independent of light from the surface for energy. A common method of nutrition for all life, prokaryotes as well as eukaryotes, is used by **chemoheterotrophs**, which consume organic molecules for both carbon and as an energy source. If an organism obtains glucose, for example, from its environment and uses it for food and energy, it is a chemoheterotroph.

Although prokaryotes are often called "primitive" due to their simplicity, they are highly evolved in many ways. Bacteria fill every conceivable ecological niche on the planet, including bacteria that live in water that is near boiling and others that live deep in solid rock. The rapid rate of bacterial reproduction, as fast as one cell division per twenty minutes, allows bacteria to evolve extremely rapidly. In fact, some of the features of eukaryotes such as the presence of introns in genes that were once taken as signs of eukaryotic superiority may indicate that eukaryotes have evolved more slowly than prokaryotes and so retained primitive features of earlier precursor cells.

Don't forget how important nitrogen metabolism is for many bacteria. See chapter 18 for more details.

CHAPTER TWENTY-ONE

Protista

Protists, single-celled eukaryotic organisms, were the first creatures to evolve that had complex, specialized organelles within their cells. The **Endosymbiont Theory** suggests that many of the organelles present in eukaryotic cells were once their own free-living prokaryotic cells. The first protists, according to this theory, arose from a symbiosis that formed between a larger cell that had engulfed a smaller one, yet for some reason did not digest its meal. Maybe the smaller cell was a parasitic species. In any case, the ingested cell may have been able to provide something for the larger cell that it could not provide itself. The smaller bacterium may have been a cyanobacterium, capable of photosynthesis, and it could gain shelter within the larger cell while providing food for the larger cell to use. Both the larger and smaller cells would divide on their own, creating a population of cells that had smaller, photosynthetic organelles within them.

The Endosymbiont Theory might sound far-fetched were it not for the extraordinary evidence that shows how similar some organelles in eukaryotic cells today are to present-day bacterial forms. Chloroplasts and mitochondria not only contain enzymes and proteins in their inner membranes that are similar in structure to bacterial proteins, they also replicate on their own, independent of the cell they are in, and *have their own DNA and ribosomes*. Although they have lost too much DNA to be able to live freely in modern times, these organelles make their own proteins and provide ATP and sugar to the larger cells they reside within, which the larger cells cannot do on their own. The origin of the nucleus and other eukaryotic organelles, such as the endoplasmic reticulum, appears to have resulted from invaginations of the cells membrane as it pinched inward to form membrane-bound structures.

The first protists on Earth probably arose about 2 billion years ago, and life on Earth appears to have remained *unicellular* for the next 1.5 billion years until the Cambrian Explosion around 550 million years ago.

MAJOR PHYLA—CELL STRUCTURE AND MOVEMENT

There are at least 30, perhaps as many as 50, different phyla of protists. The species in this kingdom of organisms are so diverse in their characteristics that many scientists debate the usefulness of having a single kingdom for all protists—despite the fact that they are mostly single-celled eukaryotic organisms. The GRE will not ask about every phylum in every kingdom, though you should know the major distinguishing characteristics among the different protists and ways to recognize a given group. There are *three major types of protists*, grouped by their similarity to organisms in other kingdoms: **animal-like protists**, **plantlike protists**, and **funguslike protists**. In general, protists within each of these groups are further divided by their feeding habits and mechanisms by which they move.

HOW CAN YOU IDENTIFY A GROUP OF PROTISTS?

In general, animal-like protists include the well-known sarcodines (such as amoebas that move by pseudopodia); the ciliates (such as the paramecia that move by cilia); and the flagellates (such as the dino-flagellates that move by flagella). Look for phylum names such as Rhizo*poda*, Actino*poda*, or *Cili*ophora, and common names such as zoo*flagella*tes. For plant-like protists, the give-away in the phylum name is the suffix *-phyta*, as in Chryso*phyta* (the golden algae) or Phaeophyta (the brown algae). For funguslike protists, look for the suffix *-mycota*, as in Oo*mycota* (the water molds) or the Myxo*mycota* (the slime molds).

Animal-Like Protists

Also known as **protozoans**, the animal-like protists are classified mainly by how they move: pseudopods, cilia, or flagella. We have already detailed in earlier chapters the mechanism by which organisms move using these appendages. The phylum *Rhizopoda*, which includes amoebas and their relatives, is representative of pseudopod-using protists. They also use their pseudopodia to engulf prey, such as another protozoan. Amoebas and other rhizopods inhabit salty, marine habitats as well as freshwater lakes and ponds. They can also be found as parasites in humans and in other organisms, spreading through the digestive system and contaminated fecal matter.

The *Ciliophora* is the representative phylum of ciliates, such as the paramecia, that move using tiny hair-like projections called cilia that project from within the cell cytoplasm. Cilia are also used to feed, primarily by sweeping food into a mouth opening (oral groove), after which the food is encapsulated in a vacuole for storage or digestion. Although many ciliates have cilia covering over their entire cell surface, many have cilia only in certain places. *Stentor* is an elongated cell with a cone-shaped widening at one end that serves as a mouth. Cilia surround this opening to sweep food into the organism, but are not found elsewhere on the cell.

The *Zoomastigophora*, or zooflagellates, are protozoans that move by the use of one or more whip-like flagella. This group of organisms is structurally related to the euglena, which are characterized by the phylum *Euglenophyta* because of their photosynthetic capability.

Funguslike Protists: The Slime Molds

The representative fungus-like protists are found in the phyla *Myxomycota* or *Acrasiomycota*. Like fungi, slime molds are not photosynthetic, relying instead on external nutrients for food. Also, like fungi and plants, slime molds produce alternating haploid and diploid generations. Many slime molds exist as a mass called a **plasmodium**, a huge single cell that many have thousands of individual nuclei within it. As occurs in the *Myxomycota*, these colorful masses can spread out over large areas and transport nuclei and nutrients along a highly branched cytoplasm. For mating, small stalks arise from the amoeboid mass, each with a ball of spores on top. These spores germinate and develop into haploid cells, which will then become a new diploid plasmodium through fertilization if the environment is right.

Other slime molds such as the *Acrasiomycota*, function independently as single-celled organisms that will group together into a multicelled amoeboid mass if conditions around them are unfavorable. Although cellular slime molds such as these go through an alternation of generations, their plasmodia are haploid, whereas the plasmodia in *Myxomycota* are diploid.

Plantlike Protists: Chlorophyta and Other Algae

Algae, dinoflagellates, and diatoms represent the plant-like protists. All of the plantlike protists are characterized by their bright coloration, the result of pigments stored within organelles called **plastids** . They also all possess chlorophyll for photosynthesis. Chloroplasts, as their name implies, are plastids that hold chlorophyll, a green pigment. But other plastids in the plantlike protists hold **carotenoid** pigments for yellow and red coloration and **xanthophylls** for brownish coloration. Some of these protists are surrounded by hardened secretions that are shed as the cells within die. Diatom shells, made partially of silica, shine like crystals on the ocean floor. Many algae are colonial, living as masses of single cells within a group, and the largest seaweeds are really multicellular algae. Some seaweeds have specialized tissues like plants, yet they are classified with the protists because of their overall simplicity. Many algae, particularly the more complex seaweeds, undergo an alternation of generations similar to the green plants.

Dinoflagellates are responsible for poisonous "red tides" that can stifle lakes or ponds and kill many aquatic organisms. Rapid growth of dinoflagellates populations within an aquatic ecosystem is called a "bloom," and blooms can kill off many organisms due to the toxins secreted by the algae.

The *Chlorophyta* are a huge group of thousands of species of green algae, some of which were likely the ancestors of modern-day land plants. The algae that symbiotically live as part of **lichens** come from this phylum.

CHAPTER TWENTY-TWO

Fungi

The fungi are eukaryotic, usually multicellular, organisms. Once thought of as "imperfect plants" because of their *lack of photosynthetic ability*, fungi remain in their own distinct kingdom separate from plants. Fungi get their nutrients and water through absorption. They may be **saprobes** (saprophytes) that absorb nutrients from nonliving matter; parasitic, absorbing nutrients from the cells of a living host; or, mutualistic, absorbing nutrients from other organisms and providing shelter or nutrients in return. Most fungi are built from **hyphae**, long strands of cells that are attached end to end. These hyphae make large networks that can cover over huge areas in a fungal mat called a **mycelium**. The fastest-growing fungi can build up to one kilometer of hyphae per day by moving nutrients to the tips of the growing hyphae.

CLASSIFICATION AND LIFE CYCLES

Fungi are classified according to differences in reproductive structures and mechanisms. There are three general divisions.

Zygomycetes

In these fungi, reproduction occurs through the use of two different reproductive structures, one male and one female, each on a different hyphae. These reproductive parts of the hyphae are called **gametangia** and they contain several haploid nuclei each. The cytoplasms of the male and female gametangia fuse in a process known as **plasmogamy**, and the haploid nuclei contained within the hyphae pair off and fuse. The fusion of these haploid nuclei in a process called **karyogamy** occurs only when conditions are favorable and starts the creation of new hyphae through mitotic

Taxonomists use the term *division* rather than *phylum* when classifying plants and fungi. These terms are essentially the same, however.

cell division. Many of the fungi in this division form **mycorrhizae**, associations with plants that occur as the fungi take up residence on the plant roots and help the plants absorb water and nutrients or nitrogen.

Ascomycetes

Also known as sac fungi, members of this division include the single-celled yeasts among a variety of multicellular cousins. Reproduction here is similar to that of the Zygomycetes; however, the female gametangia receive haploid nuclei from the male. When they possess male and female nuclei, the female gametangia start to rapidly grow into hyphae, which have closed compartments at their ends called asci (singular: **ascus**). It is in these asci that karyogamy, or fusion, of the male and female nuclei takes place. These fused, diploid, nuclei now divide by mitosis, filling the asci with diploid cells known as **ascospores.** These ascospores will be released to develop into new hyphae. An alternative means of reproduction for fungi in this division is to use asexual spores called **conidia.**

Basidiomycetes

These fungi are the commonly known mushrooms, otherwise known as club fungi, and recognized by their multiple gill structure and large, umbrella-shaped reproductive bodies called **basiocarps.** Asexual spores are held by the hundreds of millions on the underside of these basiocarps, which are the mushroom stalks we see and eat. However, for every mushroom that pops up above the ground, there may be upwards of one mile of hyphae stretched out underground, making club fungi perhaps the largest continuous organisms in the world. Reproduction, again, is similar to the other fungi: Haploid hyphae meet after spores released from mushrooms settle into the ground. These newly formed **dikaryotic hyphae** (dikaryotic = having nuclei from two different parents) will grow into new hyphae and new mushrooms complete with spores.

Don't be confused by the terminology here. The "big picture" in fungi reproduction is that two haploid hyphae meet and exchange nuclei. These nuclei fuse to form spores, which are released to grow into new hyphae again. There is clearly an alternation of generations that takes place in all fungi between the temporary diploid hyphae that are formed when haploid parent hyphae meet and the haploid hyphae that undergo the meeting in the first place.

Do not forget about molds and yeasts! Molds are rapidly growing fungi that reproduce only asexually. They are often used to add flavors and colors to many foods, such as blue cheese. Yeasts are unicellular fungi that inhabit moist or aquatic habitats. Most reproduce asexually by budding, which allows the formation of a new yeast cell directly off the body of a parent cell.

LICHENS

Lichens, which are a combination of a fungus and an algae living together in one organism, can be found growing on rocks or trees in relatively unpolluted environments. The algal cells produce sugar and nutrients for themselves and for the fungus, while the fungus absorbs water and houses the algae. They are often the first colonizers of landscapes that have been destroyed by fire and are extremely hardy. They are also extremely sensitive to man-made pollution and will die quickly in polluted areas. Most lichens grow very slowly and are able to shut down their metabolisms for long periods of time if unfavorable conditions persist. They reproduce both sexually, using spores and an alternation of generations similar to other fungi, and asexually, often because parts of the lichen will break off as it dries out. These parts can grow into another lichen after being blown by the wind or swept to another location.

CHAPTER TWENTY-THREE

Animalia

The kingdom *Animalia* is made up almost exclusively of multicellular, heterotrophic organisms. There are some exceptions, such as the tiger flatworm that possesses chloroplasts and photosynthetic capability. However, as a rule, the animals have no walls around their cells, no chloroplasts in their cells, and no large central vacuoles for water storage. They have centrioles to aid in spindle formation for cell division and store excess sugars as glycogen rather than as starch. Animals also possess nervous and muscle tissue for coordination of behavior and movement. This chapter looks at the variety of animals that exist on Earth, their characteristics, and their evolutionary relationships.

Don't get bogged down in memorizing every characteristic of every phylum. For the exam, learn the main vocabulary and the differences among key groups, especially the largest ones (Arthropoda, Cnidaria, Annelida, Mollusca, and Chordata).

PHYLOGENETIC RELATIONSHIPS

In the animal kingdom, there are about 35 phyla, 34 of which contain **invertebrate** organisms that have no backbone. Despite being the largest and most easily recognizable organisms on Earth, the **vertebrates** are simply a tiny subphylum of the phylum *Chordata*, which contains invertebrates and vertebrates possessing both a dorsal nerve cord and a rigid rod of supporting tissue known as the notochord. The origin of all animal phyla, whether invertebrate or vertebrate, seems to have occurred in the **Cambrian Explosion**, about 550 million years ago. Most scientists agree that the entire kingdom (all 35+ phyla) is **monophyletic**, meaning that every group arose from the same common ancestor, most likely some sort of colonial protist.

Make sure to distinguish between the words **monophyletic, polyphyletic** (refers to a group of organisms composed of members that do not share a unique common ancestor), and **paraphyletic** (refers to a portion of a group that contains members having the same common ancestor but that does not include every descendant species of that common ancestor).

One of the earliest phylogenetic branches off of the evolutionary tree that depicts the kingdom Animalia is the splitting off of the phylum *Porifera* from all other animal phyla. The *Porifera*, or sponges, are distinct from all other groups of animals because they lack "true" tissues. Their cells are not specialized into particular tissues (i.e., muscle, vascular) and they lack organs. The rest of the animal kingdom is divided primarily by differences in external symmetry and internal body cavities. After the branching off of the sponges, all animal phyla are initially distinguished from each other based on their outer symmetry: whether their bodies are **radially symmetric** or **bilaterally symmetric**. Radially symmetric animals have bodies stretching out in equal dimensions from a central point (like we see in jellyfish, anemones, or other round organisms). Bilaterally symmetric animals are organized along one vertical or horizontal axis and tend to be long and thin. Worms, fishes, insects, and all vertebrates are among the bilaterally symmetric animals. They can be divided into two equal parts by cutting in only one direction, while radially symmetric animals can be divided equally in an infinite number of ways as long as the cut goes through the center of the body.

MAJOR PHYLA AND DISTINGUISHING CHARACTERISTICS

Bilaterally symmetric animals are further divided based on their internal anatomy, specifically whether or not they possess body cavities (called coeloms) and blood vessel systems among these cavities. A **coelom** is a body cavity in addition to the digestive tract. Think of your body as a tube within a tube—your outer skin and muscles form an outer tube, while your mouth, digestive tract, and anus form an internal tube right through you. Then think of a fluid-filled space between your outer and inner tubes, which serves to cushion and nourish your digestive tract. This space is your coelom.

Bilaterally symmetric animals have both **dorsal** (back) and **ventral** (stomach) surfaces, as well as **anterior** (head) and **posterior** (tail) ends.

The diagram below divides key animal phyla according to their symmetry and possession of a coelom. Each phylum is specified as **acoelomate** (having no body cavity between the gut and outer wall), **pseudocoelomate** (having a body cavity that is lined by muscles and blood vessels only on the outside surface of the cavity), or **coelomate** (having a body cavity lined by muscle tissue and blood vessels both on the outer surface and on the inner one, surrounding the entire digestive tract).

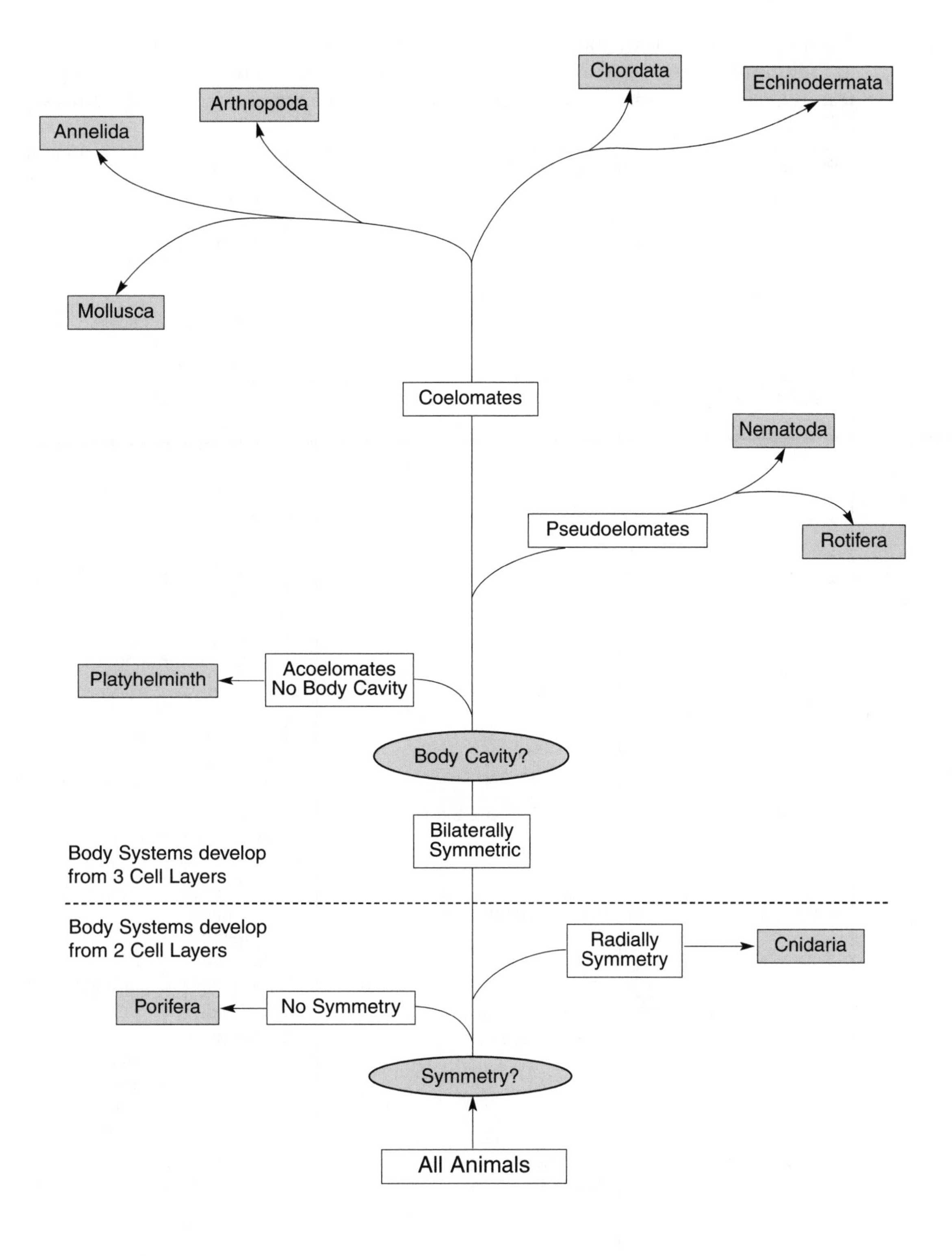
Chordata
Echinodermata
Arthropoda
Annelida
Mollusca
Coelomates
Nematoda
Pseudoelomates
Rotifera
Platyhelminth
Acoelomates
No Body Cavity
Body Cavity?
Bilaterally
Symmetric
Body Systems develop
from 3 Cell Layers
Body Systems develop
from 2 Cell Layers
Radially
Symmetry
Cnidaria
Porifera
No Symmetry
Symmetry?
All Animals

The chart below sums up the animal phyla you should be familiar with for the GRE, as well as the approximate number of species within each phylum and the phylum's distinguishing features. Don't memorize it—just look for patterns and learn one or two key features from each phylum.

Phylum	Representative Organisms	Features
Porifera (4,200)	Sponges	Body in two cell layers. Skeletons made of silica or calcium secretions. Asymmetric.
Cnidaria (9,200)	Jellyfish, Corals, Hydra	Body in two cell layers. "Baglike" with tentacles. Radially symmetric. "Stinging cells"
Platyhelminthes (6,000)	Flatworms, Tapeworms	Body flat and ribbonlike. Many forms are parasitic. Lack skeletal, circulatory, and respiratory systems. Dorsal and ventral nerves. Acoelomate. Bilaterally symmetric.
Aschelminthes (12,000)	Roundworms, Rotifers	Body slender and elongate. Possess internal organs. Many forms parasitic. Pseudoceolomate. Bilaterally symmetric.
Annelida (6,500)	Earthworms, Leeches	Body internally / externally segmented. Many internal organs repeated throughout. Coelomate / Bilaterally symmetric.
Mollusca (7,000)	Clams, Oysters, Snails	Mainly hard-shelled. All organ systems present. No segmentation. Coelomate / Bilaterally symmetric.
Arthropoda (1,000,000+)	Insects, Crabs, Shrimp	Exoskeleton. Segmentation. Ventral nerve cord. Coelomate/ Bilaterally symmetric.
Echinodermata (13,000)	Sea Stars, Sea Urchins, Sea Cucumbers	Coelomate. Deuterostome. Adults radially symmetric. Larvae bilateral symmetry. Water vascular system Tube feet for feeding and movement
Chordata (2,000+)	Sea Squirts, Lancelets ALL Vertebrates	Notochord present at some time during development, often replaced by vertebral column Paired gill slits. Dorsal nerve cord. Coelomate / Bilaterally symmetric.

One of the important differences among the many coelomate phyla lies in how the embryo develops as cleavage (mitotic division of the fertilized egg) takes place. In many coelomates (including the arthropods, mollusks, and annelid worms), embryonic cells have their developmental fate determined extremely early on in their growth. The reason for this is thought to be the *selective loss of DNA* within these embryonic cells, so that they can develop only along certain tissue lines. If you remove some of these early embryonic cells from arthropods, mollusks, or annelids, they will not be able to develop into genetically identical, separate, organisms. Arthropods and other coelomates that undergo this type of **determinate cleavage** as zygotes are known as **protostomes.** In contrast, the other coelomates (including the chordates and the echinoderms—sea stars, sea urchins) are **deuterostomes**, whose zygotes undergo **indeterminate cleavage** and whose embryonic cells retain the ability to split off and develop into new, complete embryos long into development. It is this type of cleavage in humans and in other mammals that gives us the ability to have identical twins, two embryos created from a single sperm and egg union.

Dolly the sheep was created from the DNA of an adult sheep, which was taken from the udder of a female ewe. This DNA was used to create an embryo genetically identical to the mother. This could not have been done with protostome animals because of their early selective gene loss within different cell lines.

Two phyla warrant added mention for the GRE: The first is the arthropods (phylum *Arthropoda*), which is the most successful group of animals on Earth. The arthropods, including insects, crustaceans, and spiders, comprise more than one million species worldwide, perhaps three-quarters of which are species of beetle! All arthropods have several characteristics in common:

- **Segmentation**: Specialized appendages on the members of this phylum are modified for a variety of functions: walking, eating, copulating, sensing, and defense. These appendages are usually segmented with bending joints.
- **Hard exoskeleton**: Also known as the cuticle, this is made of the amino sugar chitin and other proteins. It provides protection and is used for attaching appendages. The shell also provides an attachment point internally for muscles.
- **Molting**: Arthropods must occasionally shed their exoskeletons as they grow, and then form a new, larger exoskeleton.
- **Open circulatory systems**: A fluid called hemolymph is pushed through the organism by a series of hearts (see also chapter 13).

The second phylum to examine is phylum *Chordata*, which includes the vertebrates: humans, mammals, birds, reptiles, amphibians, and fishes. You should note that the phylum also contains over 1,000 species of "**invertebrate chordates**," organisms like the lancelet and sea squirt that possess rigid rods of notochord tissue to support their dorsal nerve cord, but never develop a hard bony vertebral column like the vertebrates do. Chordates all share the following characteristics:

- **Dorsal, hollow nerve cord**: The nerve cord develops from pinching up of dorsal ectoderm (neural tube) during embryo growth. Nerve cords in other phyla are usually made of solid tissue and located ventrally. The nerve cord in chordates will develop into the brain and spinal cord as development proceeds.

- **Notochord**: All *Chordata* embryos have a notochord—a longitudinal, bendable rod of tissue that runs along the back of the embryo. Located between the gut and the nerve cord, it helps to support adult invertebrates that do not develop a bony vertebral column. The notochord is derived from **mesodermal** cells. Adult vertebrates keep only some of their notochord as they develop a bony spinal column. The notochord remains as soft discs of cartilage that cushion the vertebrae.

- **Muscular tail**: Almost all chordates possess at some time during their development a tail, filled with muscles for locomotion. In humans, our tailbone (which can get hurt when we sit down too hard or fall down backward) is a **vestigial** structure of fused bones that supported a tail in our ancestors.

- **Gill slits**: At some point in the development of all vertebrates, small slits exist within the throat (pharynx) and open to the outside. These slits allow water to enter the mouth and exit without going through the whole digestive tract (like in adult fishes). In many vertebrates, these close up quickly during development.

It seems likely that the vertebrates and invertebrate chordates diverged from one another more than 500 million years ago. Today, some invertebrate chordates, such as the cephalochordate *Amphioxus*, have been extensively studied because they provide an excellent model for the ancestors of all vertebrates. They are considered more primitive than vertebrates because of their lack of a brain and paired sensory organs in the head (i.e., eyes), but their resemblance to fishes and other primitive vertebrates gives them the status of "**living fossils**," modern organisms that serve as models of ancestral forms.

PRACTICE QUESTIONS FOR PART III

1. Hypertension, or high blood pressure, is a leading risk factor in heart disease and stroke. Factors that can cause hypertension are likely to include

   (A) increased renin secretion resulting in increased sodium reabsorption in the collecting ducts of nephrons.
   (B) genetic defects that decrease the transcription of genes coding for antidiuretic hormone.
   (C) increased manufacture of hemoglobin molecules within red blood cells.
   (D) decreased sensitivity of vascular smooth muscle due to genetic defects in cell membrane transport of sodium and calcium.
   (E) rapid cessation of smoking and alcohol consumption.

2. The antigen-specific cytotoxic T cell of the vertebrate immune system represents a main line of defense against invading pathogens. All of the following are characteristics of the cytotoxic T cell EXCEPT

   (A) variable, antigen-specific receptors lie on the outer surface of its cell membrane.
   (B) it is able to recognize foreign peptides complexed to MHC glycoproteins on the surfaces of other cells.
   (C) proteins capable of lysing bacterial cell membranes are released by it in order to kill infected cells.
   (D) it forms identical T cell clones after immune system stimulation.
   (E) the ability to complex directly with viral particles and bacteria before releasing cytotoxic compounds results in a high degree of accuracy in targeting.

3. Hox genes are homeotic genes that encode differences in the development of various organs and tissues of an embryo. The Hox genes of chordates, arthropods, and nematodes have been sequenced and studied to reveal a high level of similarity across these animal phyla. Most chordates, however, have four "*trans-Abd-B*" genes, known as Hox 10-13, that arthropods and nematodes do not have. These genes are expressed in the

   (A) anterior nerve ganglia that develop into the brain.
   (B) muscular post-anal tail that extends posteriorly.
   (C) tissues that develop into the ventral nerve cord.
   (D) muscle cells of the heart and blood vessels.
   (E) tissues that are segmented by joints.

4. Some species of cubomedusan jellyfish possess as many as twenty-four well formed eyes, which each have distinct lenses, corneas, pigment layers, and light-sensitive layers. Although the jellyfish have a fairly complex visual anatomy, it is LEAST likely that they are able to

   (A) orient themselves spatially in relation to nearby light sources.
   (B) move toward less deep waters in search of food.
   (C) sense approaching predators.
   (D) form images of their surroundings.
   (E) integrate light perception with movement.

5. The resting membrane potential of neurons depends upon all of the following EXCEPT
   (A) differential distribution of ions across the axon membrane.
   (B) the selective permeability of the axon membrane.
   (C) saltatory conduction via nodes of Ranvier.
   (D) negatively charged phosphate groups on nucleic acids.
   (E) Na^+/K^+ ATPase transmembrane pumps.

6. Which of the following features is found in annelids but not in echinoderms?
   (A) a true coelom
   (B) water vascular system
   (C) specialized cells and tissues
   (D) gills
   (E) segmentation

7. Exposing certain types of cells to the hormone prolactin results in
   (A) the secretion of carbohydrate-digesting disaccharidases from the pancreas.
   (B) the stimulation of milk production.
   (C) a decrease in follicle maturation in females.
   (D) an increase in blood sugar concentration.
   (E) the stimulation of bone and muscle differentiation and growth.

8. A baby monkey removed from its mother and then returned after several months fails to breast feed or associate in any physical way with the mother. The most likely reason for this behavior is that the monkey
   (A) has passed through a critical period of maternal recognition.
   (B) has been conditioned not to respond to the mother.
   (C) has habituated to the presence of the mother.
   (D) is exhibiting innate stimulus discrimination.
   (E) experiences the mother as an unconditioned stimulus.

9. The plant hormone responsible for inhibiting growth and for closing the stomata during periods of water stress is
   (A) auxin.
   (B) gibberellin.
   (C) abscisic acid.
   (D) choline.
   (E) cytokinin.

10. In the nematode worm *C. elegans*, when a sperm cell enters a uniform egg cell at one site, cortical actin filaments within the egg contract on the side opposite the location of sperm entry. Egg cytoplasm is rapidly segregated in relation to the site of the cortical reaction, a process that
    (A) causes induction of mesodermal cells.
    (B) results in the buildup of a hard shell around the egg.
    (C) establishes an axis of mitotic spindle formation so that cell division resulting in equal daughter cells can begin.
    (D) blocks DNA replication until after the second cleavage of the egg cell.
    (E) contributes to the polarity of the egg cell so that the first cleavage is asymmetric.

Questions 11–14

(A) endosperm
(B) endoderm
(C) endotherm
(D) endothelium
(E) endorphin

11. A hormone produced in the pituitary gland that blocks the perception of pain.

12. The innermost layer of cells lining the blood vessels in mammals.

13. An organism that is able to maintain a constant body temperature through the use of various metabolic processes.

14. The innermost of the three primary germ layers in animal embryos.

ANSWERS

1. (A)

Increased renin secretion and subsequent reabsorption of sodium can lead to hypertension because the reabsorption of sodium causes more water to be retained in the blood. More water in the blood means higher fluid volume within the closed circulatory system and, hence, higher blood pressure (hypertension). Choices (B), (D), and (E) would result in less fluid retention. (C) has nothing to do with the pressure within the vessel walls.

2. (E)

T cells, unlike B-cell, cannot complex directly with antigens. They are able to bind to the body's own infected cells by recognizing a combination of foreign antigens displayed on the cell surface AND a self-identity protein called MHC (major histocompatibility complex). All other answer choices are characteristics of T cells.

3. (B)

The one trait of all chordates not found in other phyla, including in Arthropoda and Annelida, is the presence of a muscular, post-anal tail (which, in humans, is merely the vestigial coccyx—a series of fused vertebrae known as the "tailbone.") Arthropods, annelids, and chordates all have bunches of nerve cells (ganglia) at their head ends (anterior ends) that develop into the brain, so (A) is wrong, and chordates do not have ventral nerve cords at all (choice C). All phyla described in the question stem have muscular hearts, even the arthropods that have an open circulatory system; thus, (D) can be eliminated. One characteristic of all arthropods is the possession of jointed appendages, such as limbs. The "tans-Abd-B" Hox genes must, therefore, code for the development of a tail.

4. (D)

Jellyfish and other cnidarians have no complex nervous system. They have no brain for processing any of the information that may come in from structures such as eyes. So despite the ability of signals received by the eyes to cause changes in movement, especially if the eyes are found near tentacles (which they are), it is extremely unlikely that jellyfish can form images of their surroundings. Choice (D) is correct here.

5. (C)

Saltatory conduction describes the ability of the nerve signal to "jump" down the length of the axon from unmyelinated section to unmyelinated section. Nodes of Ranvier are spaces between myelin coverings around the axon, and the nodes allow the movement of ions such as Na^+ or K^+ across the nerve cell membrane, whereas the myelinated areas do not. Yet Nodes of Ranvier are involved in nerve cell action potentials, not in resting potentials. Choice (C) is correct here.

6. (E)

The GRE loves to ask questions like this, so learn the basic differences between animal taxa (groups). The annelids are also known as the "segmented worms." Both echinoderms and annelids are coelomates, yet only echinoderms possess a water vascular system. Both have specialized cells and tissues and neither has gills. Thus, choice (E) is right.

7. (B)

Even if you had never seen the name of this hormone, you should be able to guess based on the letters "lact" within the name. Think "lactose sugar" and you'll be all set. Lactose is the sugar present in milk and prolactin is a hormone that acts on the mammary glands to stimulate milk production.

8. (A)

Many organisms pass through a critical period during which they'll latch onto their mothers so they can recognize the mothers over any other individuals around them. This is important evolutionarily for survival, because mothers take care of food and protection. Some animals, however, will latch onto other moving stimuli, even another type of animal, during this critical period if they are not exposed to their proper mothers. And animals not allowed to bond to a maternal figure during the critical period often show extreme developmental difficulties, as does the baby monkey presented in this question stem.

9. (C)

This is a factual recall question that tests your ability to recognize the names of the plant hormone responsible for growth inhibition. Abscisic acid is the chemical that causes plants to stop growing and become dormant, particularly in times of stress (such as during a drought). Auxins and gibberellins are growth-promoting hormones, and cytokinins retard aging and control apical growth. Choline is not a plant hormone at all, but rather a building block of the neurotransmitter acetylcholine.

10. (E)

Most eggs have a distinct polarity as they develop; that is, they do not divide equally, but rather push more cytoplasm to certain parts of the embryo than to others. This polarity is what guides the proper placement of limbs and organs as the embryo matures. Polarity starts immediately after fertilization, however, as the cell organizes itself to prepare for the first, unequal division (cleavage). Choice (E) is correct here. Mesodermal cells, (A), do not yet exist at such an early stage, and blocking DNA replication (choice D) would halt cleavage altogether.

11. (E)

Endorphins are hormones released by the pituitary gland that block the perception of pain. They act as inhibitory neurotransmitters within certain nerve pathways that transmit pain signals and allow for the modulation of pain so that injuries become rapidly less painful after they occur.

12. (D)

Not to be confused with choice (B), the endoderm, the endothelium is the inner layer of cells that line mammalian blood vessels.

13. (C)

Here, look for the ending *-therm*, meaning heat. Endotherms, otherwise known as homeotherms, are able to regulate their body temperature as the temperature around them changes. For example, humans are endotherms because we keep our body temperature around 98.6 degrees Fahrenheit. Reptiles, however, are not, and must remain in the sun or in warm areas in order to keep moving.

14. (B)

The endoderm is one of the primary embryological tissues, along with mesoderm and ectoderm, that are present at the gastrula stage of development. Endodermal cells will develop into the pancreas, the liver, the lungs, and the lining of the digestive tract.

Part IV

ECOLOGY AND EVOLUTION

Introduction to Ecology and Evolution

Theodosius Dobzhansky, the famous American geneticist, is famous for his statement that nothing in biology makes sense except in the light of evolution. **Evolution, simply put, is a change in the frequency of different alleles in a population over time.** Individuals do not evolve; populations evolve. It is populations that change their genetic makeup over time, as different individuals leave more offspring than others. This differential reproductive success is caused by a variety of environmental conditions and random mutations that allow some organisms to survive and reproduce better than others in the same population. In the following chapters, we will look at the wide variety of ecosystems in which organisms live on Earth, factors in those ecosystems which affect the survival of these organisms and the growth of their populations, and the mechanisms which lead to population change over time.

In 1908, two scientists had independently come up with a mathematical way in which to study evolution in populations. The Hardy-Weinberg Theorem, named for these two scientists, describes an "ideal" gene pool, a population that is not evolving, and it is through studying a population in a nonevolving equilibrium that we are better able to understand the factors that cause populations to actually change, or evolve, over time. Hardy and Weinberg came up with five requirements that a population needs to meet in order to remain in a nonevolving state. In reality, there are no populations on Earth that meet these five conditions. However, one might argue that because humans are able to control environments to such a high degree, we do meet these conditions and are in a nonevolving state (a philosophical argument not relevant for the GRE exam!). The conditions that must be met to be considered a nonevolving population are:

- **No emigration or immigration.** If organisms enter or leave the population, they either take genes away or bring genes in, and this will alter the relative frequency of alleles for certain traits in the population.

- **Large population size.** In small populations, even a few mutations can have a large effect on the frequency of different alleles. With a large population, these mutations are effectively diluted.
- **No net mutations.** Alleles that change into others affect overall allele frequencies. Because large populations have many copies of the same allele distributed throughout, the changing of a few alleles does not affect the overall (net) frequency.
- **No natural selection.** As with random mating, certain alleles that confer a survival advantage over other alleles will allow allele frequency to change in subsequent generations if natural selection is occurring.
- **Random mating.** Mates must be selected at random, not based on any particular characteristics. If this is not the case, certain alleles become favored over others and allele frequencies will change in subsequent generations.

The chapters in this section will examine the structure and growth of populations in nature and will examine how the principles of Hardy and Weinberg apply to the evolution of populations.

CHAPTER TWENTY-FOUR

Community Ecology

BIOMES AND HABITAT SELECTION

It is nearly impossible to discern exactly where one kind of habitat ends and another begins, since ecosystems change gradually as one moves higher or lower in altitude or latitude. At the same time, however, most species are found in specific types of environments, characterized by their differences in temperature, humidity, seasonal patterns, and location. Most of these distinct regions, or *biomes*, can be named for the kinds of vegetation that grow there, but all have characteristic organisms that have evolved adaptations allowing them to survive in those particular biomes.

Taiga (Coniferous Forest)

The taiga consists of dense forests made up of **coniferous**, or cone-bearing, trees such as pine, spruce, fir and hemlock. There is often little to no undergrowth in these cool forests, since the canopy of conifers is thick enough to block most sunlight from reaching the forest floor. If you have ever hiked in the woods of northern New England (Maine, Vermont, and New Hampshire) or the Canadian Rockies, you have likely been in a northern coniferous forest, or taiga. A blanket of fallen pine needles covers a thin, poorly fertilized (nutrient-poor) soil, which is made acidic by the slowly decomposing pine needles. At the same time, however, these forests receive a great deal of sunlight, which allows the conifers to grow tall quickly. What kinds of animals would you expect to live in the taiga? What kinds of adaptations would they need for a climate with long, cold winters, lots of snow, and a short summer with a lot of rain? Check out **table 1-1** later on in this section for the answers.

Desert

Deserts are the driest of all biomes, with very little rain, limiting the kinds of vegetation that can grow there. Plants that do grow there need to be able to store water in their leaves and stems effectively and for long periods of time. The varnishlike coating on creosote and other plants' leaves slows evaporation and helps to conserve water. Thick leaves and stems also help to hold in water. Shrubs such as acacia are common, as are large plants like the Joshua tree and the saguaro cactus, and when the brief rains hit during wet seasons, vast blooms of color from brilliant flowers and fruits can be seen across the expanse of sand and brush. Plants need to produce seeds and fruits quickly so that they can begin to grow during the rainy season, yet many plants, such as the acacia, have seeds that can delay their germination for years!

The desert is home to a large variety of insects such as beetles and cicadas, and other arthropods like tarantulas, whip scorpions, centipedes, and millipedes. There is even a variety of shrimp—the tadpole shrimp (genus *Triops*)—which emerges in pools of water during the rainy season from eggs that were buried and dormant over the long, dry periods. What other organisms live in the desert and what adaptations allow them to be there? See **table 1-1** for some answers.

Tundra

Sometimes called a frozen desert, this biome is easily distinguished by its long, harsh winters that leave vast areas of Earth's northernmost latitudes dark for up to 10 months of the year. In terms of biodiversity, the tundra has fewer species of plants and animals than other biomes. The plants that can live in such a cold place have a very short growing season, typically grow very close to the ground, and do not need much water to survive. Lichens, mosses, and short shrubs predominate. Animals must not only be able to camouflage themselves in the snow for winter, but must also be able to hide from predators in the brief summer months. They also need to be able to live over the long winter months in very cold weather and without lots of food. Check out **table 1-1** after you have thought about which animals might be found in the tundra and what adaptations would help them to survive.

The tundra is best known for its characteristic **permafrost**, a layer of subsoil that never thaws, even in the warmer summer months. Permafrost prevents plants from anchoring themselves with deep roots and from absorbing lots of water, which keeps them short and compact. Plant growth is very similar to that found in the desert, as brief bursts of growth and blooming of flowers and fruits can be seen during the few warm, summer months that have their counterpart in the desert's brief rainy season.

Chaparral and Savanna

Found in coastal areas which have long, hot summers and short, rainy winters, this biome is characterized by its abundance of **fire-resistant plants**. These are short, dense shrubs such as the chamise, manzanitas, and liveoaks, whose seeds sprout quickly and vigorously after fires that frequently sweep across the chaparral. Often the seeds are released only upon exposure to extreme heat. Moreover, many species of chaparral plants store extra starch in their thick roots to regrow very quickly after a fire destroys the plant's stems and leaves. The chaparral spreads over low coastal mountains in southern California, parts of the Mediterranean and parts of Chile in South America and Angola in southwestern Africa. Organisms that live in the chaparral include those that forage for seeds and larger scavengers, as well as certain birds of prey, that can easily hunt for these smaller creatures when they are exposed among the short shrubs. See **table 1-1** for details after you consider what organisms might fall into these categories.

The savanna biome is similar in many ways to the chaparral with its often sparse covering of short trees and lots of shrubs. But the savanna is mainly a **grassland** biome, covering vast areas of the tropics and subtropics in South America, Africa, and Australia with dense, tall grass and scattered trees. Parts of the Australian savanna are far more humid than the African or South American savanna, with patches of rain forest and even mangrove swamps, making it more similar to wetlands than grasslands. If you have ever looked at pictures of the African plains with hordes of large mammals grazing or huddled around watering holes, you are looking at characteristic savanna. Seasons are similar to chaparral seasons—hot and dry; warm and wet—but there's often an added period of cool, dry weather as well. Most plants are fire-adapted just like chaparral flora, so fires cause the release of new seeds and clear out old growth to renew the landscape. In addition to lots of insects and burrowing herbivores, which often scurry around at night searching for seeds and plant parts to eat, most of Earth's largest mammals and their predators can be found in this biome. It should be easy to guess what kinds of animals live on the savanna—take a stab at it before you look at **table 1-1**.

WHAT'S SO SPECIAL ABOUT NITROGEN?

Nitrogen limits the growth of crops more than any other major nutrient. The atmosphere is 80% nitrogen, but this must be converted to nitrate (NO_3^-) and ammonia (NH_4^+) before it can be used by plants for protein, hormone, and nucleic acids synthesis. Bacteria in the soil limit the amount of nitrogen available since they are responsible for the conversion!

Temperate Grasslands

As with other grasslands like the chaparral and savanna, temperate grasslands have both hot, dry seasons, and cool, somewhat rainy seasons. Grasslands also have periodic fires that wipe out vast areas and allow fast-growing, fire-resistant plants to recolonize within days. Large grazing mammals are common and this prevents vegetation larger than shrubs from taking hold. Temperate grasslands are the biome sung about in "America, The Beautiful" with its amber waves of grain. They are the prairies across the midwestern United States and much of northern Europe, and they are known for their fertile, *nitrogen-rich soil* that is readily fertilized and farmed or used for pasture. In addition to the large herbivores that characterize the prairie, countless small animals and invertebrates live on or in the ground of temperate grasslands, consuming vast amounts of dead animal and plant material and recycling organic compounds into the *nutrient-rich soil*. What kinds of animals would you expect to find here?

Tropical Forest

These biomes exist in much of the world where temperature and day length are fairly stable throughout the year. Temperatures are warm and mild (80°F) and the sun shines for almost 12 hours each day. The type of tropical forest that comes to mind when this biome is mentioned is the tropical rain forest, the wettest and most lush land biome on Earth. Over 75 percent of the earth's biodiversity can be found in tropical rain forests, including many plant, animal, and invertebrate species that have never been discovered or catalogued. The interesting thing to note about the tropical rain forests is that the soil is very low in nutrients, but so much can grow there because there is such a high rate of turnover of nutrients.

Because there are so many organisms living and dying in these forests at any one time, a great deal of organic material always exists that can be decomposed and recycled on the forest floor. *The soil does not hold onto these nutrients for any appreciable amount of time.* The larger problem with the soil's poor quality is that when tropical forests are cut or burned down, the forests can neither grow back at any appreciable rate nor regain anything close to their original biodiversity. Surprisingly, the undergrowth of a tropical rain forest is usually sparse, not like a thick jungle that is hard to walk through. This is because the trees that grow there are tall enough and dense enough at the forest's uppermost layer or canopy that little light actually reaches the forest floor.

WHAT IS EVAPOTRANSPIRATION?

This refers to the evaporation of water from the leaf surfaces of plants. Different environmental conditions affect transpiration rates, such as surrounding humidity, temperature, and wind velocity. Except in deserts where vegetative evapotranspiration is extremely low, most forests release large amounts of water back into the atmosphere. Rivers and streams that run through forests accumulate much less water than they might because so much water is sucked up by trees and lost via transpiration.

Often referred to as dry rain forests, **tropical dry forests** grow in warm tropical regions that do not experience the rainfall characteristic of "wet" rain forests. These are typically found in lowland regions, and they resemble a dense chaparral with their shrubs and plants that are adapted to prolonged periods without rain. In areas with moderate rainfall but that still experience a dry season, tropical **deciduous** forests are populated with taller deciduous trees and shrubs, which will often lose their leaves until the next rainy season. Whether dry, deciduous, or wet, all tropical forests serve as essential watershed areas. See **table 1-1** for a small sample of tropical forest organisms.

WHAT IS A WATERSHED AREA?

Watersheds are basinlike areas of land that hold onto a great deal of water in underground streams and nutrient-rich soil after the water has been "shed" by precipitation. Much of the water cycles back into forest vegetation and is transpired, but watersheds are abundant sources of clean water, sustaining terrestrial and aquatic life in many biomes.

Temperate Deciduous Forest

In contrast to the warm tropical deciduous forests, temperate forests experience all four seasons, winter through fall. Trees growing in temperate forests are characterized by growing cycles that cause them to lose their leaves in the colder, drier winters and to regrow them in the warmer, moister spring and summer months. If you live on the eastern coast of the United States or have been to the middle latitudes of Europe, temperate deciduous forest is everywhere. Made up of tall, leafy tress like beech, maple, oak, and birch, these forests have a great deal of species diversity and a fair amount of undergrowth compared to the tropical forests. As with tropical forests, human interference has greatly damaged temperate forests, some of which have been extensively logged and damaged. See **table 1-1** after you think about what kinds of animals you might expect to find in a temperate forest and what some of their adaptations would be.

Table 1-1

Biome	Expected Fauna	Adaptations
Taiga	Bears, wolves, small cats (lynx), squirrels, many seed-eating birds	Thick coats of fur; adaptations for seed foraging i.e. thin, pointy beaks;ability to hibernate
Desert	Seed-eaters like ants, rodents, and birds; lizards, snakes, spiders; scorpions; toads	Appendages designed for burying in sand; homeostatic systems for water retention (i.e. solid urine); thick scales or carapaces; very rapid development from eggs; large ears for cooling off
Tundra	Oxen, caribou, deer, arctic fox, wolves, owls, insects (butterflies, several different kinds of flies, wingless insects), migratory birds in brief summer months	Large, compact bodies; thick fur; plumage turns white in winter and brown in summer, thick fat deposits for warmth; lack of wings in insects to cope with high wind
Chaparral	Browsers/scavengers: seed-eating birds (roadrunners; woodpeckers) and rodents; coyotes; bobcats, mountain lions; deer; reptiles; many types of insects and arthropods	Able to live in prolonged dry periods; varied diet characteristic of scavengers; animals and plants need to tolerate some frost and even snow in higher chaparral elevations
Savanna	Large plant-eaters: buffalo; bison; gazelle; big cats; large, flightless birds (e.g., emu); varied insects; rodents (nocturnal); large snakes and lizards; giraffe; zebra; etc.	Burrowing abilities for smaller animals; migratory/social group interaction; able to withstand long periods of dryness; agility and speed for large animals
Temperate Grasslands	Large, grazing herbivores—zebra; horses; buffalo; insects (ants); arachnids; birds like starlings, sparrows, doves; opossums; raccoons; coyotes	Digestive system adaptations for grass/plant consumption—second stomach; longer intestines; fur for warmth; herd behavior/migration of large mammals
Tropical Forests	Perhaps a million different species of beetles, insects, and other arthropods; snakes; great apes; monkeys; large, herbivorous mammals; rodents; amphibians; fish; worms; many birds	Climbing ability—claws or gripping appendages; jumping ability (tree to tree); flying and large wingspan; ability to trap water from rain (esp. plants)
Temperate Forests	Many rodents, birds, mammals; great diversity	Similar to tropical forest above; more fur for cooler temps; hibernation

ECOLOGICAL SUCCESSION

Occasionally, communities of organisms are completely restructured by environmental disturbances that result either in the creation of a new landscape where life has not previously existed or in the sweeping away of an existing community's flora and fauna (plant and animal life). The first case is known as **primary succession**, and the second case is called **secondary succession**.

Primary Succession

Approximately one million years ago, the Galapagos Islands, 600 miles off the northwest coast of South America, were created from violent undersea volcanoes. The eruptions left cooling lava on the surface of the Pacific Ocean which condensed to form volcanic islands of various sizes and shapes. As the islands cooled, different types of seeds blown onto the islands from the continent nearby or dropped by wayward birds had the opportunity to grow. Lizards and tortoises floated onto the islands on large rafts of seaweed and vegetation. With these birds and animals came a host of insects and other invertebrates, and over hundreds and thousands of years, the Galapagos Islands grew into lush and complex communities of organisms.

Primary succession is the gradual change of new and lifeless areas of land into thriving ecosystems as various plant life and other "colonizing" organisms arrive. The first plants to take hold include small, hardy plants such as lichens, mosses, and grasses. The first creatures to thrive may be insects, small birds, and some reptiles. Larger organisms will follow if these smaller ones are successful. It may take many attempts for life to gain a strong foothold. Primary succession occurs not only on volcanic islands, but also in places where a previously uninhabited area becomes habitable. When a glacier retreats north as the climate warms up, it leaves large areas of land to be colonized by various plants and other organisms over the course of several hundred years.

WHY WOULD ISLANDS LIKE THE GALAPAGOS HAVE UNIQUE CREATURES THAT EXIST NOWHERE ELSE?

Although their ancestors arrived from nearby landmasses, the Galapagos flora and fauna are unique because they have been isolated for thousands of years, exposed to environmental conditions that differ from island to island and from the South American mainland. Over time, genetic mutations and natural selection have caused these isolated populations to be sufficiently different in behavior and appearance, so that they would no longer mate with similar organisms on the other islands or on the mainland. Primary succession coupled with isolation can lead to the evolution of entirely new species!

Secondary Succession

Secondary succession is the rebuilding of a community of organisms that has been destroyed by a natural disaster or by human interference. This typically occurs after a forest fire, volcanic eruption, or flood, and it results in the formation of a new community. Smaller grasses, mosses, and shrubs provide a foundation for the growth of larger conifers and deciduous trees. The end (climax) community will generally differ from the community that existed before the disaster because of various factors that affect the process of succession. One of these factors is

interspecific competition (competition for sparse resources *between* populations of different species): Communities ravaged by natural disasters or human destruction may "succeed" into communities made up almost entirely of the species that were good colonizers just after the disaster. If the destruction is milder or less extensive, certain species may be able to successfully outcompete the original colonizers for space, food, and water.

ARE CLIMAX COMMUNITIES REALLY STABLE?

New evidence suggests that the species that make up the final stage of a community after succession can disappear for hundred of years and then reappear again. "Stable" forests are continually being reshaped by periodic fires and other natural changes; as such, it may not be proper to consider any stage in a community's development to be final and relatively unchanging.

How well the "**pioneer organisms**" (those that first take hold in the barren habitat) fare largely depends upon how well they can adjust to the quality of the soil, amount of sunlight and precipitation, temperature fluctuations, and availability of resources like water and nutrients. In addition, other organisms that are colonizing in the first stages of the new community may act to inhibit other pioneers through competition for resources and space; at the same time, though, many pioneer organisms facilitate the growth of others. The main area in which **facilitation** aids in the growth of new species is when a particular plant species improves the quality of the soil such that another species of plant can then develop and grow successfully. For instance, plants that are able to extensively fix atmospheric nitrogen using mycorrhizal fungi in their roots can add nitrates to the soil, thereby facilitating the growth of new species that could not have grown there without a nitrogen-rich soil.

Pine trees grow tall and quickly in a successional ecosystem, yet their saplings need a lot of sunlight. As they grow, they eclipse their own saplings so new pines do not grow as well. However, falling pine needles and other organic material from the conifers add enough nutrients to the soil to facilitate the growth of oak and hickory and other hardwood trees, which grow in the place of new pines. Thus, pine trees facilitate new species which inhibit their own growth! In another example of **inhibition**, monoterpenes (chemicals found in pine resins) leach into soils from fallen pine needles and pine cones. Monoterpenes inhibit nitrogen-fixing bacteria in the soil and block "**nitrification**"—specifically the conversion of ammonia to nitrate. Overall, the processes of inhibition and facilitation are crucial in determining what type of climax community will result after succession.

INTERSPECIFIC RELATIONSHIPS—COMMUNITY INTERACTIONS

All of the interactions that exist between organisms of different species in an ecosystem fall into the category of *symbiosis*, a word meaning "to live together." These relationships include **competition**, **parasitism**, **commensalism**, and **mutualism**.

Competition

The previous section introduced one of the most fundamental interactions that takes place between two or more species in the same habitat: competition. Competition can arise in several ways and manifest itself when species physically fight to gain access to resources (**interference competition**), or when they simply compete by consuming the same resources (**exploitative competition**). It has been demonstrated in the lab that two species competing for the same limited resources will result in one of the species being driven to extinction. This principle is known as the **competitive exclusion principle**: It was developed in the early 1930s by scientists working with cultures of *Paramecia* after they noted that two species of Paramecia in the same culture cannot coexist if they depend on the same food source. One species will outcompete the other, which will die off quickly as its access to food is diminished. It is understandable that this principle would exist in nature to some extent, yet natural ecosystems often provide enough options and flexibility that organisms excluded from one food source can still survive and flourish.

A discussion of competition would not be complete without an explanation of the term **niche**. The niche of an organism is often thought of as the physical area in which the organism lives and what resources it needs to survive. But a niche is more than that: A niche is an organism's role in the ecosystem in which it lives, its job so to speak. In other words, the niche explains what it eats, what eats it (where it fits into the food web), whether it is **diurnal** or **nocturnal**, which other species in its community it interacts with and in what way, and many other things. Just because two species require the same nutrients or food source does not mean that they share the same niche. Yet if the niches of two species are too similar, then those species almost certainly could not coexist in the same community.

When speaking about niches, many ecologists refer to a species' **fundamental niche** as compared to its **realized niche**, two terms you need to be familiar with for the GRE. The fundamental niche is comprised of all the potential resources that a species can use in its environment, while the realized niche is comprised of the resources the species actually uses. Why are members of a species necessarily limited to a certain subset of all the possible resources available? Competition with other species, disease, predation, and the location of the other resources limit what is accessible.

Parasitism

In this form of predation, one organism lives on or within another host organism in order to derive energy and nutrients for itself. In this type of interaction, the parasite is strengthened and the host harmed; therefore, this type of interspecific relationship is considered to be a "+/–" interaction, where one organism is helped and the other hurt. Parasites can be **ectoparasites**, which live "ecto-" to their hosts (on the host's surface), such as mosquitoes, leeches, lice, and ticks. Parasites can also be **endoparasites**, such as pinworms, trypanosome protozoans which cause African Sleeping Sickness, and hundreds of species of bacteria and viruses, which live inside their hosts and gain extra protection while feeding off of host nutrients. Some forms of parasitism, called "**brood parasitism**," do not involve actual feeding, but rather fooling a host organism into caring for the parasite's offspring in place of its own: Certain bird species lay eggs in the nests of other bird species and lead the host parents to neglect their own eggs while raising the young of the parasitic species.

The most effective parasites do not kill their hosts, To do so would mean the death of the parasite itself. That is why most endo- and ectoparasites simply weaken their hosts for their own gain

Commensalism

Commensals are organisms that interact in a way resulting in one organism benefiting from the relationship, while the other neither benefits nor is harmed. This "**+/0**" interaction may actually involve some as yet undiscovered benefit to the "neutral" organism, but commensalism often takes place between species when one organism hitchhikes or scavenges using another organism for locomotion and protection. Sharks are often followed by groups of remoras, small fish that can attach to the underside of the shark using a suction-cup like apparatus on the dorsal surface of their heads. This allows the remoras to ride along with the sharks with minimal energy use, if any, and **scavenge** bits of food left over from the often sloppy and explosive eating habits of the host sharks. It is unlikely that the sharks are affected in any way, positively or negatively, by this interaction, but the remoras benefit a great deal from the free ride and the free meal. Other commensals include the mouse that lives behind your kitchen sink foraging for scraps of food left over from a recent meal, and the clown fish that picks up food floating by as the fish hides in the tentacles of a large, stinging sea anemone for protection.

Mutualism

Although Darwin himself referred to interactions in nature as being "red in tooth and claw," the vast majority of interactions in ecosystems are cooperative ones, where species interact for mutual benefit in a "+/+" manner. Often evolutionary changes that occur in one species involved in a mutualistic relationship will drive evolutionary change in the other species, because the survival of the two species is so closely linked. The bobtail squid is able to glow and emit light from its body to illuminate its surroundings because of the luminescent bacterium, *Vibrio fischeri*, that lives in its gut. The bacteria that live on the surface of our skin and mucous membranes outcompete foreign, potentially disease-causing bacteria. And the bacteria that live in the guts of cattle allow the cattle to break down the cellulose in the grass they eat. There are countless other examples of mutualism in the natural world, and the structure of ecosystems absolutely depends upon the stability and continuity of these interactions.

What Is Coevolution?

Two organisms are said to coevolve when adaptations of one species act in a selective manner to guide changes in the other species. Coevolution means that evolutionary changes in one of the species in a mutualistic relationship act to select certain traits in the other species of the relationship. Some insects, for instance, have evolved some adaptations to combat certain plant toxins, allowing them to feed on plants that cannot otherwise be touched by other insects. As those plant toxins change in structure, certain mutations in the insects that feed on those plants may be favored and selected so that the insects evolve resistance to the new plant toxin over a few generations.

COMMUNITY STRUCTURE AND DIVERSITY

As the number of species in an ecosystem increases and the population size of each of species grows, the greater the number of symbiotic interactions are possible and the more complex the ecosystem will be. It is important to note that the number of organism types in an ecosystem is very different from the number of organisms of each species in an ecosystem, yet both factors are important in considering the complexity and diversity of a given community within an ecosystem.

The number of different kinds of organisms that a community contains is referred to as **species richness**, while the **relative abundance** refers to how many members of a particular species appear in a given area. Both of these concepts together add up to the **biodiversity** that is present in a community. A desert community which has 10 different species of cactus, with two of those species occuring in much greater abundance than any of the other eight, would appear to have less biodiversity than a comparable environment where all 10 species of cactus grew in the same relative abundance. In fact, the first desert community described might have a harder time adjusting to climate or other environmental changes, especially if these changes affected the two dominant cactus species in a significant way. If the major species in an environment are adversely affected by certain changes that take place, the entire community may collapse.

The ability of a community to survive through various environmental changes is called resilience; ecologists distinguish a community's **resilience** from its **stability**, the ability for a community to regain its original biodiversity and complexity after some environmental change or destruction. A resilient community may grow back quickly after a widespread drought ends, but the species that take hold may not be representative in richness or relative abundance to the species that were there before the drought. A stable community can regrow into its original state after environmental change or destruction, such as a pine forest that regrows its spruce, hemlock, and fir trees in the same relative abundance and with similar species diversity after an extensive fire.

Food webs are not always as simplistic as basic biology textbook illustrations suggest. As ecologists study food webs in a community, the complex roles played by various predators and producers can become clearer. As is often the case, one or more species may be **keystone species**, which if removed would cause widespread havoc and destruction in a community.

A keystone species controls the growth or spread of certain other species and maintains a balance in the community. When keystone species disappear, habitats can change dramatically and many resident species may die off. As species rapidly die off, other species can move in and change the biodiversity of the community. African elephants seem to be a keystone species in subSaharan grasslands: With the disappearance of the elephants, the grasslands actually begin to grow greater numbers of tall, woody plants, and to resemble forests or shrublands. Organisms dependent on the environment and on the species diversity of the grasslands can die out quickly as the landscape changes.

FOOD WEBS AND PRODUCTIVITY

All ecosystems are comprised of three basic "classes" of organism: **producers**, which can harness light or chemical energy and turn it into sugars and other biomolecules; **consumers**, which take energy built by producers and use it to grow and reproduce; and, **decomposers**, a type of consumer, which are responsible for breaking down the bodies and cells of dead producers and consumers. Although an ecosystem may be able to function with only producers and decomposers, it is the vast array of consumers that have evolved in every ecosystem which has helped to fill Earth with its extensive biodiversity.

Producers, Consumers, and Decomposers

Producers use light (**photosynthetic autotrophs**) or chemicals (**chemosynthetic autotrophs**) along with simple molecules (carbon dioxide, water and minerals) to drive the **biosynthesis** of their own proteins, carbohydrates, and lipids. The energy a producer (such as a plant) gets from the sun is stored in chemical bonds in the biological molecules it produces. Producers use some of the energy they synthesize to drive their own respiration and biosynthesis; they store the rest of the energy they produce in carbohydrates or oils. Producers form the foundation of any community, passing on their stored energy to other organisms. In a terrestrial environment, green plants, photosynthetic bacteria, or mosses are producers, using the energy of sunlight to produce biosynthetic energy through photosynthesis. In marine environments, green plants or algae are the main producers. Marine ecosystems at deep ocean geothermal vents are home to chemosynthetic bacteria that use the energy of inorganic molecules released from the volcanic vents to drive biosynthesis.

Consumers get the energy to drive their own biosynthesis and to maintain life by ingesting and oxidizing the complex molecules synthesized by other organisms. Since they get their energy by consuming other organisms, they are called **heterotrophs. Herbivores** (plant eaters), **carnivores** (meat eaters), and **omnivores** (eating both plants and animals) are all consumers. The adaptations of each consumer depend on the type of food it eats. Herbivores tend to have flattened teeth for grinding and long digestive tracts that allow for the growth of symbiotic bacteria to digest cellulose found in plants. Carnivores are more likely to have pointed, fang-like teeth for catching and tearing prey and shorter digestive tracts than herbivores. **Primary consumers** like the cow, the grasshopper, and the elephant are animals that consume producers such as green plants. **Secondary consumers** are carnivorous and

consume primary consumers (such as frogs, tigers, and dragonflies). Finally, **tertiary consumers** feed on secondary consumers, and would include such organisms as snakes that eat frogs.

Decomposers, also known as **saprophytes** or **detritivores**, are heterotrophs, since they derive their energy from oxidizing complex biological molecules, but do not consume living organisms. Decay organisms get energy from the biological organic molecules they encounter left as waste by producers and consumers, or the debris of dead organisms. They perform respiration to derive energy, and return carbon dioxide, nitrogen, phosphorous and other inorganic compounds to the environment to renew the cycles of these materials between the biotic and physical environments. Bacteria and fungi are the primary examples of decay organisms. **Scavengers** such as hyenas or vultures play a similar role, living on the stored chemical energy found in dead organisms.

Energy Flow in Communities

The term food chain is often used to describe a community, depicting a simple linear relationship between a series of species, with one eating the other. One food chain might contain grass as the producer, mice as the primary consumer, snakes as the secondary consumer, and hawks as the tertiary consumer. The different levels in the food chain, such as producers and primary consumers, are sometimes called **trophic levels**. A more realistic depiction of the relationships within the community is a food web, in which every population interacts not with one other population, but with several other populations. An animal in an ecosystem is often preyed upon by several different predators, and predators commonly have a diet of several different prey, not just one. The greater the number of potential interactions in a community food web, the more stable the system will be, and the more able it will be to withstand and rebound from external pressures such as disease or weather.

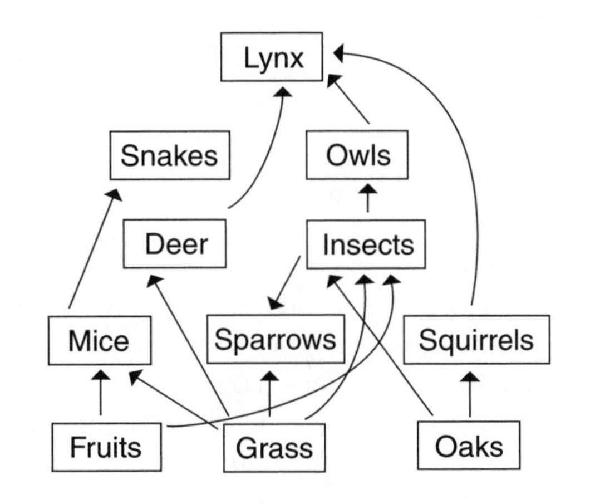

Why Don't Food Webs Have 10–15 Trophic Levels or More?

Because only a fraction of energy is available as one moves up the food chain, organisms at higher trophic levels must consume more than those at lower levels to get the energy they need to survive. With so many trophic levels, there would not be enough energy to sustain viable populations of animals at those higher levels.

Each trophic level in a food web contains different quantities of stored chemical energy in the populations it contains. When consumers eat producers, and secondary consumers eat primary consumers, *some energy is lost in each transfer from one level to another.*

As producers get energy from the sun, not all of the energy is converted into stored energy in chemical bonds. Some of the energy is lost at that level to the metabolic needs of organisms. Plants consume some of the energy they produce in respiration to support their own metabolic activities. The total chemical energy generated by producers in a given area is that ecosystem's **gross primary productivity**, and the total productivity with losses from respiration and other energy use by plants subtracted is the **net primary productivity**.

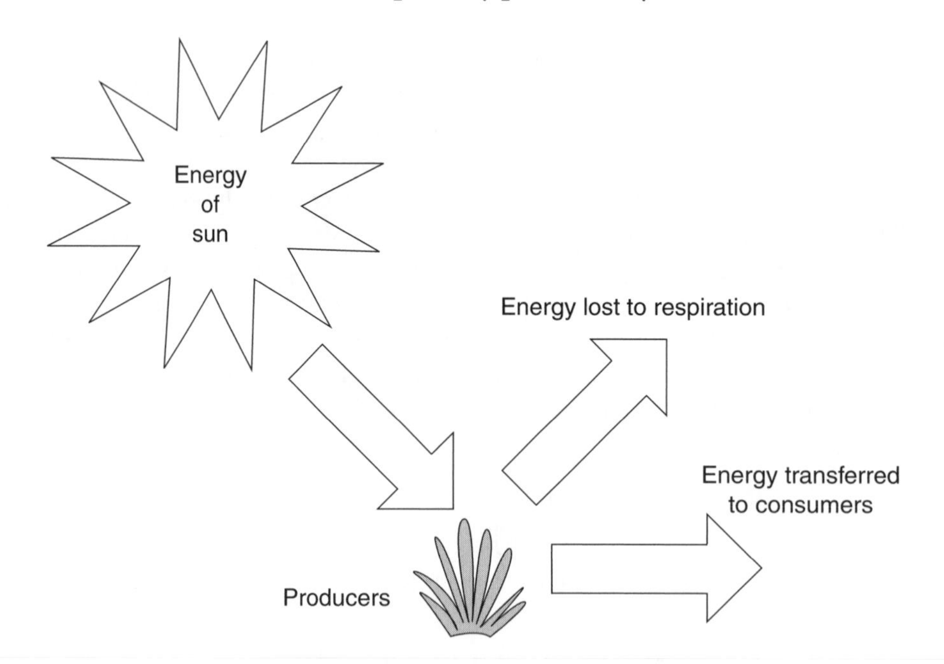

Only about 10 percent of the stored chemical energy is present in the next higher trophic level at every stage, although this number is very approximate and depends greatly on the habitat and the organisms involved. The energy contained in a community can be visualized as a pyramid (see figure below). Although the efficiency of **energy transfer** between levels can differ greatly, the pyramid will always have the *most energy in the producer level*, with less in each level of consumers. A similar pyramid is usually observed if one compares the **biomass**, or numbers of individuals in a community, because each successive level contains about 10 percent of the biomass in the level beneath it. In terms of numbers, the shape of the pyramid can often vary, with a single large producer like a tree supporting a large number of primary consumers like birds or insects. Pyramids of biomass are not always bottom-heavy. In many aquatic ecosystems, where the birth and death rates of the primary producers (algae and other plankton) are so high, a smaller biomass of producers can support a much larger biomass of primary consumers. The high turnover of producers makes many aquatic biomass pyramids have a smaller foundation and larger second level.

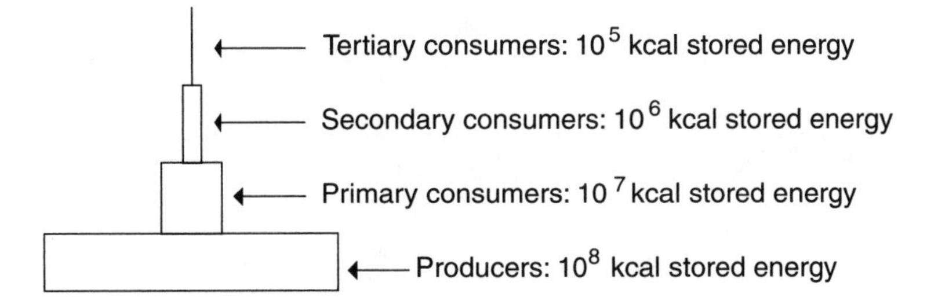

CHAPTER TWENTY-FIVE

Population Size and Its Determinants

A **population** is defined as a group of organisms from a particular species that inhabit the same area. This chapter addresses the concepts of genetic drift and gene flow, which can readily alter the gene pools of local populations, as well as the factors that help determine population density and distribution. Keep in mind that natural selection acts to change populations, not individuals. So while population structure and survival depend upon the interactions of the individual organisms with each other and with the environment, the population itself—not the individuals—undergoes genetic change over time.

EXPONENTIAL VERSUS LOGISTIC MODELS OF POPULATION GROWTH

Unrestrained and unregulated growth of populations cannot and does not occur in nature for very long, due to **limiting factors** such as climate and the supply of resources. Although we will discuss birth and death rates in more detail later, these rates can be used to predict a population's growth over time.

Exponential Population Growth

Most populations experience some sort of unlimited, rapid growth over a brief period of time. In this so-called **exponential growth** period, resources are plentiful and all organisms in the population can reproduce at their maximum rate. This maximum rate of increase in a population is known by the mathematical term r_{max} where r represents the difference between the average birth rate of the population (b) and the average death rate (d) of the population: $r = b - d$. The terms b and d are decimals, calculated by studying a population's number of births and deaths *per individual*. If there are eight births for every 100 organisms in a population each year, the average birth rate (the *per capita* birth rate) would equal 8/100, or 0.08. If d for the same population were 0.05 per year, that would mean that for every 100 organisms in the population, on average five of them would die that year. Using these numbers, r for this population could be found using the equation above: $r = b - d$, or $r = 0.08 - 0.05 = 0.03$. A positive value for r represents an *increase* in population size over time, in this case by three organisms for every 100 (or 30 for every 1,000, etc.) every year. A negative value for r means that the population is decreasing in size due to a greater number of organisms dying than being born.

Because r is simply a rate of growth, the term r must be multiplied by the number of organisms in a given population for it to take on any relevance in predicting the growth of a particular population. Therefore, the change in the population size over a given period of time can be represented by rN, where N = the number of organisms in the population being studied. Changing this last sentence entirely into a mathematical equation gives us:

$$\frac{\Delta N}{\Delta t} = rN$$

where ΔN is the change in population size and Δt is the amount of time over which the population is studied. For a population undergoing maximal growth, exponential growth, this equation is modified slightly to:

$$\frac{\Delta N}{\Delta t} = r_{max}N$$

If you were to plot exponential population growth on a curve, you would see J-shaped curves, such as the ones shown below, that start off rather flat for a short period of time and then very rapidly increase:

For which line would you predict a greater value of *r*? The steeper the line, the greater the value of *r*, because a higher *r* corresponds to faster exponential growth. The population whose growth is shown by the dotted line has a higher *r* than the population depicted by the solid line!

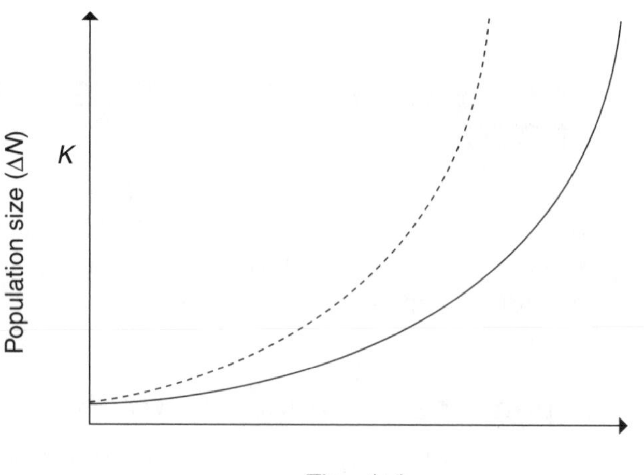

If you see J-shaped curves on the GRE, think exponential growth! What kinds of populations most often undergo this kind of explosive growth? Only those that are small enough and in areas of plentiful resources—perhaps populations that are filling in a new area such as a newly formed volcanic island or an area recently destroyed by fire. In fact, you should connect both primary and secondary succession with exponential growth and high r values.

Logistic Population Growth

In contrast to exponential growth, the **logistic model** describes the more realistic pattern of growth in a population, in which the population grows exponentially for a period of time and levels out in number as it stops growing. This leveling occurs as the population reaches what is known as its **carrying capacity**, the maximum size at which the population can stably interact with its environment for a long period of time. In population equations, as introduced above, carrying capacity is represented by the letter *K*. Using *K*, we can rewrite the above equation as:

$$\frac{\Delta N}{\Delta t} = r_{max} N \left(\frac{K - N}{K}\right)$$

where $(K - N)$ represents the difference between the current population size and the carrying capacity, the maximum attainable size. $\left(\frac{K-N}{K}\right)$ is, therefore, the fraction of the carrying capacity that the current population is at. By plugging in some numbers, you would see that, as *N* increases, *r* decreases: In other words, as the population nears its carrying capacity, the rate of growth quickly slows down. This is reflected in the S-shaped logistic growth curve characteristic of most populations after growing for a certain length of time in a particular environment:

What Is the Difference between K-Selected Species and R-Selected Species?

K-selected species live at a population density near their carrying capacity. R-selected ones have population densities which change rapidly and often.

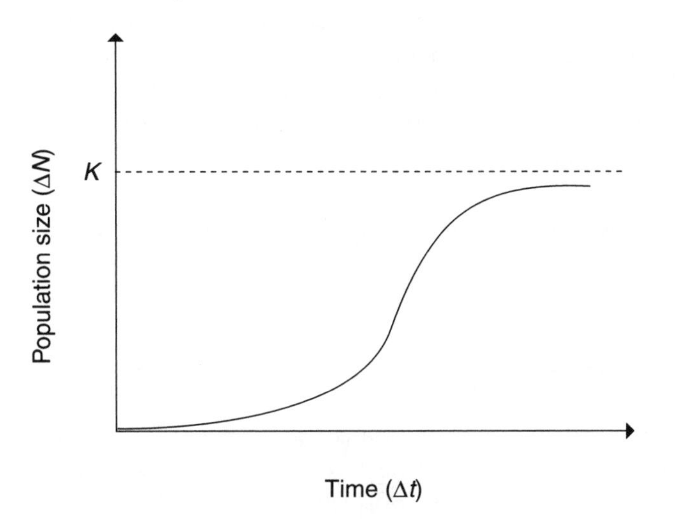

Carrying Capacity and Limiting Factors

Again, *carrying capacity* refers to the number of organisms of a particular species that a given area (habitat) can hold indefinitely *without damaging the habitat or impairing the habitat's productivity*. Although carrying capacity may be reached in a gradual manner by many populations, some groups of organisms in a given area may reproduce and spread so rapidly that their carrying capacity, *K*, is actually exceeded for some period of time. At this point, the death rate (*d*) will become greater than the birth rate (*b*) until the population levels out at the *equilibrium imposed by the carry capacity* of the ecosystem. If the population has exceeded *K* by a large amount, it may, in fact, dip significantly below the carrying capacity before rebounding. This can be seen in the graph below of an imaginary population that has exceeded *K* and takes some period of time before regaining equilibrium at *K*:

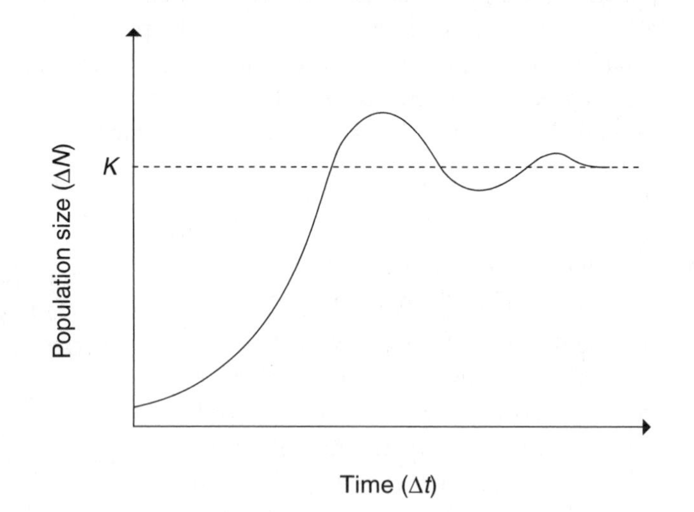

What environmental factors determine a population's carry capacity? These factors, known as **limiting factors**, include food and water, predators, disease, sunlight, and climate patterns. Some limiting factors are **density-dependent**; that is, they affect a population's growth to a greater degree the larger the population gets. Limited food resources and limited space are both density-dependent because they affect a population more as it gets larger, and less as it gets smaller. Denser populations run out of food and space more quickly! **Density-independent** factors include drought, an early frost, or natural disasters like forest fires and storms. These events will likely affect a population's growth regardless of the size of the population or how well-adapted the organisms are to their environment.

LIFE HISTORY STRATEGIES

While it may not be quite applicable to humans, the purpose of life is to pass on one's DNA in as effective a manner as possible to the next generation; that is, to insure that one's genetic information is reliably passed to one's offspring and that they, in turn, can pass on this information to their offspring. To this end, all organisms have adopted certain strategies, called **life history strategies**, that enable them to pass on their DNA efficiently and to breed successfully according to the environment around them. Some species are adapted to produce lots of offspring in multiple births over the course of many years, while others have very few offspring—perhaps one or two every few years. Some reproduce and then die immediately after giving birth.

It stands to reason, then, that a species' life history strategy can be deduced by examining the following four areas:

(a) How often does the species breed?

(b) How many offspring are born to each individual?

(c) At what age is reproduction first initiated?

(d) At what age does the organism begin to grow old and die?

There are as many combinations of these four factors that make up a life history as there are species on Earth. To complicate matters even further, some species change their life history strategies according to the season or other changes in their environment.

Take, for instance, one type of water python, *Liasis fuscus*, which shows differences in its reproductive timing simply as a result of small differences in its nest-site location. Female pythons use either nest sites with cooler temperatures and frequent temperature variations, or nest sites with higher and more stable temperatures. Those in the cool nests (e.g., hollows within root systems near water) take longer to reproduce and as a consequence, must wait up to two years before being able to reproduce again. Those in the warmer nests (e.g., underground burrows at slightly higher altitudes) reproduce more quickly and can reproduce again one year later. Yet, pythons laying eggs in the cooler nests remain with their eggs till hatching, whereas pythons in the warmer nest-sites generally desert their eggs within a few days of laying them. So while the **cost of reproduction** is much greater for cool nest-site pythons (since they have to remain to incubate the eggs), cool nest-sites afford a higher chance that the eggs will survive till birth. Eggs at the warm nest-sites fall easy prey to those predators searching for a quick egg meal—a good reason that the trade-off of laying eggs in warm sites and being able to desert them requires females to reproduce again only one year later (rather than two years later like the cool nest-site pythons).

As you can see, life history strategies are all about trade-offs for a species: Do you spend lots of time nurturing your eggs or offspring with the consequence that you have less energy to invest in the next reproduction and more time to wait to reproduce again? Or do you invest little energy and reproduce often, even though many of your offspring will never reach maturity? Perhaps something in between?

For the GRE, these two generalizations will help you solidify your understanding of life history strategies with what you know about *K*-selected and *r*-selected species:

***K*-Selected Species**	***r*-Selected Species**
Produce a few, large, well-developed young; often parental care; slow growth; delayed maturation and hence delayed reproduction of offspring; long gestation	Lots of offspring at once; often little or no parental care; fast growth; young able to reproduce relatively soon after birth; short gestation

Keep in mind that these characteristics are only generalizations. The vast majority or organisms are not completely *K*- or *r*-selected.

DEMOGRAPHY—POPULATION BIRTH AND DEATH RATES

The term *demography* refers to the study of the vital statistics of a society or population: the birth rate, the death rate, the time it takes between offspring being born and then bearing their own offspring (generation time), the age structure (What portion of the population is old? young? of reproductive age? nearing reproductive age?), and gender ratios (What portion of the population is male? female?). The age and gender structure of a population is absolutely critical to how fast a population grows and can be used to predict future growth many years down the road.

When demographers study human societies and nations, they use pyramid-like structures to depict the age and gender structures of certain populations. A typical pyramid might look something like the one below. This imaginary population has a large number of individuals at or below reproductive age (approximately mid-teens to mid-40s)—this is a young population overall.

By looking at the demographic table below, it can be concluded that this population should experience fairly rapid growth for the next decade or two, while birth rate outpaces death rate. It should also be evident that, if we were to look at the demographic table of this population in several decades, there would be many more individuals in the upper age ranges (e.g., 50–90 years old), as the population that is now reproducing ages.

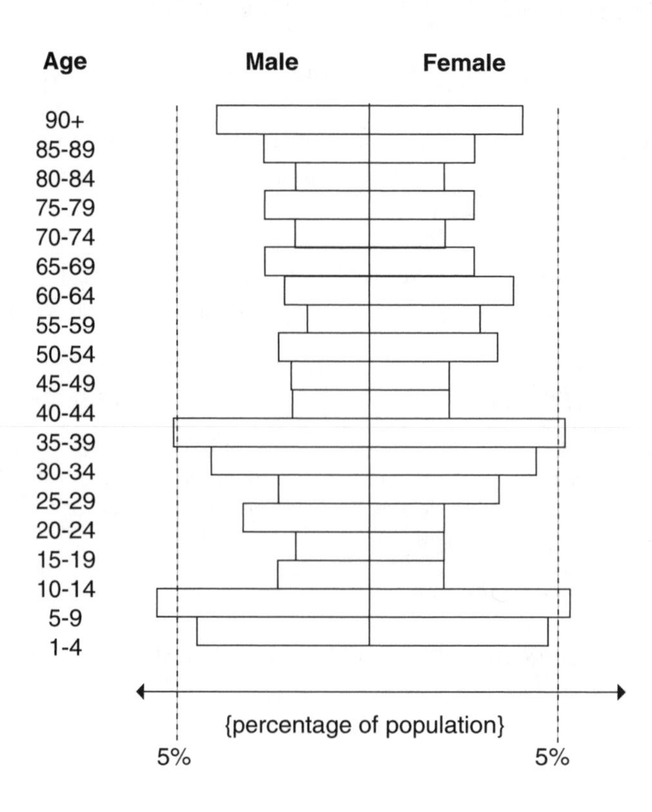

CHAPTER TWENTY-SIX

Mutation and Genetic Variability

Populations evolve through changes in their gene pools over time. These changes occur as the relative frequencies of certain alleles vary from generation to generation. There are some processes that have the ability to add new alleles into a population's gene pool, such as gene flow and mutation. In contrast, other processes remove alleles from the gene pool. Genetic drift, natural selection, and sexual selection are examples of evolutionary events which work toward stabilizing the genetic variability in a population. This chapter focuses on those events which increase and maintain a population's genetic diversity, phenomena essential for mechanisms such as natural selection to operate.

ORIGINS OF GENETIC DIFFERENCES BETWEEN ORGANISMS

The main process that leads to genetic variance within a population is **mutation**, already covered in some detail in the first part of the book. Although mutations have the ability to *create new alleles* from preexisting ones, it is the process of **genetic recombination** through meiosis and sexual reproduction that has the capacity to "shuffle" existing alleles into *new combinations* of alleles. These two processes together lead to the formation of offspring whose DNA incorporates novel alleles and allelic combinations not found in the offspring's parents and perhaps not in any other organism in the population.

Mutation

Because only a small fraction an organism's DNA actually codes for proteins, the vast majority of mutations are considered *neutral* in that they neither harm nor help the organism's offspring when passed on. Because there are many instances of **gene duplication** in most organisms, mutations occurring in duplicate copies of genes generally will not affect an organism's ability to produce enough of the protein that gene codes for. In fact, successive mutations within duplicate copies of genes may allow the formation of novel alleles without compromising an organism's ability to produce a key protein or phenotype. In addition, mutations within many genes may have little to no effect on the final protein product, especially if the mutations are point mutations or if they occur toward the very end of the coding section of a gene.

It's important to remember that the only mutations that can be passed to offspring are "germ-line" mutations–those that occur in cells leading to the production of gametes (sperm and egg)! Somatic mutations are never passed on.

Mutations that tend to have a beneficial effect (**beneficial mutations**) on the phenotypes expressed in offspring can arise from a variety of genetic change. Point mutations may alter a single amino acid in a protein, allowing cells with that protein on their surface to bind with more affinity to another cell. Gene duplications may lead, for example, to an insect that is able to produce twice as much of a particular insecticide-digesting enzyme. A **gene translocation**, which moves a gene from one region of a chromosome nearer to a more active promoter region, could lead to increased activation of the gene and result in some beneficial phenotype. These mutations or translocations just described could more easily have a negative impact (**deleterious mutations**), which is usually the case.

Keep in mind the following additional points about mutations for the GRE:

- Mutations limit the rate of evolution—beneficial mutations need to achieve a high enough frequency over the course of several generations that they "fix" in the population.

- Changes in the environment can quite suddenly lead to certain allelic variants (and therefore phenotypes) becoming favorable. This occurs independent of how fast mutations are occurring within the population, as already established variants (mutants) within the population gain a fairly sudden advantage over others in the population. In a population of bacteria exposed to antibiotics, some of these bacteria could have naturally occurring mutations that allow them to be resistant to the antibiotic. The mutations conferring resistance had occurred in previous generations, perhaps as neutral mutations or changes in duplicate genes, yet become advantageous as the environment around the bacteria changes.

- Mutant alleles may increase in frequency in a population if they are simply linked to some beneficial allele (i.e., found on the same chromosome); these mutant alleles are passed on to offspring literally as hitchhikers. In cases of hitchhiking, only recombination can lead to separation of the linked alleles, possibly decreasing the frequency of one of the alleles or selecting the allele out of the population altogether.

Recombination

Although mutations are the ultimate source of new alleles, genetic recombination randomly combines parental alleles so that offspring are truly unique mixes of their parents' genes. Even without recombination, the number of possible genetic combinations between two individuals alone can be amazingly high. Because of simple Mendelian independent assortment alone, every human can build gametes with over 8 million possible combinations of his or her chromosomes! *This number does not begin to take into account the additional variation arising from random genetic recombination*, or **crossing over** as it is also known. Sexual recombination is essential to producing the genetic variability seen in almost all populations of sexually reproducing organisms.

Single-celled organisms such as bacteria generate genetic diversity almost entirely through mutations and not via recombination, which makes sense since they reproduce asexually. But don't forget that many single-celled organisms can, in fact, recombine their DNA whether they reproduce sexually or asexually! Paramecia, for example, exchange DNA between parents through a sexual process known as **conjugation**. Among organisms that reproduce asexually, bacteria can exchange bits and pieces of DNA through the movement of plasmids, small circular pieces of DNA additional to the main circular chromosome that are readily shipped from bacterium to bacterium.

POLYMORPHISMS AND LINKAGE DISEQUILIBRIUM

Polymorphism

We have seen the term polymorphism earlier in section II, where **RFLPs** were discussed in the genetics chapters. Recall that RFLPs are discrete, single-base changes in the nucleotide structure of genes from one individual to another. Because there are a certain number of varieties or **morphs** of a particular gene in every population, studying restriction fragment length polymorphisms can help to characterize an individual's specific genotype. Likewise, on a more macro level, many traits within a population vary from individual to individual in a **noncontinuous** manner, where two or more forms of the particular trait are present in the population but *intermediate forms do not exist.*

Perhaps the most commonly cited example of a polymorphism in the human population is the ABO blood proteins common to all red blood cells. Humans have four distinct polymorphisms of blood type: type A proteins on their red blood cells; type B; type AB; or type O (neither A nor B proteins present). There are no intermediate forms. You can have a blood type that is only one of four variants. In contrast, some traits vary in a **continuous** manner throughout a population and do not have discrete forms. These traits are complex and result from the interactions of multiple genes. Examples include height and skin color, which differ along some continuum, as seen below, though a certain range may predominate in a given population (e.g., the average Pygmy is shorter then the *average* American):

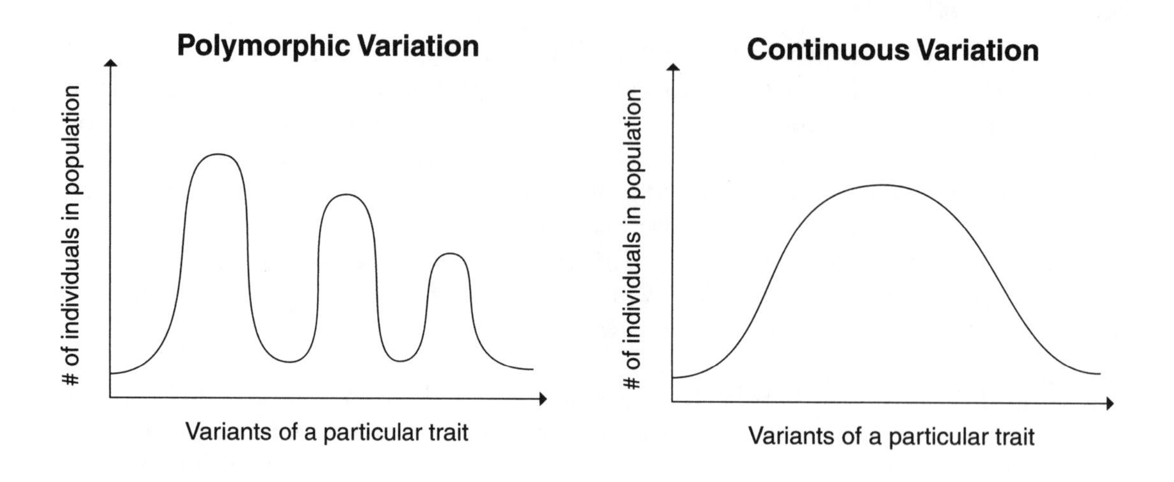

Balanced Polymorphisms

Polymorphisms are referred to as balanced when two or more morphs of a particular trait exist side by side within a population for an extended period of time or perhaps permanently. Very often, a balanced polymorphism can be maintained because organisms in the population will choose only mates who have similar characteristics. Some bird populations, for instance, include birds that exhibit either large, sturdy beaks or small, thin beaks. Because of differences in food present in the environment, large-billed birds and soft-billed birds can both be successful, yet medium-billed birds will not be. They are out-competed for food by both beak extremes. As the birds in the population continue to mate with like-beaked partners, this polymorphism (the two discrete beak forms) will be maintained or balanced. The trait of beak size never varies continuously in this particular group of birds.

Two mechanisms act to maintain polymorphisms in a population:

- **Heterozygote advantage**: Scientists have discovered several lethal or harmful alleles which are kept at a relatively high frequency in many human populations. The most cited of these, the allele causing misshapen hemoglobin protein and subsequent sickle-cell anemia, allows those who have one normal copy of the hemoglobin-producing allele (S) and one mutant, sickle-cell copy (s) to be more resistant to the parasite that causes malaria. Because the malaria parasite, *Plasmodium falciparum*, lives within red blood cells, those who produce at least some deformed red blood cells (due to misshapen hemoglobin molecules within) cause the parasite to die off more quickly as it has fewer cells within the body in which to live and reproduce. The heterozygote individuals (Ss) have an advantage over either of their homozygous counterparts, particularly in environments infested with malaria-carrying mosquitoes. Homozygous dominant individuals (SS), producing all normal hemoglobin, fall victim to malaria quite easily, and homozygous recessive individuals (ss) fall victim to the debilitating effects of malformed hemoglobin in every one of their red blood cells. *If the most advantageous genotype in a population* *is the heterozygote genotype (in this case, Ss), it allows for the continued birth of homozygous dominant and homozygous recessive genotypes in subsequent generations, even though these genotypes have a survival disadvantage.* This heterozygous advantage maintains both the S allele and the s allele in the population and, therefore, the three different possible genotypes in the population—HH, Hh, and hh—a balanced polymorphism! Without it, the population might be made up of only the HH genotype.

- **Frequency-dependent selection**: This occurs when a particular variant, or morph, loses its survival advantage as it becomes more frequent within the population. As a simple example, when resistance of a group of organisms to a particular pathogen increases, with more and more of the organisms' offspring expressing resistant phenotypes, genes in the pathogen that tend to make it more virulent or allow it to reproduce faster will be strongly favored. Because it is so much more lethal, this resistant phenotype will become less and less successful the more frequent it becomes in the population. A higher frequency of resistance correlates with a higher virulence of the pathogen. With low frequencies of the resistant phenotype, there is a much smaller selective push toward increased virulence in the pathogen. In such a way, *frequency-dependent selection, the maintenance of small "pockets" of particular variants in a population, tends to preserve the balanced polymorphism of alleles for a particular trait.*

Linkage Disequilibrium

Among the terms to be familiar with when studying mutation and recombination is linkage disequilibrium, which you may find on the GRE written as **gametic disequilibrium**. This complicated sounding term refers to the fact that certain alleles at different genetic loci may be linked so that the traits coded for by these genes appear together much more frequently than would be expected if the alleles were assorting independent of one another. The term disequilibrium comes from the fact that, due to gene linkage, the relative frequency of certain phenotypes in a population differs from what would have been predicted by the Hardy-Weinberg *Equilibrium* equations.

The stronger the linkage disequilibrium between two alleles (or two genetic loci), the closer the alleles are assumed to be on a particular chromosome. In addition, recombination through crossing over tends to disrupt linkage disequilibrium by separating these linked genes. Therefore, sexual reproduction reduces linkage disequilibrium.

How Does Mutation Affect Linkage Disequilibrium?

Mutation often decreases disequilibrium because it tends to increase the amount of polymorphism in a population. As polymorphism increases, the chances decrease that even two closely linked genetic loci will remain unaffected and constant over many generations.

SPATIAL PATTERNS IN GENETIC VARIATIONS

It is quite common to find a graded series of changes in phenotypes expressed by a species that is widespread across a particular geographic region with various habitats and local climate differences. For example, house sparrows show a continual geographic variation in body size across their range, which covers almost the entire United States. House sparrows with the largest body size are found in the northern-most latitudes, while the smaller sparrows are found in the warmer, more southern climates. The smallest sparrows are found in the San Francisco area, with average body size gradually increasing in local populations extending eastward from California. The larger size of the birds living in northern latitudes most likely helps them retain their body heat in the colder climates (recall: surface area to volume ratios, where larger size decreases the surface area: volume ratio, and smaller animals lose more heat per unit size than do larger animals). This graded variation across a diversity of climates and other environmental conditions is called a **cline**.

In some cases, local populations have significant phenotypic differences from other nearby or neighboring populations. Although still able to interbreed, these populations do not always show a smoothly-graded transition from one population to another; yet one may readily find hybrids at the borders between these populations that resemble blends of the distinct phenotypes. The *Ensatina* genus of salamander (see Introduction), living throughout a variety of habitats in California, shows this type of cline, with neighboring populations often differing greatly from one another in body patterns (size and coloration of spots or banding) due to significant changes in local climate across the variety of coastal and inland habitats where one can find these salamanders. The local populations themselves, the distinct "forms" of salamander, are called **ecotypes**: *locally adapted variants of an organism, differing genetically from other local forms.* While the environment clearly plays an important role in how certain phenotypes get expressed, e.g., certain coloration in certain areas, genetics seem to play an essential role in clinal variation as well. Ecotypes moved into new habitats will develop much of the same coloration and many of the same traits that they would have expressed in the *original* habitat. In other words, keep in mind that much of the variation seen across a cline is *genetically* encoded and is not dependent upon environmental influences.

Biogeography and Global Clines

Biogeography is the study of the past and present ranges of species throughout the world. Scientists look at environmental data as well as resource availability for a given region over time to learn about population shifts and changes in a given population's range. Species dispersal and the limits of a given species are studied by **transplantation experiments**; for these, a small, founder population of a species is moved to a new area where the population has not previously spread, in order to see whether the organisms take hold there.

Species richness and distribution change as one moves in latitude or longitude across the earth, and biogeographers attempt to study these global clines, or patterns of gradual and sudden species distribution change. The greatest species richness overall occurs both at the equator and deep under water near marine deep sea vents.

CHAPTER TWENTY-SEVEN

Speciation and Gene Flow

Mount Graham rises 10,000 feet above the Sonoran Desert in southwestern Arizona, surrounded by desert and grasslands. Near the top of this mountain at the southern end of the Pinaleno mountain range exists a cool, moist forest of spruce and fir, made of trees hundreds of years old and far more characteristic of a northern Canadian forest than any desert habitat. In this old-growth spruce forest live at least two dozen species of plants and animals that exist nowhere else on Earth. They were isolated on the top of Mt. Graham as the glaciers of the last ice age retreated and left the Pinaleno Mountains surrounded by desert about 11,000 years ago. The aptly named Mt. Graham Red Squirrel is one of best known of these endangered species, separated from neighboring red squirrel populations by the shifting climate. Numbering around 600, the Mt. Graham squirrels cannot spread out into and mate with other genetically similar red squirrel populations, nor can they incorporate individuals from these other populations into their own: The Sonoran desert is too formidable a barrier to any movement off Mt. Graham.

In contrast, salamanders of the genus *Ensatina*, which inhabit the moist areas of coastal California, form an almost continuous population of salamanders up and down the coast. They display a range of body colors and patterns that seem to change gradually as one moves from the northernmost salamander populations to the southernmost ones. Despite the lack of isolation among particular groups of salamanders and neighboring groups to the north and south, in most cases salamander populations with a given color pattern will mate only with others of the same pattern and coloration. As these salamanders have spread up and down the coastline of California, they have adapted to new environments with novel color and pattern schemes; these changes have created a barrier to mating despite the proximity of other salamander populations which are nearly identical genetically. The *Ensatina* salamander populations are **isolated** not **geographically** (like the squirrels above), but **reproductively**, from each other.

Across the planet, we see varying degrees of species isolation as a result of environmental factors or human interference. As mentioned earlier, a nonevolving population of organisms has, as one of its characteristics, isolation from other populations. This isolation means that genes from one population—that population's **gene pool**—cannot mix with genes from another population and affect the relative frequency of alleles for certain traits. *Remember that evolution is defined as a change in the frequency of alleles over time.* Isolation alone, however, does not qualify a population for "nonevolving" status, since the population would also have to meet the other four Hardy-Weinberg conditions.

At the same time, we will soon learn that isolation from other populations is absolutely essential for new species to develop from existing ones. For evolution to occur as a result of species isolation, however, **genetic mutation** and **natural selection** must also be present.

SPECIATION

Most populations inhabit ranges that are continuous and stable for long periods of time. Yet it is possible for certain populations of organisms to become isolated from others: When that happens, one population of a particular species changes over time in ways that are distinct from the changes taking place in other populations of the same species. In turn, this can lead to the formation of a new species, a process known as **speciation.** One of the five tenets of the Hardy-Weinberg Law is that organisms from one population of a given species must not emigrate out of or immigrate into another population of that species. Otherwise, frequencies of certain alleles in the original population will change over time, and evolution will take place.

Organisms can become isolated from others in their population in a variety of ways. Whether sudden or gradual, this isolation is equivalent to an emigration of alleles out of the original population. When this happens, different genes may be selected over time in the newly isolated population and new mutations can occur. As days, months, or years pass, the isolated population can acquire traits that do not exist in the original population, as the two groups of once-similar organisms diverge. In most cases, if one were to "reacquaint" these two divergent populations, their members would no longer be able to reproduce with each other. In this way, the process of speciation will have occurred. A species is loosely defined as a group of interbreeding organisms that share common attributes.

Speciation Requires Isolation

Just because two populations of organisms are geographically separated and do not mate does not preclude their mating should they encounter each other. Thus the formation of a new species must involve a change in genetic makeup or in gene expression, such that the potential to interbreed is significantly reduced or eliminated altogether. It is important to understand that isolation can lead to speciation if the isolation leads to a reproductive barrier between two groups of organisms. This can only occur if the isolation is accompanied by sufficient genetic changes over time that the two groups become reproductively incompatible. One would think that geographically isolating a group of organisms would be the right way to start the process of speciation, but **reproductive isolation**, believe it or not, can also occur right in the midst of the original population of organisms! Let us look at each of these speciation mechanisms in turn.

TYPES OF REPRODUCTIVE ISOLATION

1. *Prezygotic*: factors prohibiting the union of gametes, e.g., different reproductive cycles/periods (*temporal isolation*); differences in courtship patterns, such as bird songs or mating while flying through the air versus on the ground (*behavioral isolation*); and, the physical inability to copulate, due to size differences or anatomical differences, including the binding of sperm cells to the egg cell membrane (*mechanical isolation*).

2. *Postzygotic*: factors prohibiting survival or birth of the offspring AFTER successful fertilization, e.g., abortion of zygote formed between two species or frail, underdeveloped offspring (*hybrid inviability*); hybrids are born sterile and cannot produce normal gametes (*hybrid sterility*); and, fertile hybrids occur in the first generation but successive generations are infertile or inviable (*hybrid breakdown*).

Allopatric Speciation

The term "allopatric" comes from Greek and Latin roots meaning "other land." This type of speciation occurs when a physical barrier geographically separates populations. The Mount Graham Red Squirrel discussed at the start of this chapter is an example of this process.

The creatures that Charles Darwin observed on the Galapagos Islands are another example of the ability for geographic isolation to lead to the formation of new species. Birds, reptiles, and plant seeds were blown over to the islands by wind or swept over the ocean on rafts of vegetation from the South American mainland. They became isolated on these newly formed volcanic islands. Over time, genetic changes occurred in each of these island populations, creating new species of birds, reptiles, and plants—related to the original South American species—which could not reproduce with members of the mainland species or, in many cases, even with similar populations on the other islands. **Geographical isolation** can occur slowly, such as when mountain ranges gradually form and divide populations; it can also occur quite quickly after natural disasters like earthquakes or volcanoes, which can rapidly divide a landscape and isolate populations. It is now thought that even the simple building of a two-lane highway through an ecosystem can serve as a geographical barrier to populations of organisms left on either side of the road—enough so that these populations will be left isolated from each other and no longer able to interbreed.

Sympatric Speciation

In this type of speciation, a group of organisms becomes reproductively isolated within the parent population—hence the prefix *sym-* which is Greek for "together." This occurs *without geographic isolation*. It is most frequently found in plant species, though there are some instances in animals. Many plant species suffer from cell division errors during reproduction, where offspring end up with extra sets of chromosomes. This process would be like forming human sperm and egg cells that are each diploid—when fertilization occurred, a zygote would be formed with 46 pairs of chromosomes rather than the conventional 23 pairs. Although humans would miscarry zygotes formed in this manner, because many plant species can self-fertilize, the likelihood of this happening successfully in a group of plants is high. Assuming these new **polyploid** offspring are viable, they can continue to grow and self-fertilize.

Imagine a field in which several groups of polyploids spring up among the original plants. Because the polyploids would not be able to form viable offspring with plants from the original population (due to such great differences in chromosome number), a new and genetically distinct plant species will have cropped up in the midst of the species of which it was once part! This process is known as **autopolyploidy** (the act of becoming polyploid by *self*-fertilization).

A much more common instance of sympatric speciation in plants occurs when two genetically similar plants, each with a slightly different number of chromosomes, mate and form hybrids—just like a donkey and horse mating to create a mule. However, as with mules, the hybrid plant offspring are usually nonfertile. But plants can do something mules and most other animals cannot: reproduce asexually! If these plant hybrids are viable and can reproduce asexually, a new fertile, polyploid hybrid species can arise, again right in the midst of the original species. This process is known as **allopolyploidy** (the act of becoming polyploid by combining your chromosomes with those of another species). Plant species that have arisen through sympatric speciation include potatoes, tobacco, and certain types of wheat.

Sympatric speciation in animals might occur in a species that has two distinct characteristics (morphs) for the same trait. In a beaver population that is dimorphic for fur pattern (where there are beavers with completely dark coats of fur, and beavers with a large amount of white patterning within the dark fur, but very few beavers in between), speciation would likely occur between the two morphs if beavers began to preferentially mate only with like-patterned mate.

GENE FLOW AND GENETIC DRIFT

The phenomenon of **genetic drift** can often play a major role in evolution. Genetic drift results from what is known as **sampling error**. Let's say you could put the gene pool of a population (all the dominant and recessive alleles) into a large sac and pull randomly from the sac to "create" new offspring. If you created only a few offspring, there is no way that their gene pool will come close to the diversity of alleles that were present in the original sac. The more you pull alleles from the sac and the more offspring you create, the more the new population will come to resemble the old population in terms of the frequency of different alleles. It's the same phenomenon when you toss a penny: You expect 50 percent heads and 50 percent tails, but tossing the penny only a few times will not necessarily give you anything close to that ratio.

Therefore, the smaller a population, the greater the effect of small changes in allele frequency or of mutations on the next generation's gene pool. In larger populations, the effects of a mutation in a certain allele might be drowned out by the fact that there are so many of the normal allele in the new population that the overall frequency of the allele does not change. However, in smaller populations, if mutations make it through to the next generation (if they are picked out of that hypothetical sac), then they can make a big difference in the frequency of a particular trait or allele in the next generation.

For the GRE, you should know that *genetic drift* describes the changes that occur in a population's gene pool (sac of alleles) from generation to generation due to random chance, not due to some sort of selection for certain traits. Although habitat destruction by humans has drastically decreased the population sizes of various species across the world, there are natural situations that result in populations small enough for genetic drift to occur: the **founder effect** and the **bottleneck effect**.

The Founder Effect

When a small group of individuals colonizes an island or a new, isolated habitat, a founder effect results. Because this group represents only a small portion of the alleles that existed in the parent population, subsequent generations will be formed from only this subset of genes, and the isolated "founder" population can diverge quite rapidly from the original one. Any mutations in this small subset of alleles and in this small group of organisms are likely to be disproportionately represented in subsequent generations. In addition, genetic drift is influenced by how many offspring each set of parents has. The more offspring a given set of parents has, the more their alleles are represented in their offspring (since, due to meiotic shuffling of alleles, each offspring will contain only a portion of the parents' total alleles). Therefore, the greater the number of offspring, the greater the chance that the offspring will carry forward the full complement of the parents' alleles into the next generation. This is genetic drift on a very small (single-family) scale, but it clearly has wider effects on the population as a whole.

What Long-Term Result Can Arise from Genetic Drift?

Because each generation is an independent genetic event, random changes in the gene pool due to genetic drift may result in the frequency of a particular dominant or recessive allele reaching 1.00 and the frequency of its corresponding allele reaching 0.00. In this case, the population has become homozygous for that trait and no further change is possible! In the absence of genetic drift, or with a large population, the frequencies of genes will almost always be same from one generation to the next.

The Bottleneck Effect

This relatively sudden narrowing of genetic diversity in a population can occur when a portion of a population is wiped out from a natural disaster, such as a flood or fire. Because these kinds of disasters tend to kill off organisms quite randomly, the alleles that "survive" through to the next generation will have done so based on no merit of their own. In other words, the remaining population will contain a random subset of the original genetic diversity, not a subset that resulted from certain alleles conferring a greater survival benefit than others. The problems that result from a population's genetic bottleneck can be quite severe. Unfortunately, once genetic drift starts, it can often proceed without restraint toward a situation where one allele (i.e., the dominant or recessive allele) for a particular trait becomes far more prevalent in the population. Thus, populations that remain small and in which genetic drift has begun can rapidly diverge from the gene pool of the parental population.

Gene Flow

The term *gene flow* relates to population genetics in that it refers to the movement of genes from one population to another. Alleles for different traits literally "flow" from one group of interbreeding organisms to another as individuals emigrate out of and immigrate into neighboring populations. When gene flow is occurring at high rate, the gene pools of neighboring populations can clearly become much more uniform, and small differences between these populations can become "diluted" by the flow of genes from other populations. It is evident that this phenomenon is occurring at high rates in the human population as races and ethnicities increasingly interbreed.

CHAPTER TWENTY-EIGHT

Natural Selection and Other Mechanisms of Evolution

We have already reviewed two important mechanisms that result in significant genetic change in populations over time: the bottleneck effect and the founder effect. Both of these mechanisms have an ability to select certain alleles to be passed onto the next generation. In addition, **mass extinction** events, perhaps initiated by some natural disaster (as in the probable meteorite collision that wiped out the dinosaurs and 50 percent of other families of organisms 65 million years ago), have an enormous selective power despite their relative infrequency. This chapter deals with mechanisms of evolution and selection, selection at different levels (populations, genes, etc.), and the concept of fitness.

NATURAL SELECTION AND FITNESS

The concept of natural selection is a simple one summed up by three words: *differential reproductive success*. Remember these words for the GRE. The process of natural selection is one in which groups of organisms compete for finite resources in a limited amount of space: Those organisms possessing the genes (and therefore phenotypes) allowing them to *reproduce* most successfully in that environment will pass on their genes *more effectively* than other organisms around them. In such a way, the organisms' bodies can be considered as nothing more than vehicles for successful DNA passage from one generation to the next; genes that build bodies able to survive and reproduce well tend to increase their frequency (occurrence) in subsequent generations. Evolution is, quite simply, the increase or decrease of certain alleles within a population over time due to environmental factors that "select" certain phenotypes to be more adept than other phenotypes at producing viable offspring. Keep in mind that certain potentially lethal alleles or harmful mutations can likely be passed on from generation to generation because they are part of an organism whose *overall* genetic makeup allows it to have a high level of reproductive success, despite the presence of some deleterious alleles.

Fitness

Make sure to understand the concept of fitness. Sometimes written as **reproductive fitness**, this term always refers to how successful an organism is at reproducing. In a population, organisms with a high degree of fitness have more *surviving* offspring than those with low fitness. Natural selection works to increase the overall fitness of organisms in a population, because it increases the frequency of highly fit organisms, those with reproductively successful genotypes. It is also important to understand why the phrase "survival of the fittest" is not an appropriate definition of natural selection. First, it is potentially confusing with regard to what fitness means (and it does not mean strength, power, speed, etc.), and second, *natural selection is not about survival, it is about reproductive success.* While being able to survive in order to reproduce is essential, organisms that live very long lives are not considered fit in an evolutionary sense unless they have also left lots of surviving offspring!

Modes of Natural Selection

Because natural selection changes the relative frequency of certain alleles in a population over time, it is necessary to understand the ways in which this process acts on a population. There are three main outcomes or "modes" of natural selection, which you should understand well for the GRE.

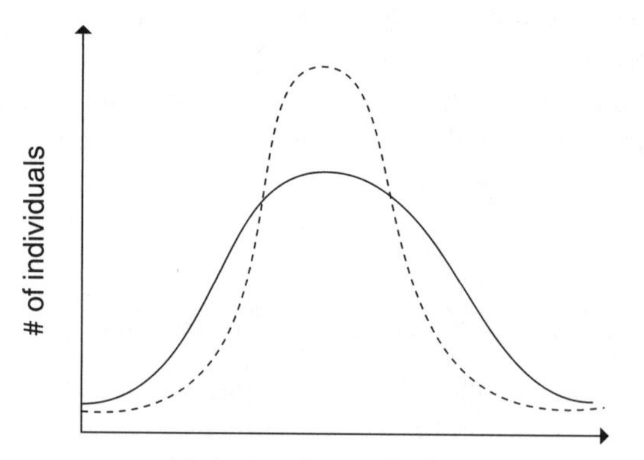

The first mode is known as *stabilizing selection*, where natural selection acts to remove "extreme" variants of a particular trait from the population. In other words, this mode of selection narrows the range of values for a particular trait. Let's say a population of moths ranged from totally white to totally dark in a certain habitat. The population would show stabilization if moths with body color somewhere in the middle of the two color extremes (grey perhaps) were most fit for a particular habitat while white and black extremes died off without reproducing much at all (perhaps because they were picked off by predators more easily than grayish moths). Over time, a population undergoing stabilizing selection for certain traits would show reduced variation in phenotypes and a greater expression in a narrower range for those traits. Notice in the graph above how the population of these moths after several generations (dotted line) demonstrates a stabilizing selection of the original population (solid line).

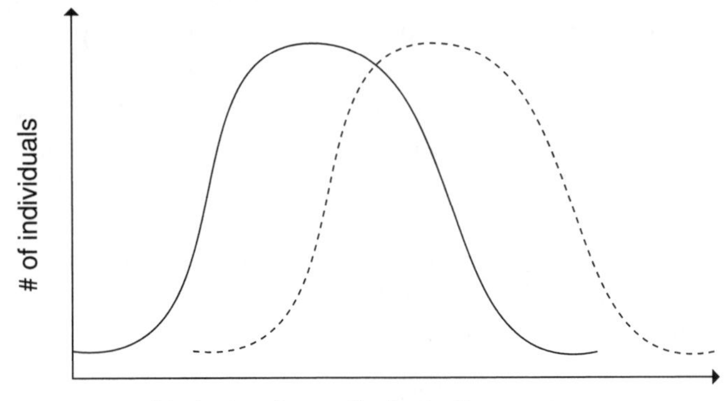

The second mode is known as *directional selection*, where natural selection acts to increase the frequency of expression of certain variants at one extreme of the phenotypic range. In other words, this mode of selection takes the range of values for a particular trait and moves it so that the phenotypes of more and more offspring over time show one extreme of a trait that had previously covered a particular range. If the same population of moths as above ranged from totally white to totally dark in their habitat, the population would show directional selection if moths with a body color toward one of the two color extremes, e.g., white or black, were most fit for a particular habitat, and the other extreme died off without reproducing much at all. Over time, a population undergoing directional selection for certain traits would show the elimination of one phenotypic extreme while increasing expression of the other extreme. Notice in the graph above how the population of these moths after several generations (dotted line) demonstrates a directional selection of the original population (solid line).

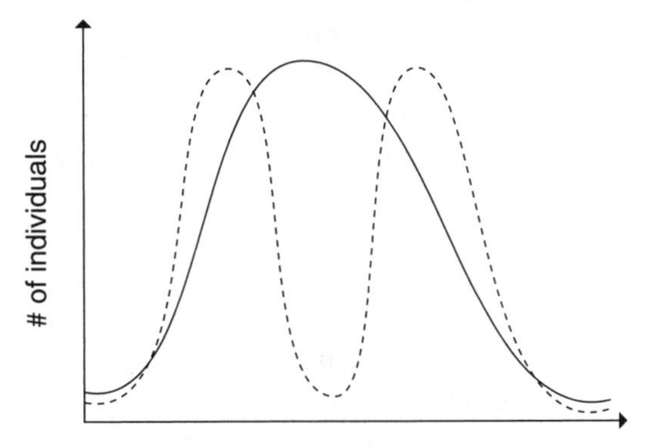

The final mode is known as *diversifying* or *disruptive selection*, where natural selection acts to favor the extreme variants over intermediates. If the environment of the moths described above contained two different features (e.g., differences in tree bark color—some light, some dark), two extreme phenotypes have a selective advantage as intermediates would be noticed first by predators. Gradually, natural selection would increase the frequency of expression of variants at both extremes of the phenotypic range. Over time, a population undergoing diversifying selection for a certain trait would show the elimination of intermediate phenotypes while increasing expression the extremes. Notice in the graph how the population of these moths after several generations (dotted line) demonstrates a diversifying selection of the original population (solid line).

Overall, keep in mind that *diversifying and directional selection occur most often when some aspect of the environment undergoes significant change,* while *stabilizing selection occurs more often in a more stable and unchanging environment.*

LEVELS OF SELECTION

There continues to be great debate among biologists regarding the level or levels at which natural selection acts. Most agree that natural selection does its selecting at the level of the individual organisms, i.e., gene frequencies change from generation to generation according to the differential reproductive success of *individual organisms.* Some scientists maintain that natural selection acts most powerfully at the level of the gene itself, while others maintain that it acts at the larger level of the group/population.

Group Selection

While we have discussed the selection of individuals and the "individual-as-genetic-vehicle" idea above, the term group selection refers to the understanding that a trait might be expected to increase in the population as a whole if the rate of extinction of the group is lowered by a high frequency of that trait. Under group selection, the increase in a trait such as a particular cooperative behavior might enhance the proliferation of new populations carrying that trait. Keep in mind, though, that the notion of organisms being "altruistic" and cooperative raises troubling questions, since it is likely that what we view as **altruism** or cooperation is nothing more than individuals acting to maximize their own benefits in a particular environment.

For the GRE, focus only on the *general concepts* concerning levels of selection, rather than specifics, since the test makers will not likely probe such a hotly debated area.

The idea that organisms act for the good of their group is decades old: It assumes that there are population-level traits that somehow help organisms to limit their rates of reproduction or resource use. In turn, the population is able to control itself, more or less, in a changing environment.

Kin Selection

Kin Selection Theory attempts to explain the evolution of some of the more complex aspects of animal behavior. In particular, it looks at altruistic and cooperative behaviors, as well as social grouping and mating systems. As mentioned earlier, there has been much debate as to whether these behaviors are simply individual survival mechanisms masquerading as something more complex. Kin selection in many ways overlaps ideas of group selection as discussed above.

The basic idea behind kin selection is that of **inclusive fitness**, a term you may see on the GRE referring to the overall fitness of a group of organisms due to cooperative behaviors within the group. A group's overall fitness is increased when some of its members behave in such a way that helps their offspring survive and reproduce. These behaviors are considered altruistic: The fitness of the members initiating the behaviors is often diminished, while their offspring's fitness is often increased. An adult prairie dog may make loud cries to notify offspring and related kin of approaching predators. Or it may may face off against a clearly superior predator in order to distract the predator from any nearby offspring. In both cases, the actions help the survival of the offspring and relatives, despite putting the adult prairie dogs in grave danger. Thus, kin selection may be selected for because helping a relative helps your own alleles in the long run.

Selection at the Gene Level

The body of an individual organism has been referred to in the literature as an "**interactor**," since it is the body that directly interacts with the environment. One's genes are referred to as "**replicators**," since they possess the information needed for replication (reproduction). Many biologists have proposed that genes themselves are the units of natural selection, since the genes as a whole, not the organism's body, are passed onto the next generation. However, because it is the organism's body that interacts with the environment, and it is the environment that does the selecting according to the theory of natural selection, it seems reasonable to conclude, as most biologists do, that the individual is the unit of selection.

All you need to understand for the GRE is that *natural selection may indeed act on many different "levels," from the chromosome to the organism to the group.* It goes without saying that all these levels are absolutely interconnected and dependent upon changes at any other level. Be aware of the terms introduced in this section, however, since you are likely to encounter them in some form on the exam.

ARTIFICIAL SELECTION

Artificial selection or **selective breeding** is the process through which humans carefully pick organisms with particular traits, and use them to breed lines of offspring expressing those traits. It has been practiced for as long as civilizations have existed on Earth. In contrast to natural selection, humans do the selecting; as such, organisms' differential reproductive success is based not on traits that help them survive and reproduce, but rather on traits that confer on them a utility to humans. The types of features that might be desired are high-yield milk or beef production in cattle; resistance to disease in plants; or large size and brilliant color in flowers and fruits.

You will not likely be asked about artificial selection on the GRE. It is presented here to round out the discussion of the forces that shape genetic changes in organism populations over time.

Artificial selection is not necessarily an easy practice; often, these breeds need specific environments. They might require particular temperatures and humidities in which to grow, antibiotics, or other controversial "safety measures" to keep breeds healthy over many generations. Given these conditions, ethical considerations abound.

With the continued progress of biotechnology and genetic engineering, artificial selection has taken a bold step: Newly designed genes or DNA from other animal or plant species will be used to introduce traits never before possessed by certain species. Plants will be able to fend off parasites or predators because they can produce a bacterial toxin from transplanted bacterial DNA; plants will be able to produce various types of natural plastic, such as polystyrenes; and pigs will be able to express human antigens on their cells' surfaces, allowing the potential for pig-to-human organ transplant without rejection.

OTHER MECHANISMS OF SELECTION

Natural selection is not a random process. Although the environmental changes that select species, mutants, or varieties are more or less random (as are the mutations themselves), the process itself is not random. In addition to natural selection, there are other processes that have a great deal of power to shape the gene pools of populations over time; some of these processes, such as mass extinction, "select" *in a random fashion* those organisms that will pass alleles successfully onto the next generation.

Mass Extinction

Similarly, mass extinction events, which have occurred quite rarely but regularly throughout history, have enormous potential to reshape or eliminate altogether the gene pools of many populations. Mass extinction events, such as occurred 245 million years ago in the Permian and 65 million years ago in the Cretaceous, cause the dying off of many local populations and even entire species (especially ones not geographically widespread). At the same time, these events leave open many *available niches* for the rapid growth and *adaptive radiation* of existing species. The dinosaur extinction 65 million years ago is widely credited with having opened up niches for the development of larger and larger diurnal mammals, eventually leading to the evolution of primates, including ourselves.

Adaptive radiation is defined as the emergence of many species from a common ancestor after this ancestral population has been introduced into an environment with diverse conditions and open niches. Examples of speciation through adaptive radiation abound, with the most commonly cited one being Darwin's Finches on the Galapagos Islands, where an ancestral species of coastal, mainland finch migrated to settle in small populations on many of these volcanic islands. As time progressed, these subpopulations diverged from one another in size, nesting, food preferences, beak sizes, and coloration. Random mutations and genetic drift provided the raw material for natural selection to mold these subgroups in different ways to the diversity of environments presented on the islands. Today, over a dozen species of finch exist in the Galapagos from a single South American ancestor.

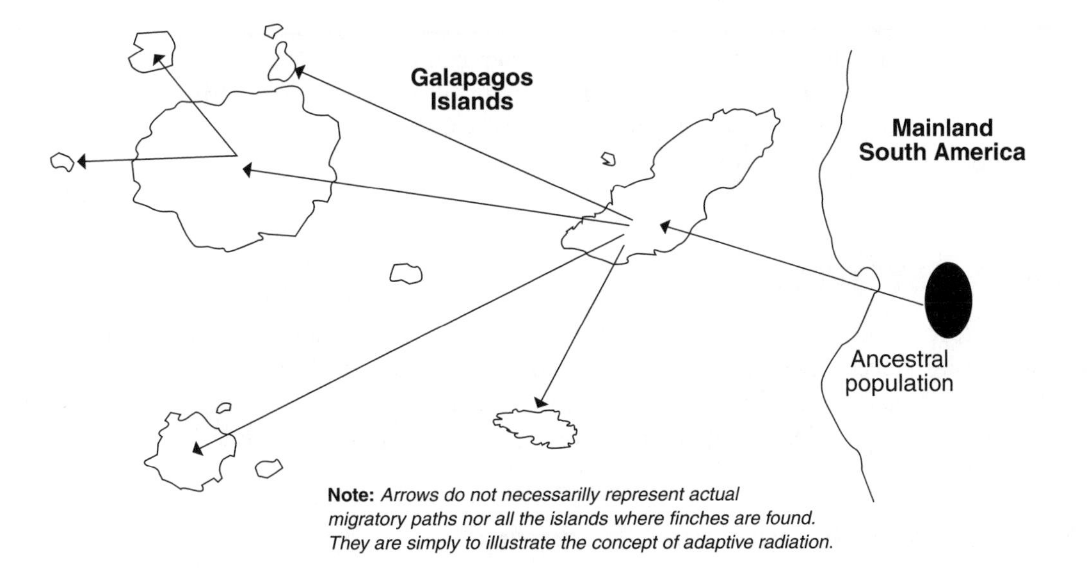

Sexual Selection

Although usually considered a subset of natural selection, this form of selection requires a separate mention. Sexual selection relies not on environmental factors for determining advantageous genotypes, but rather on the relative attractiveness of one potential mate over another. Most mates are drawn to one another by certain coloration patterns, songs, or oversized appendages (e.g., a peacock's tail). There are probably no sexually reproducing species on Earth that mate in a totally random manner, and this kind of selection guarantees that allele frequencies will depart from Hardy-Weinberg predictions. For the GRE, remember that the process of sexual selection also means that there will be *fewer heterozygotes in any given population* than expected by Hardy-Weinberg equilibrium laws: Nonrandom mating drives "purebred" homozygotes (both recessive and dominant) to mate with like purebreds, especially if one of these genotypes (homozygous dominant or homozygous recessive) confers a definite advantage in the population's current habitat. The reason for this is simple: Heterozygous mating allows the possibility of offspring being born with the unfavorable genotype, thereby decreasing the overall fitness of the population.

Keep in mind, of course, that there is a trade-off implicit in an organism making itself so attractive that the attractiveness becomes a hindrance to survival! A peacock's tail must be large enough to maximally attract, yet lightweight enough to allow the peacock to escape from predators. Mating songs need to be short enough to attract mates without making the caller an easy homing beacon for predators nearby. You may see this on the GRE in the context of the **runaway-selection model**, proposed by R. A. Fisher, whose idea describes the potential for positive reinforcement of an "attractive trait," whereby the trait's frequency and expression in the population increase in a runaway fashion until an equilibrium is reached between the mating benefit and the survival risk of expressing the trait to such a high degree.

Note: Genetic drift and gene flow, described earlier, can affect the changes in a population's gene pool that take place independent of natural selection. In addition, speciation mechanisms such as auto- and allopolyploidy can cause new populations and, hence, new gene pools to arise in the midst of established groups.

MATING SYSTEMS

In some form, sexual selection is present in all sexually reproducing species. Yet not all of these species have highly ornamented (decorated) males, and some are impossible to differentiate (male versus female) from a simple external view. It is evident, then, that the runaway-selection model (described previously) does not apply in many cases.

Monogamy, Polygamy, and Polyandry

Monogamous species are characterized by males that pair with one (and only one) female during the breeding season. In most cases, these male-female mating pairs remain together for their entire lives. Many reasons have been proposed for why a monogamous relationship forms between two organisms:

- The costs of *not* remaining monogamous are too great given the habitat. Perhaps there is a low density of females, so it is very costly in terms of energy use (resource use) to search for a mate every season. Or, overall fitness of males is much greater if they remain at a single nest-site to defend eggs and rear the young.

- The male's fitness is increased if it can guard against insemination of females by rival males, since the females will be carrying only eggs fertilized by the one male.

- If not competing against rival males for the available females, a male can more easily defend its territory.

Depending on the habitat, some males and females opt for a **polygynous** relationship, where males "keep" and mate with multiple females, either all at the same time or in some sort of order as the males move through the group (called **sequential polygyny**). Given the fact that so many species allocate very little time and few resources to raising their young, polygyny seems like the logical choice for most males: They are able to fertilize many more females and leave more young than they could in a monogamous relationship.

The more resources available to males (food for energy, space for territory, and females for mating), the more likely it is that the males will form a polygynous mating system.

Polyandry is a mating system in which females have several male partners. This can occur in sequentially hermaphroditic species as described above, or in instances where females are in high density and males are not. Pipefish are elongated and small, saltwater fishes that live in long beds of undersea grass: Each female mates with many males during a very short breeding season, with the most highly ornamental and colorful females attracting the greatest number of males.

There is a great deal of literature on mating systems and sexual selection, and many hypotheses abound for why males and females pair up the way they do. For the GRE, it is unlikely that you will see anything else on this topic other than the three basic systems described above.

Polyandry requires a low density of males, and high levels of resources for females.

SOCIAL SYSTEMS

A massive amount of material has been written on social structure within animal populations, ranging from well-known works such as Jane Goodall's studies of primates to Edward O. Wilson's research on ants. Suffice it to say that for the GRE you should know the "big" idea; that is, many animal species are organized by complex dominance hierarchies and social systems that allow certain dominant animals to reproduce more often and more successfully than less dominant ones. Social structures tend to maximize the overall fitness of a population.

CHAPTER TWENTY-NINE

The Hardy-Weinberg Law and Its Use

BACKGROUND AND APPLICATIONS

The *Hardy-Weinberg Equilibrium Law* was named for two scientists, G.H. Hardy and W. Weinberg, in 1908. The Law states that sexual recombination alone cannot change the relative frequencies of alleles or genotypes in a population over the course of repeated matings. In other words, meiosis, which acts to rearrange alleles every generation, has no effect on the overall gene pool of a population. The only way the allelic frequencies of a population can be changed is if the population fails to *meet one or more of the five conditions* specified below.

The Law characterizes a *nonevolving population* of individuals as one in which the following occur:

1. No emigration or immigration
2. Large population size
3. No net mutations
4. Random mating
5. No natural selection

In reality, this type of population could never be found in nature, so in an indirect fashion, the Hardy-Weinberg Law tells us about the major phenomena which drive evolution. If the frequencies of alleles that we *expect* a population to show are not the frequencies that are actually shown, we can conclude that the population is evolving.

Imagine a population in which 80 percent of the individuals have the ***A*** allele at one gene locus on a chromosome, and 20 percent have the ***a*** allele in the same locus. In a population set up with the conditions above, how will sexual recombination affect the allele frequency? The probability of an offspring getting two ***A*** alleles (one from each parent) would be $0.8 \times 0.8 = 0.64$ (or 64%). The probability of an offspring getting two ***a*** alleles would be $0.2 \times 0.2 = .04$ (or 4%). The probability of creating a heterozygous (***Aa***) individual would be $0.8 \times 0.2 = 0.16$, multiplied by two because both the sperm and egg can donate *either* the ***A*** or the ***a***, so the probability here is 0.32 (32%).

Hardy and Weinberg labeled the *frequency* of the dominant allele **p** and the frequency of the recessive allele **q**. Therefore, in our parent generation, $p = 0.8$ and $q = 0.2$. Since the frequency of all possible alleles put together must equal 1 (or 100%), $\mathbf{p + q = 1}$.

Using the math from above, the probability (from above) of generating an *AA* zygote is p^2 (0.64). The frequency in the offspring of an *aa* zygote would be q^2 (0.04), and the frequency of heterozygotes in the population would be $2pq = 0.32$ (from above). *Because all offspring frequencies must add up to 1, we get for the offspring of p and q parents:*

$$p^2 + 2pq + q^2 = 1 \ (0.64 + 0.32 + 0.04 = 1, \text{ in this case})$$

As you can see from this equation, the frequency of the dominant allele (p) in the *next set* of offspring would be $0.64 + (0.5)(0.32) = 0.8$, which is the same frequency of the p allele in the original parents. The (0.5)(0.32) term came from the fact that *half* of each heterozygote's genotype is made up of the dominant allele.

Therefore, without evolutionary pressures in an ideal population, genotype and allele frequencies remain in equilibrium—hence, the term Hardy-Weinberg equilibrium!

Summary of Hardy-Weinberg Equations for the GRE
$p + q = 1$ (Used for frequencies of single alleles) $p^2 + 2pq + q^2$ (Used for frequencies of individuals expressing a particular trait)

So why do we need this math, and what does the law tell us? It allows us to calculate frequencies of alleles in gene pools if we know the frequencies of genotypes and vice versa. This is important in studying not only population demographics, but also how populations change over time. Even if you are given only phenotypes, Hardy-Weinberg equations can be used to deduce genotypes and allele frequencies.

Note: If a question states "the frequency of an allele is _____," it means you're dealing directly with p or q. If you are asked about the frequency of a disease itself or a particular genotype, you're dealing with p^2 or q^2, because it takes two alleles to make a trait or a genotype. p^2 represents the frequency of homozygous dominant individuals, q^2 represents the frequency of homozygous recessive individuals, and 2pq represents the frequency of heterozygous individuals.

CHAPTER THIRTY

Conclusion—The Evolution of Life on Earth

Perhaps the most daunting task evolutionary biologists face is how to understand and correctly build the relationships of one species to another. In other words, how closely related are two given species and what criteria should be used to show that they are closely linked? As biotechnology gets more advanced, studies of species' DNA will help biologists understand the relationships between species on Earth better than they ever have before.

SYSTEMATICS AND PHYLOGENY

The term **classification** is a fairly broad one, meaning to put things in some sort of order. Even before Carolus Linnaeus in the 1700s, humans were attempting to piece together the diversity of life on Earth in a way that showed relationships among species.

- **Phylogeny**: The relationship of one species to another.
  You might say to someone, "The fact that Archaeopteryx had both birdlike features and reptilian features helps us understand its phylogeny, its relationship to both birds and reptiles."

- **Systematics**: The process of classifying organisms based on their phylogeny.
  The way we classify a species depends on our understanding of its phylogeny, something that changes all the time as new data (e.g., fossils) are discovered and new genetic techniques arise.

You may very well see on the test a mention of **cladistics**. Cladistics is the grouping of species determined to be related because they all share some distinguishing novel feature. For example, at a certain point in the history of life on Earth, the hard-shelled amniote egg was a new evolutionary development, something that allowed entire groups of organisms to take over and expand into certain niches. The species using this new feature become a **clade** and form a branch point on a large "tree of life" that depicts evolutionary history. In this branching **cladogram**, each branch point represents the starting point of a new group of organisms possessing a new trait that has never before arisen.

The main problem with cladistics is that the traits chosen to show relatedness between species can be quite arbitrary. Why choose an amniote egg to differentiate birds, lizards, and mammals from amphibians and fishes, when one could choose scaly or thick, waterproof skin? As the choices for the cladogram change, the cladogram can itself change.

CONVERGENT AND DIVERGENT EVOLUTION

Two mechanisms exist that result in species having external structures and internal systems similar to each other. In some cases, organisms of different species express similar characteristics because they are descended from the same common ancestor. In other cases, similar traits evolve in unrelated organisms simply because they live in comparable environments and natural selection has shaped their bodies in similar ways.

Divergent evolution refers to the process in which many species arise over time from a common ancestor. Because these species share the *same genetic foundation*, their appendages and body systems often share the same underlying structures, depending upon how much they have diverged from the **common ancestor**. This process of divergent evolution is exactly how adaptive radiation works (described earlier), and results in related species sharing what are called **homologous structures**. Homologous structures are characteristics that are structurally related to each other yet serve different purposes, because the species being compared live in different kinds of habitats. An often cited example is the flipper of a dolphin and the hand of a primate: Both have nearly the same underlying bone structure, because both dolphins and primates are mammals arising from a common ancestor, yet they serve vastly different purposes and are able to perform very distinct functions.

Convergent evolution occurs when species with *no recent common ancestor* and not much genetic similarity evolve nearly identical structures because they happen to live in very similar environments. This is extremely strong evidence to suggest that nature *has a limited number of solutions for particular "problems"* faced by organisms in their habitats, and that if a variation is present due to the right chance mutations, nature will select the same types of variants even across unrelated and geographically distant species. Both whales and fish have fins and streamlined bodies for maximal speed, yet they share only the most distant ancestor. Their environment selected these similar traits because others simply would not work in that setting: Long, thin bodies generally have fins and all are streamlined. The alternative would be a smaller, radially symmetric body. Structures shared due to convergent evolution (so-called because unrelated organisms have "converged" on similar traits) are called **analogous structures.**

Current work in biotechnology has been focusing more on homologies of protein structure across species than on macro structures like limbs and fins. Analysis of **protein homology** gives a much clearer indication of species relatedness than any comparison of overall anatomy.

PRACTICE QUESTIONS FOR PART IV

1. In some populations of organisms, the number of adults in the population may be greater than the number of adults actually contributing genes to the next generation. This situation would

   (A) increase the calculated genetic drift of the population because the size of the population is effectively smaller than it seems.
   (B) increase the number of heterozygotes in the population and, thus, the amount of variation within the population.
   (C) contribute to disruptive selection within the population and the formation of two separate subspecies.
   (D) decrease the probability that the frequency of certain alleles possessed by nonmating members of the population would drift toward zero.
   (E) result from monogamous relationships between males and females of the population.

2. Among certain populations of salamander in California exist subspecies that have overlapping geographic ranges and frequently encounter each other, although mating between the populations does not always occur. These subspecies would be considered

   (A) allopatric.
   (B) sympatric.
   (C) parapatric.
   (D) hybrids.
   (E) sexually isolated.

3. When organisms that are homozygous for a particular beneficial allele gradually replace those who possess more harmful alleles within a population, this is known as

   (A) disruptive selection.
   (B) Hardy-Weinberg equilibrium.
   (C) stabilizing selection.
   (D) unstable equilibrium.
   (E) directional selection.

4. Inbreeding can sometimes bring an inbred population to a genetic equilibrium in which its fitness is increased over the fitness that existed before inbreeding began. A reasonable explanation for this is that

   (A) inbreeding causes a initial decline in fitness due to the loss of heterozygosity in the affected population which then rises again as heterozygotes increase in number.
   (B) self-fertilization increases the mean fitness of many organisms because it results in the loss of variety among the phenotypic features of members of the population.
   (C) new allele frequencies after the population stabilizes will reflect only successful genotypes.
   (D) deleterious alleles masked by heterozygosity in earlier populations will be exposed and eliminated due to inbreeding.
   (E) most homozygous inbred populations are less fit than other, more variable populations.

5. Both insects and birds use wings to fly, yet at best they share a distant evolutionary ancestor. In reality, the evolutionary lineages of insects and birds diverged perhaps as long as 400,000,000 years ago. The feature of wings that both birds and insects share is considered

    (A) homologous.
    (B) analogous.
    (C) divergent.
    (D) a mutation.
    (E) polygnous.

6. A biome that experiences all four seasons—winter, spring, summer, and fall—and is characterized by warm, moist springs and summers is the

    (A) tropical deciduous forest.
    (B) taiga.
    (C) temperate deciduous forest.
    (D) chapparal.
    (E) tundra.

7. Photoperiod, or day length, is a critical factor for many plants and animals, as it regulates metabolic and growth processes. In northern alpine forests, many species show a variation in their metabolic response to changes in day length. Northern species are genetically programmed to slow down their metabolism more quickly than southern species as day length shortens and winter approaches. This geographic variation is known as

    (A) a cline.
    (B) gametic disequilibrium.
    (C) genetic drift.
    (D) selective photoperiodism.
    (E) a balanced polymorphism.

8. The move onto land by animals and their success in a variety of biomes could be accomplished only in stages as certain novel traits arose. These traits necessary for the proliferation of life on land included all of the following EXCEPT

    (A) muscular limbs to resist gravity.
    (B) the amniote egg for embryonic development.
    (C) homeothermy for effective temperature regulation.
    (D) moist skin for diffusion and gas exchange.
    (E) fur for added warmth and protection.

Questions 9–10

Consider the following blood group data taken from a population in Hardy-Weinberg equilibrium with respect to the alleles responsible for different blood factors. All individuals in the population possess two different blood factors, each coded for by a dominant allele and a recessive allele. For the first blood factor, allele R is dominant to allele r, so both RR and Rr individuals test as blood type R, while rr individuals test as blood type r. For the second blood factor, allele F is dominant to allele f, so both FF and Ff individuals test as blood type F, while ff individuals test as blood type f. The frequencies observed for blood type in the population are as follows:

Type	Frequency
RF	0.60
Rf	0.15
rF	0.24
rf	0.01

9. The frequency of the r allele in this population is _____, while the frequency of the F allele in the population is _____.

    (A) 0.60, 0.50
    (B) 0.25, 0.25
    (C) 0.50, 0.60
    (D) 0.25, 0.84
    (E) 0.50, 0.40

10. Given the information above, all of the following statements concerning this population are true EXCEPT

    (A) there is no mutation between the blood factor alleles.
    (B) there is significant migration between this population and others.
    (C) the population is large in size.
    (D) there is no positive selection for the R allele.
    (E) mating is random.

11. Parthenogenesis differs from hermaphroditism as a means of reproduction because

    (A) parthenogenesis is a form of sexual reproduction while hermaphroditism is a form of asexual reproduction.
    (B) parthenogenesis usually involves the production of haploid offspring from unfertilized eggs while hermaphroditism produces diploid offspring from a sperm and egg union.
    (C) hermaphroditism cannot occur without meiosis, while parthenogenesis can.
    (D) hermaphroditism is advantageous in more stable environments where mates can be easily found, while parthenogenesis is advantageous in environments where mates are not easily found.
    (E) homeotic genes are not involved in parthenogenetic development, yet play a crucial role in the development of hermaphroditic young.

Questions 12–15

(A) clade
(B) hybrid zone
(C) cline
(D) ecological niche
(E) phylogeny

12. The history of the descent of a group of organisms from common ancestors.

13. A region in which genetically distinct populations come into contact with each other and produce some offspring of mixed ancestry.

14. The combination of all relevant environmental variables in which a species or population lives.

15. The set of species descended from a particular ancestral species.

ANSWERS

1. (A)

In some populations, one male may mate with many females, especially when members of the population live as harems made of a dominant male and several females. Because only the dominant males end up passing on their genes to the next generation, the population size is effectively smaller than it would be if all males contributed their genes to the gene pool. Thus, genetic drift increases as it typically does with small populations. Genetic drift tends to decrease the number of heterozygotes in a population because it reduces the genetic variation within a population, so choice (B) is incorrect. Because genetic drift weeds out heterozygotes, it allows the population to drift toward the more successful homozygous dominant individuals (called directional selection), so (C) is also wrong. And (D) is out because many alleles possessed primarily by nonmating members will drift toward a frequency of zero within the population over several generations, since these members are not passing on these alleles. Monogamous relationships in (E) do not usually result in a situation where only a few males pass on the genes for the next generation.

2. (B)

Sympatric species are those species whose geographic ranges overlap and who frequently encounter each other. Think of *sym-* meaning "same" and *-patric* meaning "land"—these species occupy the same land. Allopatric species, choice (A), live in distant habitats and will not meet, whereas parapatric species, choice (C), overlap geographic ranges only at their borders, where hybrid species can sometimes be formed.

3. (E)

Movement in one phenotypic direction over time is defined as directional selection, where one extreme phenotype is favored over all other possible phenotypes given a particular environment. Disruptive selection, choice (A), weeds out intermediate phenotypes, favoring extremes; (C) stabilizing selection, weeds out extreme phenotypes allowing the majority of members within the population to converge on the intermediate physical feature (i.e., very light and very dark fur color may be less advantageous than a grayish, intermediate color).

4. (D)

Just like the alternation of generations in plant life cycles may help to weed out unfit haploid genotypes, inbreeding depression (as it is called when the fitness of a population declines due to inbreeding) may wipe out the most unfit alleles, allowing the population to rebuild with a stronger overall gene pool. Inbreeding, however, is generally bad for a population because it greatly decreases the genetic variability of the population, making its members susceptible to even minor changes in the environment.

5. (B)

Analogous traits are defined as characteristics that have a similar structure and function in two different organisms that are not evolutionarily related. Choice (A), homologous structures, refers to traits similar in structure because they derive from common ancestry (e.g., a chicken's wing and a human hand).

6. (C)

The characteristics mentioned in the question stem describe a temperate deciduous forest, such as that which exists in the northern United States. These biomes experience a spring and summer growing season, cold winters, and plenty of moisture. All other answer choices are incorrect as they include biomes with climates and seasonal variability (markedly different) from the biome described in the question.

7. (A)

A cline can be defined as a gradual change in the expression of a trait over a particular geographic range. In this case, northernmost species slow down their metabolisms more rapidly than southernmost species as day length shortens. This change as one moves from the south to the north is defined as a cline.

8. (D)

Although moist skin is necessary for diffusion of nutrients and gases, it was not a novel (new) trait for land animals. Land animals do struggle to maintain moist membranes in their lungs and on their body surface; yet the gills of fishes and mollusks (as well as the tissues of a hydra) must be kept moist for effective diffusion. All other traits mentioned in the answer choices describe novel adaptations to life on land that had not previously been present in aquatic life.

9. (C)

This is a fairly difficult question with some trick answer choices. You might have been tempted to choose (D) if you simply added the frequency of individuals expressing blood type *r* (0.24 + 0.01) and did the same for blood type *F* (0.24 + 0.60). This type of addition works for the recessive *r* allele, but it does *not* work for the *F* allele. The frequency of the *r* trait, q^2, is (*rF* + *rf*), or 0.24 + 0.01 = 0.25; thus, the frequency of the *r* allele, *q*, is 0.5. However, individuals with blood type F can be either homozygous or heterozygous; therefore, the frequency of *RF* individuals in the question includes both *FF* individuals as well as *Ff* ones. So you cannot simply add 0.60, the *RF* frequency, to the *rF* frequency to calculate the frequency of the *F* allele. You must first find the frequency of the recessive *f* allele and then subtract that from 1.00 in order to find the frequency of the *F* allele. The frequency of the *f* trait is 0.15 + 0.01 = 0.16 so the frequency of the *f* allele is 0.4. The allele frequency of *F* is 1.00 - freq(*f*) = 0.60. So choice (C) is correct.

10. (B)

Hardy-Weinberg equilibrium requires no mutation, large population size, no natural selection, and random (panmictic) mating—but no migration from other populations with different allele frequencies. Thus, (B) is the correct answer.

11. (B)

Parthenogenesis is the production of offspring from unfertilized eggs. Offspring produced in this manner are usually haploid and form their own eggs by mitosis, not by meiosis. While parthenogenesis is an asexual form of reproduction, hermaphroditism is a sexual form of reproduction. Even though both gametes can, though rarely do, come from the same parent, the sperm and egg cells have each been produced in separate meiotic events. Therefore, hermaphroditism involves the genetic contributions of two parents, even when a hermaphrodite mates with itself. All other answer choices are incorrect regarding differences between these modes of reproduction.

12. (E)

The evolutionary history of an organism or species is its phylogeny. The phylogeny of a species traces its ancestry back in time.

13. (B)

Offspring of mixed ancestry are known as hybrids, and a hybrid zone is the area of overlapping geographic ranges between two subspecies where viable offspring can sometimes be produced from interbreeding populations.

14. (D)

Also known as the "role" each organism plays in its habitat, an organism's niche is the combination of how the organism interacts with the important environmental factors surrounding the organism (e.g., predators, prey, climate, resource availability, etc.)

15. (A)

A cladogram is an evolutionary tree that groups related organisms by shared features, so a clade is a set of species that descends from the same common ancestor and shares similar features and body structures.

Part V

PRACTICE TEST AND EXPLANATIONS

This section contains a full-length practice test. Familiarize yourself with the directions, then go on to take the practice test. Please note that the format of this practice test varies slightly from that of the actual GRE Biology exam.

In addition, note that some questions will introduce terminology or vocabulary that was not introduced in this study guide. These new terms are explained in detail in the answer explanations. Part of the design of this practice test is to teach you to be able to conceptually reason through questions based on what you have learned.

Before taking this test, find a quiet room where you can work uninterrupted for three hours. Make sure you have a comfortable desk, several No. 2 pencils, and a watch, so that you can time yourself. Use the answer grid provided to record your answers (You can cut it out or photocopy it.)

Once you start to take the test, don't stop until you've finished. You have two hours and 50 minutes. You'll find the answer key and explanations following the test.

Answer Sheet

1 A B C D E
2 A B C D E
3 A B C D E
4 A B C D E
5 A B C D E
6 A B C D E
7 A B C D E
8 A B C D E
9 A B C D E
10 A B C D E
11 A B C D E
12 A B C D E
13 A B C D E
14 A B C D E
15 A B C D E
16 A B C D E
17 A B C D E
18 A B C D E
19 A B C D E
20 A B C D E
21 A B C D E
22 A B C D E
23 A B C D E
24 A B C D E
25 A B C D E
26 A B C D E
27 A B C D E
28 A B C D E
29 A B C D E
30 A B C D E
31 A B C D E
32 A B C D E
33 A B C D E
34 A B C D E
35 A B C D E
36 A B C D E
37 A B C D E
38 A B C D E
39 A B C D E
40 A B C D E
41 A B C D E
42 A B C D E
43 A B C D E
44 A B C D E
45 A B C D E
46 A B C D E
47 A B C D E
48 A B C D E
49 A B C D E
50 A B C D E
51 A B C D E
52 A B C D E
53 A B C D E
54 A B C D E
55 A B C D E
56 A B C D E
57 A B C D E
58 A B C D E
59 A B C D E
60 A B C D E
61 A B C D E
62 A B C D E
63 A B C D E
64 A B C D E
65 A B C D E
66 A B C D E
67 A B C D E
68 A B C D E
69 A B C D E
70 A B C D E
71 A B C D E
72 A B C D E
73 A B C D E
74 A B C D E
75 A B C D E
76 A B C D E
77 A B C D E
78 A B C D E
79 A B C D E
80 A B C D E
81 A B C D E
82 A B C D E
83 A B C D E
84 A B C D E
85 A B C D E
86 A B C D E
87 A B C D E
88 A B C D E
89 A B C D E
90 A B C D E
91 A B C D E
92 A B C D E
93 A B C D E
94 A B C D E
95 A B C D E
96 A B C D E
97 A B C D E
98 A B C D E
99 A B C D E
100 A B C D E

101	Ⓐ Ⓑ Ⓒ Ⓓ Ⓔ	126	Ⓐ Ⓑ Ⓒ Ⓓ Ⓔ	151	Ⓐ Ⓑ Ⓒ Ⓓ Ⓔ	176	Ⓐ Ⓑ Ⓒ Ⓓ Ⓔ
102	Ⓐ Ⓑ Ⓒ Ⓓ Ⓔ	127	Ⓐ Ⓑ Ⓒ Ⓓ Ⓔ	152	Ⓐ Ⓑ Ⓒ Ⓓ Ⓔ	177	Ⓐ Ⓑ Ⓒ Ⓓ Ⓔ
103	Ⓐ Ⓑ Ⓒ Ⓓ Ⓔ	128	Ⓐ Ⓑ Ⓒ Ⓓ Ⓔ	153	Ⓐ Ⓑ Ⓒ Ⓓ Ⓔ	178	Ⓐ Ⓑ Ⓒ Ⓓ Ⓔ
104	Ⓐ Ⓑ Ⓒ Ⓓ Ⓔ	129	Ⓐ Ⓑ Ⓒ Ⓓ Ⓔ	154	Ⓐ Ⓑ Ⓒ Ⓓ Ⓔ	179	Ⓐ Ⓑ Ⓒ Ⓓ Ⓔ
105	Ⓐ Ⓑ Ⓒ Ⓓ Ⓔ	130	Ⓐ Ⓑ Ⓒ Ⓓ Ⓔ	155	Ⓐ Ⓑ Ⓒ Ⓓ Ⓔ	180	Ⓐ Ⓑ Ⓒ Ⓓ Ⓔ
106	Ⓐ Ⓑ Ⓒ Ⓓ Ⓔ	131	Ⓐ Ⓑ Ⓒ Ⓓ Ⓔ	156	Ⓐ Ⓑ Ⓒ Ⓓ Ⓔ	181	Ⓐ Ⓑ Ⓒ Ⓓ Ⓔ
107	Ⓐ Ⓑ Ⓒ Ⓓ Ⓔ	132	Ⓐ Ⓑ Ⓒ Ⓓ Ⓔ	157	Ⓐ Ⓑ Ⓒ Ⓓ Ⓔ	182	Ⓐ Ⓑ Ⓒ Ⓓ Ⓔ
108	Ⓐ Ⓑ Ⓒ Ⓓ Ⓔ	133	Ⓐ Ⓑ Ⓒ Ⓓ Ⓔ	158	Ⓐ Ⓑ Ⓒ Ⓓ Ⓔ	183	Ⓐ Ⓑ Ⓒ Ⓓ Ⓔ
109	Ⓐ Ⓑ Ⓒ Ⓓ Ⓔ	134	Ⓐ Ⓑ Ⓒ Ⓓ Ⓔ	159	Ⓐ Ⓑ Ⓒ Ⓓ Ⓔ	184	Ⓐ Ⓑ Ⓒ Ⓓ Ⓔ
110	Ⓐ Ⓑ Ⓒ Ⓓ Ⓔ	135	Ⓐ Ⓑ Ⓒ Ⓓ Ⓔ	160	Ⓐ Ⓑ Ⓒ Ⓓ Ⓔ	185	Ⓐ Ⓑ Ⓒ Ⓓ Ⓔ
111	Ⓐ Ⓑ Ⓒ Ⓓ Ⓔ	136	Ⓐ Ⓑ Ⓒ Ⓓ Ⓔ	161	Ⓐ Ⓑ Ⓒ Ⓓ Ⓔ	186	Ⓐ Ⓑ Ⓒ Ⓓ Ⓔ
112	Ⓐ Ⓑ Ⓒ Ⓓ Ⓔ	137	Ⓐ Ⓑ Ⓒ Ⓓ Ⓔ	162	Ⓐ Ⓑ Ⓒ Ⓓ Ⓔ	187	Ⓐ Ⓑ Ⓒ Ⓓ Ⓔ
113	Ⓐ Ⓑ Ⓒ Ⓓ Ⓔ	138	Ⓐ Ⓑ Ⓒ Ⓓ Ⓔ	163	Ⓐ Ⓑ Ⓒ Ⓓ Ⓔ	188	Ⓐ Ⓑ Ⓒ Ⓓ Ⓔ
114	Ⓐ Ⓑ Ⓒ Ⓓ Ⓔ	139	Ⓐ Ⓑ Ⓒ Ⓓ Ⓔ	164	Ⓐ Ⓑ Ⓒ Ⓓ Ⓔ	189	Ⓐ Ⓑ Ⓒ Ⓓ Ⓔ
115	Ⓐ Ⓑ Ⓒ Ⓓ Ⓔ	140	Ⓐ Ⓑ Ⓒ Ⓓ Ⓔ	165	Ⓐ Ⓑ Ⓒ Ⓓ Ⓔ	190	Ⓐ Ⓑ Ⓒ Ⓓ Ⓔ
116	Ⓐ Ⓑ Ⓒ Ⓓ Ⓔ	141	Ⓐ Ⓑ Ⓒ Ⓓ Ⓔ	166	Ⓐ Ⓑ Ⓒ Ⓓ Ⓔ	191	Ⓐ Ⓑ Ⓒ Ⓓ Ⓔ
117	Ⓐ Ⓑ Ⓒ Ⓓ Ⓔ	142	Ⓐ Ⓑ Ⓒ Ⓓ Ⓔ	167	Ⓐ Ⓑ Ⓒ Ⓓ Ⓔ	192	Ⓐ Ⓑ Ⓒ Ⓓ Ⓔ
118	Ⓐ Ⓑ Ⓒ Ⓓ Ⓔ	143	Ⓐ Ⓑ Ⓒ Ⓓ Ⓔ	168	Ⓐ Ⓑ Ⓒ Ⓓ Ⓔ	193	Ⓐ Ⓑ Ⓒ Ⓓ Ⓔ
119	Ⓐ Ⓑ Ⓒ Ⓓ Ⓔ	144	Ⓐ Ⓑ Ⓒ Ⓓ Ⓔ	169	Ⓐ Ⓑ Ⓒ Ⓓ Ⓔ	194	Ⓐ Ⓑ Ⓒ Ⓓ Ⓔ
120	Ⓐ Ⓑ Ⓒ Ⓓ Ⓔ	145	Ⓐ Ⓑ Ⓒ Ⓓ Ⓔ	170	Ⓐ Ⓑ Ⓒ Ⓓ Ⓔ	195	Ⓐ Ⓑ Ⓒ Ⓓ Ⓔ
121	Ⓐ Ⓑ Ⓒ Ⓓ Ⓔ	146	Ⓐ Ⓑ Ⓒ Ⓓ Ⓔ	171	Ⓐ Ⓑ Ⓒ Ⓓ Ⓔ	196	Ⓐ Ⓑ Ⓒ Ⓓ Ⓔ
122	Ⓐ Ⓑ Ⓒ Ⓓ Ⓔ	147	Ⓐ Ⓑ Ⓒ Ⓓ Ⓔ	172	Ⓐ Ⓑ Ⓒ Ⓓ Ⓔ	197	Ⓐ Ⓑ Ⓒ Ⓓ Ⓔ
123	Ⓐ Ⓑ Ⓒ Ⓓ Ⓔ	148	Ⓐ Ⓑ Ⓒ Ⓓ Ⓔ	173	Ⓐ Ⓑ Ⓒ Ⓓ Ⓔ	198	Ⓐ Ⓑ Ⓒ Ⓓ Ⓔ
124	Ⓐ Ⓑ Ⓒ Ⓓ Ⓔ	149	Ⓐ Ⓑ Ⓒ Ⓓ Ⓔ	174	Ⓐ Ⓑ Ⓒ Ⓓ Ⓔ	199	Ⓐ Ⓑ Ⓒ Ⓓ Ⓔ
125	Ⓐ Ⓑ Ⓒ Ⓓ Ⓔ	150	Ⓐ Ⓑ Ⓒ Ⓓ Ⓔ	175	Ⓐ Ⓑ Ⓒ Ⓓ Ⓔ	200	Ⓐ Ⓑ Ⓒ Ⓓ Ⓔ

GRE Biology Test

Time—170 minutes

200 Questions

Directions: Each question or incomplete statement below is followed by five possible answer choices or completions. Select the correct answer and fill in the corresponding oval on the answer grid.

1. Which of the following INCORRECTLY pairs a metabolic process with its site of occurrence?

   (A) Glycolysis—cytosol
   (B) Citric Acid cycle—mitochondrial membrane
   (C) Electron transport chain—mitochondrial membrane
   (D) ATP phosphorylation—mitochondria
   (E) Oxidative decarboxylation of pyruvate—mitochondria

2. Consider a biochemical reaction A → B, which is catalyzed by the human enzyme AB dehydrogenase. Which of the following statements is true of this reaction?

   (A) The reaction will proceed until the enzyme concentration decreases.
   (B) The reaction will be more favorable at 0°C.
   (C) A component of the enzyme is transferred from A to B.
   (D) The free energy change (ΔG) of the catalyzed reaction is the same as the free energy change for the uncatalyzed reaction.
   (E) AB dehydrogenase will change the equilibrium but not the rate of the reaction to form A from B.

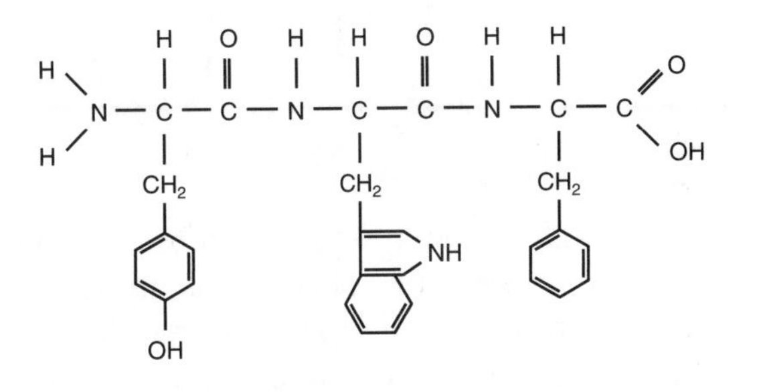

3. The molecule pictured above can be considered a(n)

   (A) Oligonucleotide
   (B) Triglyceride
   (C) Amino acid
   (D) Trisaccharide
   (E) Polypeptide

CONTINUE ON THE NEXT PAGE

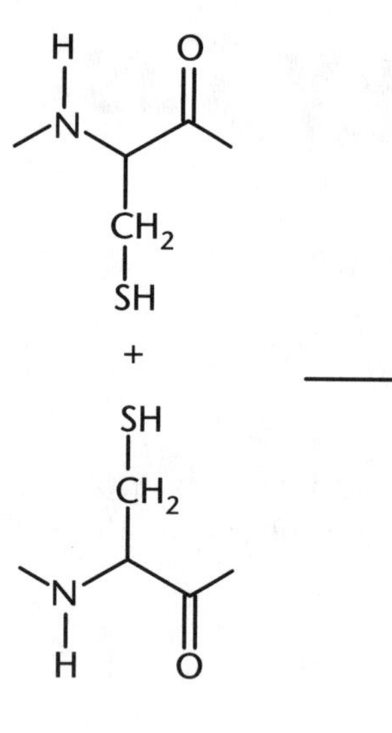

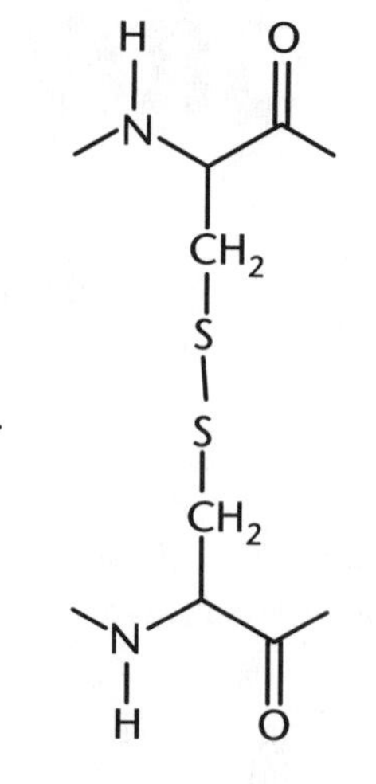

4. The diagram above depicts a common bond formed between two nearby cysteine amino acids within a protein. The two cysteine molecules become oxidized to form cystine. The bond formed between them, consisting of two covalently bonded sulfur atoms, would most likely play a role in a protein's
   (A) primary structure.
   (B) secondary structure.
   (C) tertiary structure.
   (D) quaternary structure.
   (E) enzymatic capabilities.

5. The cancer drug Taxol interferes with the separation of chromosomes during anaphase of mitosis. The likely mechanism of Taxol's action is
   (A) interference with the synthesis of cyclin-Cdk complexes.
   (B) prevention of kinetochore microtubule breakdown.
   (C) excitation of inhibitory transcription factors.
   (D) methylation of select areas of DNA on metaphase chromosomes.
   (E) blockage of spindle formation from microtubule organizing centers.

6. In tabby cats, black coloration is caused by the presence of a particular allele on the X chromosome. A different allele at the same locus can result in orange color. The heterozygote calico cat, however, with splotches of both black fur and orange fur, is due to
   (A) polar body formation and inactivation.
   (B) crossing over during gamete formation.
   (C) the formation of thymine dimers.
   (D) segregation of alleles in different cell lines.
   (E) Barr body formation.

7. The DNA of a mouse is analyzed to reveal the presence of alleles resulting in brown fur color. The mouse is heterozygous, Bb, yet is completely colorless in phenotype. This can be explained as an occurrence of
   (A) Epigenesis
   (B) Epistasis
   (C) Incomplete dominance
   (D) Pleiotropy
   (E) Codominance

8. On a chromosome, genes A and W are five map units apart, W and G are 15 units apart, E and G are 12 units apart, and W and E are three units apart. Between which two genes would you expect the highest recombination frequency?

   (A) A and W
   (B) W and G
   (C) E and G
   (D) W and E
   (E) A and G

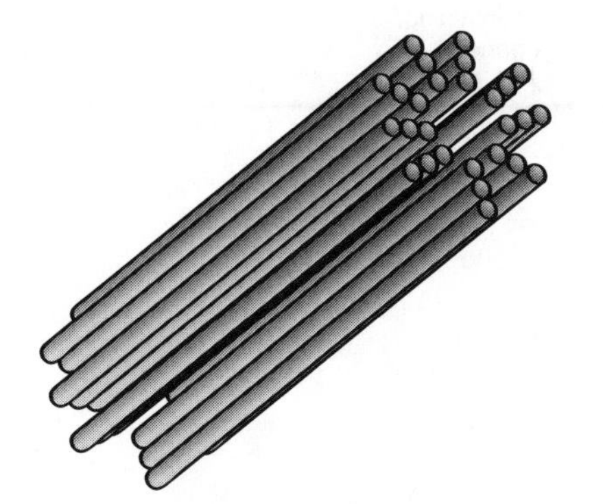

9. The structure pictured above shows the arrangement of microtubules that can be found within

   (A) Cilia
   (B) Flagella
   (C) Basal bodies
   (D) Sarcomeres
   (E) Motor kinesins

10. Calmodulin is an intracellular signaling molecule that complexes with calcium ions in order to activate certain protein kinases. These protein kinases can be activated by the direct binding of Ca^{2+} / calmodulin complexes or by autophosphorylation of the kinase subunit. In the unphosphorylated state, one would expect the kinase subunit to

    (A) be constitutively active.
    (B) phosphorylate target proteins that are nearby.
    (C) be inactive.
    (D) be unable to bind calmodulin.
    (E) act as a transducer of calcium signals.

11. The sex-determination pathway in *Drosophila* results in the formation of either a male or female because of the presence or absence of key proteins coded for by a single gene known as *double-sex*. The *double-sex* gene codes for the protein dsx^f in order to build a female fruit fly, but it codes for the related protein dsx^m in order to build a male fruit fly. It is therefore appropriate to conclude that

    (A) the two protein products of the double-sex gene are created by alternative RNA splicing in male and female fruit flies.
    (B) the protein products differ in males and females because females have two copies of the *double-sex* gene and males only have one.
    (C) RNA polymerase in females reads the *double-sex* gene differently than it does in males, altering the transcription of the sex-determining gene.
    (D) the dsx^f and dsx^m proteins must have different signal sequences, allowing them to go through alternative post-translational modifications.
    (E) the proteins coded for by the *double-sex* gene have little to do with the presence of X or Y chromosomes in the cells of *Drosophila*.

CONTINUE ON THE NEXT PAGE

12. Which of the following compounds are both products of the light reaction in photosynthesis and reactants in the Calvin cycle?

    (A) Pyruvate and acetyl-CoA
    (B) NADPH and O_2
    (C) $NADP^+$ and CO_2
    (D) $NADP^+$, ATP, and CO_2
    (E) NADPH and ATP

13. The primary function of fermentation is to

    (A) generate ATP for the cell.
    (B) synthesize glucose.
    (C) regenerate NAD^+.
    (D) synthesize ethanol or lactic acid.
    (E) add to the total amount of ATP produced by cellular respiration.

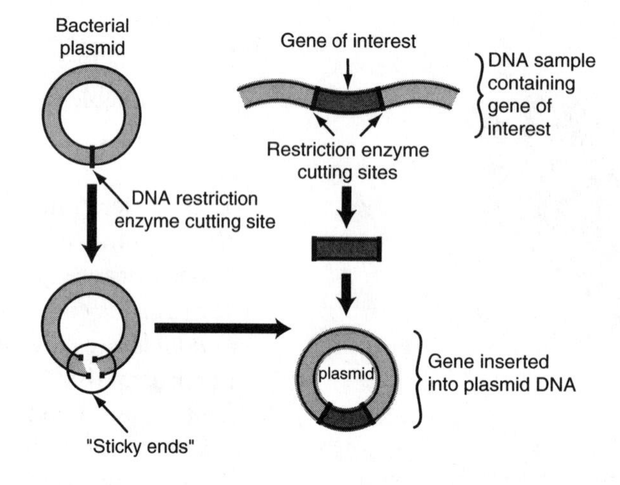

14. The diagram above shows the process used to insert a foreign gene into a bacterial plasmid, thereby transforming the bacteria into an organism that produces proteins typically produced in other organisms. This technique may fail unless the scientist

    (A) uses the entire mammalian stretch of DNA that contains the gene of interest, including the introns and exons that are part of that DNA.
    (B) uses cDNA made from an RNA template of the gene of interest.
    (C) inserts a reporter gene along with the gene of interest in order to measure the success of transformation.
    (D) selects for antibiotic resistance before mixing the transformed plasmid with the bacterial cells.
    (E) uses RNA nucleotides that can be directly converted into protein by the bacteria without first being transcribed.

15. All of the following are considered post-transcriptional modifications that occur in the nucleus EXCEPT
    (A) 5' capping with methylated guanines.
    (B) the addition of a 3' poly-adenine tail.
    (C) the excision of introns from mRNA via spliceosome formation.
    (D) mRNA attachment to polyribosomes.
    (E) stabilization of mRNA by snRNPs.

16. "DNA fingerprinting" uses RFLPs and gel electrophoresis in order to map the genetic differences between two individuals based upon
    (A) residue left from their fingerprints at a crime scene.
    (B) slight differences in the lengths of pieces of their DNA after restriction enzyme digestion.
    (C) single-nucleotide differences within genes for similar traits.
    (D) differential gene expression within the cells of the two individuals.
    (E) the existence of tissue-specific promoter sequences within short segments of their DNA.

17. In the cross between two individuals heterozygous for two different traits, it is expected that a 9:3:3:1 ratio of phenotypes will occur in the offspring. Yet, a ratio of 6:6:2:2 might indicate
    (A) Pleiotropy
    (B) Genomic imprinting
    (C) Incomplete dominance
    (D) X-linkage
    (E) Gene linkage

18. The LDL receptor is a cell-surface protein receptor that allows certain cells to pull lipoproteins containing cholesterol out of the bloodstream into their cytoplasm. All of the following are likely true concerning the LDL receptor EXCEPT
    (A) it matures in the Golgi before being transported to the cell membrane within secretory vesicles.
    (B) its production is upregulated in the presence of higher levels of cholesterol in the blood.
    (C) it has polar and non-polar amino acid regions for placement within the cell membrane.
    (D) the receptor protein contains ER-retention signals in order to be pulled back into the cell once bound to blood lipoproteins.
    (E) mutations in the receptor can result in receptors that are synthesized correctly in the ER but fail to migrate to the Golgi.

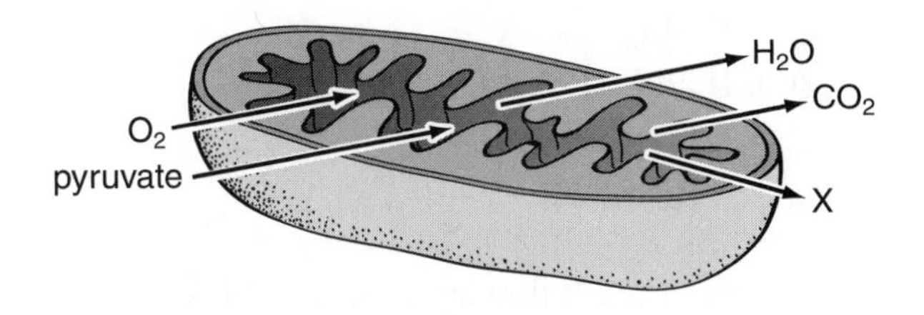

19. In the diagram pictured above, the letter X represents
    (A) Glucose
    (B) $NADP^+$
    (C) ATP
    (D) ADP
    (E) NADH

CONTINUE ON THE NEXT PAGE

20. The compound tosyl-L-phenylalaninechloromethyl ketone (TPCK) has a phenyl group that allows it to fit into the active site of the chymotrypsin and bind covalently to the enzyme. TPCK would be considered a(n)

    (A) Allosteric effector
    (B) Covalent modifier
    (C) Irreversible inhibitor
    (D) Noncompetitive inhibitor
    (E) Competitive inhibitor

21. Rotifers are tiny, pseudocoelomate animals that have complete digestive systems and other specialized organ systems. Which of the following statements are also true of rotifers?

    I. They possess both a mouth and an anus.
    II. Their internal body cavity is entirely enclosed by mesoderm.
    III. Their dorsal nerve cord runs superior to their notochord.

    (A) I only
    (B) II only
    (C) II and III only
    (D) I and III only
    (E) I, II, and III

22. Which of the following would be LEAST able to make bacteria more virulent and infectious?

    (A) the ability to evade nonspecific and specific body defenses
    (B) the secretion of lipoteichoic acid, which contributes to septic shock
    (C) increasing the rate of bacterial replication
    (D) increasing the production of proteins allowing bacterial pili to extend and contact other pili
    (E) the sudden ability of the bacteria to infect several different kinds of species rather than only one

23. Nitrogen-fixation can be carried out by free-living bacteria or by bacterial symbionts living in the roots of plants. The bacteria *E. coli* can use nitrate as an electron acceptor in the electron transport chain that is part of its cell membrane. The nitrate, therefore, serves the same purpose as which molecule in eukaryotes?

    (A) Oxygen
    (B) NADH
    (C) Carbon dioxide
    (D) Hydrogen
    (E) NAD^+

24. Flourescence in situ hybridization (FISH) can be used to detect differences in rRNA sequences among individuals in a population of bacteria. To search for particular rRNA sequences, one would want to use which of the following molecular probes?

    (A) An rRNA probe complementary to the sequence being searched for
    (B) An mRNA probe complementary to the rRNA sequence
    (C) An rDNA probe made of single-stranded genes for rRNA that can hybridize to the rRNA sequences
    (D) An rRNA probe identical to the sequence being searched for
    (E) A double-stranded DNA probe containing the rRNA genes that code for the rRNA sequences being searched for

25. Skin-associated lymphatic tissue (SALT), which is located just underneath the epidermal layer of the skin, is associated with which of the following organ systems?

    (A) Digestive
    (B) Endocrine
    (C) Immune
    (D) Secretory
    (E) Nervous

26. The average distance between genes in *E. coli* is only 120 base pairs. Scientists have assigned functions to approximately 3,500 of the 4,000 genes that exist in this organism. Which of the following is also true of the *E. coli* genome?

    (A) The coding sequences are continuous and lack noncoding introns.
    (B) The genome consists of a series of linear chromosomes.
    (C) RNA genes would be nonexistent as RNA is built solely at the ribosomes.
    (D) There are multiple origins of DNA replication built into the genome.
    (E) The genome contains a centromeric region necessary for chromosome segregation.

27. All of the following statements about transmission along neurons are correct EXCEPT

    (A) The rate of transmission of a nerve impulse is directly related to the diameter of the axon.
    (B) The intensity of a nerve impulse is directly related to the size of the voltage change.
    (C) A stimulus that affects the nerve cell membrane's permeability to ions can either depolarize or hyperpolarize the membrane.
    (D) Once initiated, local threshold depolarization stimulates the propagation of an action potentials down the axon.
    (E) The resting potential of a neuron is maintained by differential ion permeabilities and by the Na-K ATPase pumps.

28. Neurotransmitters characterized as inhibitory would not be expected to

    (A) open K^+ channels.
    (B) open Na^+ channels.
    (C) bind to receptor sites on the post-synaptic membrane.
    (D) open Cl^- channels.
    (E) hyperpolarize the neuron membrane.

29. The molecule pictured below could be best described as

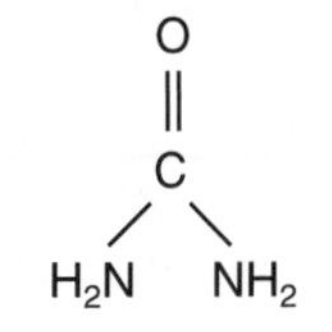

    (A) a building block of a protein.
    (B) an amino acid.
    (C) a fatty acid.
    (D) a waste product of amino acid metabolism.
    (E) the amino portion of an amino acid.

30. The production of nitric oxide (NO) in the body is regulated by a complex of genes that includes the bNOS isoform. A knock-out mouse with a mutant bNOS gene was generated by recombinant molecular techniques. The mutant mouse produced normal bNOS protein except for the identity of amino acid 675 in the protein. Here, the amino acid cysteine was replaced with the amino acid tryptophan, rendering the bNOS protein nonfunctional. Which of the following mutations was responsible for this change?

    (A) Frameshift
    (B) Single base-pair deletion
    (C) Point
    (D) Nonsense
    (E) Antisense

CONTINUE ON THE NEXT PAGE

31. Exposure to high levels of radiation in humans has been demonstrated to cause anemia. The most likely explanation for this is that the radiation damages the

    (A) Blood vessels
    (B) Spleen
    (C) Liver
    (D) Thymus
    (E) Bone marrow

32. Tetrodotoxin, an extremely potent poison produced by the puffer fish, binds tightly to voltage-gated sodium channels and blocks the flow of sodium ions but does not affect either potassium or chloride ion channels. Tetrodotoxin directly blocks which phase of the action potential?

    (A) Depolarization
    (B) Repolarization
    (C) Hyperpolarization
    (D) Neurotransmitter release
    (E) The refractory period

33. Oversecretion of gastric HCl can be treated by severing the vagus nerve in a procedure called a vagotomy, which reduces parasympathetic activity. Which of the following effects is LEAST likely to be caused by a vagotomy?

    (A) Decreased heart rate
    (B) Decreased gastric motility
    (C) Decreased HCl production
    (D) Increased blood pressure
    (E) Increased dilation of the pupils of the eye

34. The respiratory chain electron carrier ubiquinone can accept electrons not only from NADH, but also from a Krebs cycle intermediate called succinate. The presence of succinate in the mitochondrial matrix will

    (A) increase the likelihood that alternative sources of energy, such as fatty acids, will be used to supply the ETC with electrons.
    (B) lead to an increase in the number of hydrogen ions that are transported from the mitochondrial matrix to the intermembrane space.
    (C) lead to an increase in the amount of reduced NADH in the mitochondrial matrix.
    (D) have no effect on the amount of ATP produced.
    (E) decrease the rate of ATP synthesis.

35. Based on the chemiosmotic model, which of the following will occur when isolated mitochondria that have been maintained in a solution of pH 8 until saturation are suddenly transferred into a second solution at pH 4 that contains ADP and P_i (inorganic phosphate)?

    (A) No ATP will be produced.
    (B) There will be a burst of ATP synthesis.
    (C) The rate of ATP production will be unchanged.
    (D) ATP will be converted to ADP and P_i.
    (E) Products of glycolysis will accumulate within the mitochondria.

36. Sarcomere shortening in muscle fibrils requires
    (A) the rapid influx of sodium ions into the cytoplasm after release from the sarcoplasmic reticulum.
    (B) T-tubule shortening after calcium ion release by the sarcoplasmic reticulum.
    (C) ATP release from the sarcoplasmic reticulum and subsequent myosin attachment to actin filaments.
    (D) conformational modifications of the tropomyosin-troponin complex within muscle fibers.
    (E) an increase in the transcription of muscle regulatory proteins such as troponin and actin.

37. In I-cell disease, high concentrations of the hydrolytic enzymes normally found within lysosomes are directed out of the cell into the surrounding extracellular fluid. What statement below best explains the cause of I-cell disease?
    (A) The ribosomal enzymes that are used by the cell to "read" mRNA sequences and build correct enzymes are defective.
    (B) A defect in the ER causes certain sugar residues not to be attached to certain proteins, allowing incorrect targeting of the lysosomal enzymes.
    (C) Sugars on the lysosomes bind to the inner leaflet of the cell membrane and cause the lysosomes to release their enzymes into the surrounding fluid .
    (D) The lysosomal enzymes have been altered during synthesis so that they possess a nuclear localization signal.
    (E) The lysosomal enzymes are built as steroids that can pass easily through the lysosomal and cell membranes.

38. Which of the following is an example of passive immunity?
    (A) A nurse gets stuck with a needle containing blood from a patient infected with tuberculosis (TB), gets a brief flu-like illness, and a few years later tests positive for anti-TB antibodies.
    (B) A child receives a vaccination for polio consisting of inactivated polio virus.
    (C) An adult exposed to a certain influenza strain will not become sick again because he was exposed to that strain as a child.
    (D) A baby born to a woman who has antibodies to hepatitis may be temporarily resistant to the disease.
    (E) A farmer exposed to a virus infecting some of his farm animals does not get ill when exposed to a related virus infecting his family.

39. Which of the following contribute(s) to gas exchange in the alveoli?
    I. Low partial pressure of O_2 in the pulmonary capillaries when compared to the inhaled air
    II. Low partial pressure of CO_2 in the pulmonary capillaries when compared to inhaled air
    III. Presence of surfactant
    (A) I only
    (B) II only
    (C) II and III only
    (D) I and III only
    (E) I, II, and III

CONTINUE ON THE NEXT PAGE

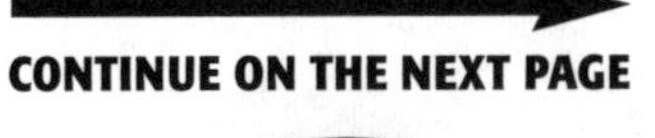

40. All of the following can explain why the removal or rearrangement of groups of cells at pregastrula stages in vertebrate embryos can be tolerated by the embryos, and allow the embryos to develop normally, EXCEPT

    (A) Destruction of small numbers of cells is not harmful because other cells nearby can replace them.
    (B) Rearrangement of cells within a group has no effect because all cells in the pregastrula stage are equivalent in competence.
    (C) Moving cells to other areas of the embryo is tolerated because cells can adapt to their surroundings without being in the presence of an inducer.
    (D) Moving cells that will become organizing inducers to other regions will simply allow cells in that new location to respond according to their distance from the inducer.
    (E) The selective gene loss occurring within the pregastrula cells allows certain groups of cells to be removed without detrimental effect.

41. Junctions between animal cells that resist shearing forces and attach cells through linker proteins are known as

    (A) Tight junctions
    (B) Plasmodesmata
    (C) Anchoring junctions
    (D) Occluding junctions
    (E) Gap Junctions

42. Red blood cells of humans serve essentially the same function as which of the following in an insect?

    (A) Malphigian tubules
    (B) Digestive system
    (C) Tracheal tubes
    (D) Hemolymph
    (E) Open circulatory system

43. Lichens are a mutualistic combination of which of the following organisms?

    (A) Fungus and bacteria
    (B) Bacteria and algae
    (C) Green plant and fungus
    (D) Bacteria and green plant
    (E) Fungus and algae

44. All of the following are true of restriction endonculeases EXCEPT

    (A) They can be used to create recombinant DNA molecules.
    (B) They often cut DNA strands at complementary palindromic sequences.
    (C) The "incisions" left by the enzymes can cause the formation of sticky, overhanging ends.
    (D) They can be used in nuclear transfer techniques for cloning purposes.
    (E) Most are named according to bacteria from which they have been isolated.

45. The haploid structures of fungi through which sexual reproduction occurs are known as

    (A) Gametangia
    (B) Mycelia
    (C) Haustoria
    (D) Mycorrhizae
    (E) Dikaryotic hyphae

46. Which of the following characteristics could most readily distinguish a marsupial from a placental mammal?
    (A) Egg-laying ability
    (B) Fusion of the lower jaw bones in the marsupial
    (C) The presence of a layer of insulating hair in the placental
    (D) Whether or not the young are nourished by their mother's milk
    (E) The degree of fetal development at the time of birth

47. The "multiregionalist" hypothesis of Homo sapiens evolution asserts that modern-day humans evolved in parallel from three separate groups of Homo erectus. Which of the following pieces of evidence would best refute this theory?
    (A) In 1959, Mary Leakey found Australopithecus robustus fossils in the same fossil-bearing strata as Australopithecus boisei.
    (B) In 1925, Raymond Dart found evidence to suggest the existence of two separate Australopithecines, robustus and africanus.
    (C) Fossils of Homo erectus in Asia carbon dated to about 40,000 years ago placed them as living at the same time as the first Homo sapiens in Asia.
    (D) It was established that up to 10 species of Homo may have co-existed together by about 100,000 years ago, and many more hominid species may have existed 2–3 million years ago.
    (E) It was established that homo sapiens have a much larger brain than Homo erectus did, as well as a great reduction in body hair.

48. The Rh factor is a cell surface protein on red blood cells similar to the type-A and type-B proteins that also exist on the surface. A dominant gene codes for the presence of Rh protein; therefore, being Rh^+ (positive) is dominant over being Rh^- (negative). A woman who has type-A blood and is Rh^+ has an O^+ daughter and a son who is B^-. Which of the following is a possible genotype for the mother?
    (A) I^AI^ARR
    (B) I^AI^ARr
    (C) $I^Ai\ rr$
    (D) I^AiRr
    (E) I^AiRR

49. Members of the phylum Bacillariophyta are yellow or brown eukaryotic algae. They have unique, glasslike walls made of silica and the walls are built in two sections that fit together like a box lid and a box. They are also known as
    (A) Dinoflagellates
    (B) Zygomycetes
    (C) Diatoms
    (D) Green algae
    (E) Volvox

50. A bilaterally symmetric deuterostome might be classified in the phylum
    (A) Annelida
    (B) Chordata
    (C) Platyhelminthes
    (D) Porifera
    (E) Arthropoda

CONTINUE ON THE NEXT PAGE

51. Which of the following statements is not consistent with Darwin's theory of natural selection?

    (A) Individuals in a population exhibit variations, some of which can be passed along to offspring.
    (B) Organisms change during their lifespans to better fit their environment, and these changes can be passed along to offspring.
    (C) Natural selection can lead to speciation.
    (D) Individuals that reproduce most successfully are more likely to have offspring that also reproduce successfully if the environment remains stable.
    (E) Certain organisms have a higher rate of reproductive success than other organisms due to a variety of environmental factors.

52. A community is made up of

    (A) the factors that create an organism's niche.
    (B) the interaction of several ecosystems.
    (C) populations of organisms and their abiotic environment.
    (D) several different populations of organisms living together in the same habitat.
    (E) a group of organisms that interact in a mutualistic fashion.

53. Two plants growing together in the same pot are separated and planted in different pots. One plant dies while the other grows much higher. The plants most likely had what kind of relationship?

    (A) Parasitic
    (B) Commensalism
    (C) Mutualistic
    (D) Mycorrhizal
    (E) Competitive

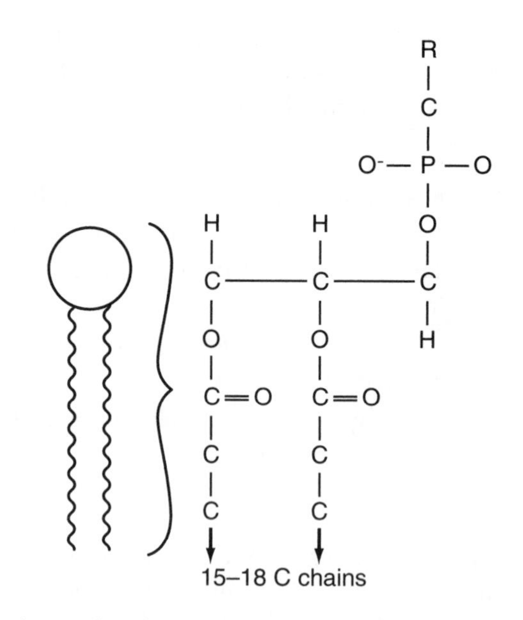

54. The molecule pictured above could be found in all the following places EXCEPT

    (A) the nuclear membrane.
    (B) cell walls in fungi.
    (C) Golgi vesicles.
    (D) secretory vesicles.
    (E) an axon terminal.

55. Rotavirus is an encapsulated virus possessing a double-stranded RNA genome, and it is the major cause of diarrhea in infants. Rotavirus probably causes the diarrhea by

    (A) decreasing the absorption of sodium and water from the intestinal lumen due to damaged intestinal villi.
    (B) secreting RNA-encoded proteins that pull water from intestinal cells into the intestinal lumen.
    (C) increasing the absorption of water into damaged intestinal villi.
    (D) causing damage to renal glomeruli so that excess water and salts are allowed to filter out of the bloodstream into the excretory system.
    (E) absorbing food particles in the intestines so that less food and more water reach the large intestine.

56. *Ras* is one of a family of genes that have been found in tumor virus genomes and seem responsible for the ability of the viruses to cause tumors in cells they infect. The cellular counterpart to this viral cancer-causing gene is known as a

   (A) Second messenger
   (B) Oncogene
   (C) Proto-oncogene
   (D) Tumor promoter
   (E) Metastatic growth factor

57. A hormone able to exert negative feedback on both the hypothalamus and pituitary glands during the menstrual cycle is

   (A) FSH
   (B) LH
   (C) hCG
   (D) GnRH
   (E) Estrogen

58. A major difference between ectotherms and endotherms is that

   (A) as ambient temperature rises, ectotherms maintain nearly constant body temperature.
   (B) endotherms receive most of their body heat from their surroundings.
   (C) endotherms derive body heat from metabolic reactions and use energy derived from metabolic reactions to cool their bodies.
   (D) ectotherms maintain their body at lower temperatures than do endotherms, therefore leading to the term "cold-blooded."
   (E) ectotherms cannot live on land because of temperature fluctuations that can damage their organ systems.

59. Benefits of asexual reproduction include all of the following EXCEPT

   (A) it often allows for the production of many more offspring at the same time.
   (B) it is advantageous in changing environments in which population variety is the key to successful propagation of a species.
   (C) it is easier in certain environments to have offspring without searching for a mate.
   (D) structuring the social organization of certain species of social insects, in which certain members of the species are produced asexually through parthenogenesis.
   (E) allowing the conservation of resources otherwise allocated to finding mates and performing ritualized courtship.

60. The bark of a tree is made up mainly of

   (A) Xylem
   (B) Pulp
   (C) Vascular cambium
   (D) Tracheid cells
   (E) Phloem

61. Which of the following is true regarding phage lambda, a virus that infects bacteria?

   (A) In the lytic cycle, the bacterial host replicates viral DNA, passing it on to daughter cells during binary fission.
   (B) In the lysogenic cycle, the bacterial host replicates viral DNA, passing it on to daughter cells during binary fission.
   (C) In the lytic cycle, viral DNA is integrated into the host genome.
   (D) In the lysogenic cycle, the host bacterial cell bursts, releasing phages.
   (E) Phage lambda is able to replicate without entering a bacterial host.

CONTINUE ON THE NEXT PAGE

62. The covalent binding of a molecule other than the substrate to the active site of an enzyme often results in
    (A) noncompetitive inhibition.
    (B) feedback inhibition.
    (C) irreversible inhibition.
    (D) increasing the energy of activation.
    (E) modifying the free energy change of the reaction.

63. In a food chain that consists of grass → grasshoppers → spiders → mice → snakes → hawks, the organism(s) that possess the most biomass within the community is (are) the
    (A) Grass
    (B) Grasshoppers
    (C) Mice
    (D) Snakes
    (E) Hawks

64. Avascular plants that alternate between haploid and diploid forms and have flagellated sperm are members of the division
    (A) Rhizopoda
    (B) Bryophyta
    (C) Pterophyta
    (D) Coniferophyta
    (E) Anthophyta (the angiosperms)

65. Which of the following is not a characteristic of monocots?
    (A) One cotyledon in each seed
    (B) Parallel leaf veins
    (C) Fibrous root system
    (D) Vascular bundles of xylem and phloem complexly arranged
    (E) Petals in multiples of four or five

66. Some members of this phylum of invertebrates undergo torsion during early embryonic development, resulting in a mantle cavity and anus that are above the head in adults:
    (A) Annelida
    (B) Rotifera
    (C) Chordata
    (D) Mollusca
    (E) Cephalochordata

67. The earliest forms of life were most likely
    (A) Unicellular autotrophs
    (B) Multicellular autotrophs
    (C) Unicellular heterotrophs
    (D) Multicellular heterotrophs
    (E) Photoautotrophs

68. The punctuated equilibrium hypothesis claims that
    (A) cataclysmic events (e.g., asteroid strikes) have shaped the history of life on earth.
    (B) most speciation occurs sympatrically.
    (C) species go through long periods of time during which they do not change markedly in genotype or phenotype.
    (D) new species arise through mutations that have large effects on phenotype.
    (E) the boundaries between fossil species are largely arbitrary.

69. On average, a new mutant allele takes $4*N_e$ generations to become fixed in a population if only drift is operating (N_e is the effective population size). Assume the effective population size of humans is 10,000, and they have 500 different alleles that can occur at a particular locus. If these alleles are all neutral, then they have descended from a single allele in an individual who lived approximately

    (A) 40,000 generations ago.
    (B) 5,000,000 generations ago.
    (C) 40,000 years ago.
    (D) 80 generations ago.
    (E) 500,000 generations ago.

70. Suppose the frequency of an allele *A* in a population is 0.97. It mutates to allele *a* (that starts at a frequency of 0.03) at a rate of 0.004 mutations per generation. However, allele *a* mutates to allele *A* at a rate of 0.001 mutations per generation. Considering only these mutation rates, the equilibrium frequencies of alleles *A* and *a*, respectively, are

    (A) 0.970 and 0.030
    (B) 0.966 and 0.031
    (C) 0.20 and 0.80
    (D) 0.50 and 0.50
    (E) 0.00 and 1.00

71. The males of chimpanzee troop 1 wipe out those of troop 2 and proceed to mate with the females of troop 2. The number of males equals the number of females, and the chimpanzees have nonoverlapping generations. The males carry allele *A* at a particular locus while all the females are homozygotes for allele *B*. From this point on, the population is under Hardy-Weinberg conditions. Assuming that no males carry allele *A* by the F_1 generation and 5/8 of the females carry allele 1 by the F_3 generation, the best explanation for this shift in the frequency of allele *A* is

    (A) Overdominance
    (B) Gametic drive
    (C) Migration
    (D) Imprinting
    (E) X-linked transmission

72. Organisms in which of the following groups lack true tissues?

    (A) Insecta
    (B) Avis
    (C) Porifera
    (D) Platyhelminthes
    (E) Cnidaria

CONTINUE ON THE NEXT PAGE

73. Use the following data regarding crossover frequencies to map the relative location of four genes linked on the same chromosome: P, Q, R, and T.

Genes	Frequency of Crossover
P and Q	35%
R and Q	20%
R and P	15%
T and Q	60%
P and T	25%

Which of the following represents a possible arrangement of genes P, Q, R, and T on the chromosome where they are linked?

(A) P-Q-R-T
(B) P-R-Q-T
(C) T-P-R-Q
(D) T-P-Q-R
(E) Q-P-T-R

74. In some organisms, features that have no function become vestigial and are ultimately lost. In many cave-dwelling animals, organs such as the eyes have been lost while other sense organs have increased in size. Which of the following hypotheses to explain the loss of nonfunctioning organs would not be considered correct?

(A) Mutations causing the reduction in size of nonfunctional traits become fixed by genetic drift.
(B) Natural selection against organs that are not used exists because the organs interfere with other, more important, functions.
(C) The development of the organ requires energy expenditures that are better spent on building other tissues or maintaining other traits.
(D) All organs are maintained or eliminated as a result of how much they are used.
(E) The organs that disappear have a negative genetic correlation with other traits.

75. Speciation by polyploidy commonly occurs in plants. This form of speciation likely occurs because of what type of isolation?

(A) Behavioral
(B) Postzygotic
(C) Prezygotic
(D) Geographic
(E) Allopatric

76. An iteroparous life history, rather than a semelparous one, may evolve in a certain species if

(A) juvenile individuals devote a great deal of energy toward reproductive efforts.
(B) environmental conditions necessitate having a large batch of offspring early on in life.
(C) young individuals devote much energy early on to self-maintenance, while saving reproductive efforts for later on in life.
(D) delayed maturation increases the chances of dying prior to reproducing.
(E) there is a high adult mortality rate in its particular habitat.

77. Male birds that leave their mates immediately after mating and go off in search of other females often benefit by fathering so many offspring. This is known as

(A) Polygamy
(B) Monogamy
(C) Polygyny
(D) Polyandry
(E) Altruism

78. Heterochrony is defined as an evolutionary change in the rate of development of a particular feature. In some cases, a feature present in adult members of a species resembles the same feature that was present in juveniles of the ancestors of that species. These underdeveloped adult characteristics are often called

    (A) Ontogenetic
    (B) Morphogenetic
    (C) Paedomorphic
    (D) Heterotropic
    (E) Homeotic

79. The pedigree below traces the incidence of the hereditary disease β-thalassemia in a particular family. β-thalassemia is a hemolytic anemia that arises from the abnormal synthesis of β-chains in hemoglobin.

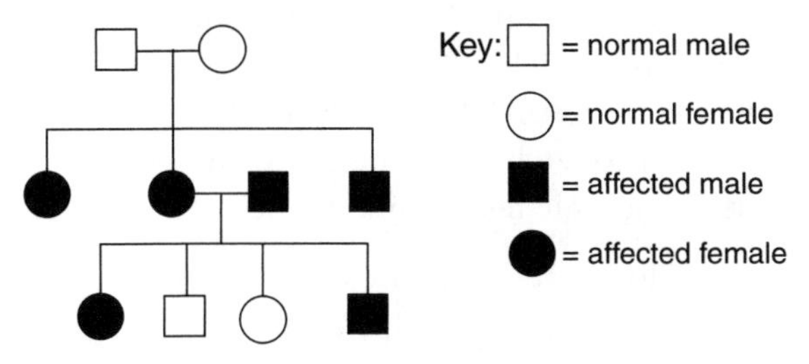

According to the pedigree, the gene for β-thalassemia is inherited as an

    (A) Autosomal recessive trait
    (B) Autosomal dominant trait
    (C) X-linked recessive trait
    (D) X-linked dominant trait
    (E) Imprinted trait

80. Which of the following is a characteristic of the hormone vasopressin?

    (A) It increases water reabsorption in the kidneys.
    (B) It increases sodium reabsorption in the kidneys.
    (C) Its secretion is regulated by the hormone ADH.
    (D) It is secreted by the anterior pituitary gland.
    (E) Its secretion is regulated by the enzyme renin.

81. Linkage disequilibrium results in a situation such that changes at certain gene loci are not independent of changes at other loci. Which of the following reasons explains why two genetic loci might remain in linkage disequilibrium?

    I. Recombination is very low or nonexistent.

    II. Nonrandom mating causes certain allelic combinations to be selected for.

    III. New mutations associated with specific alleles at other loci retain this association unless broken by recombination events.

    (A) I only
    (B) II only
    (C) II and III only
    (D) I and III only
    (E) I, II, and III

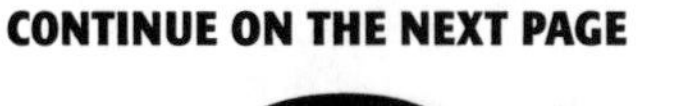

CONTINUE ON THE NEXT PAGE

82. Aneuploidy is the most common type of chromosomal disorder. The condition may arise from either a trisomy or a monosomy. The only monosomy that is generally nonlethal is known as

    (A) Turner syndrome
    (B) Klinefelter syndrome
    (C) XYY syndrome
    (D) Down syndrome
    (E) Cri-du-chat syndrome

83. Each of the following statements regarding capsules in bacteria is correct EXCEPT

    (A) Most gram-positive bacteria have capsules, whereas gram-negative ones rarely do.
    (B) Most bacterial capsules are made of polysaccharides and serve to protect the bacteria by inhibiting phagocytosis.
    (C) Bacterial capsules can vary antigenically and as a result, some bacteria have many different serologic types.
    (D) Bacterial capsules can be purified and used in vaccines against the same bacteria.
    (E) The process by which anti-capsular antibodies attach to bacterial capsules is called opsonization.

84. HIV (Human immunodeficiency virus) is classified as a retrovirus. This means that

    (A) pieces of its RNA genome are spliced into a single strand of RNA before translation.
    (B) it reverses its morphology from type C to type D as it enters cells.
    (C) it transcribes its RNA genome into a DNA genome.
    (D) the promoter regions for RNA polymerase binding are located downstream of genes rather than upstream of them.
    (E) the mechanism of the virus is to allow host cells to enter its capsid in order to take control of them, rather the typical mechanism of viral entry into a host cell.

85. Various immunologic tests are used to diagnose or aid in the treatment of a variety of medical conditions. If you wished to perform ABO blood typing on a person to whom you were about to give a blood transfusion, which of the following immunologic tests would be best?

    (A) Flourescent antibody
    (B) Radial immunodiffusion
    (C) Immunoelectrophoresis
    (D) Flowcytometry
    (E) Agglutination

86. The only type of RNA that does not leave the nucleus once it is built is

    (A) hnRNA
    (B) mRNA
    (C) rRNA
    (D) tRNA
    (E) Viral RNA

87. Which of the following must be true regarding a sex-linked recessive disorder that is 100% lethal in infancy?

    (A) Females are unable to carry the recessive allele.
    (B) The disease will cause death in both males and females.
    (C) Male children of male carriers will also be carriers.
    (D) The condition will cause death only in males.
    (E) The disease will serve to increase the potential son/daughter ratio in a cross between a female carrier and a normal male.

88. All of the following statements are true of the hormone estrogen EXCEPT

    (A) Estrogen is a steroid hormone.
    (B) Spikes in estrogen concentration in the blood bring about surges in luteinizing hormone, which causes ovulation.
    (C) Estrogen is related to testosterone in chemical structure.
    (D) Estrogen is released by the cells in the ovaries and acts only locally within the developing follicles.
    (E) Estrogen and progesterone are secreted by the corpus luteum in order to maintain the thickness of the endometrium.

89. When a muscle fiber is subjected to very frequent stimuli,

    (A) an oxygen debt is incurred.
    (B) a muscle tonus is generated.
    (C) the threshold value is reached.
    (D) the contractions combine in a process known as summation.
    (E) a simple twitch is repeatedly generated.

90. All of the following are examples of imperfections in the fossil record that make it difficult to distinguish patterns such as phyletic gradualism and punctuated equilibrium EXCEPT

    (A) Ages of most fossils can be estimated only in an imprecise manner, since most fossils themselves cannot be dated directly by radiometric techniques.
    (B) Fossils deposited over a short time interval are usually mixed together with other fossils before the sediment solidifies, such that a sample of fossils is often a time-averaged sample.
    (C) Paleozoic insect fossils are usually flattened imprints so that many character traits cannot be studied.
    (D) Mesozoic mammals often leave only a jawbone or tooth behind as their only trace.
    (E) Fossils of early Foraminifera show that the species had evolved rapidly from one relatively stable phenotype to another at the Miocene/Pliocene boundary.

CONTINUE ON THE NEXT PAGE

91. Creationists have argued that a lineage of reptiles could not be the ancestors of mammals because reptiles have only one ear bone, while mammals have three, and there are no intermediate fossils with two ear bones. The best evidence against this argument is the fact that

    (A) there are many transitional fossils between the amphibian and reptile lineages.
    (B) Mesozoic mammals are represented almost exclusively in the fossil record by jawbones and teeth, making the study of changes to their jaws very easy to accomplish.
    (C) the bones responsible for the connections of jawbone to skull in reptiles, which include the quadrate bone, remain fixed to the skull during the reptile/mammal transition.
    (D) a secondary articulation between the lower jawbone and the skull is seen to exist in early mammalian fossils.
    (E) the therapsids, considered to be the ancestors of modern-day mammals, have greatly enlarged jaw muscles and a partial separation of the breathing passage from the mouth.

92. Which of the following are characteristics of arthropods?

    I. Ectodermally secreted exoskeleton.

    II. Bilaterally symmetric and acoelomate.

    III. Water vascular system.

    (A) I only
    (B) II only
    (C) II and III only
    (D) I and III only
    (E) I, II, and III

93. Growth rings of plants, studied by dendrochronologists, consist of which of the following tissue(s)?

    (A) Leaf mesophyll
    (B) Pericycle and stele
    (C) Secondary xylem and phloem
    (D) Protoderm and procambium
    (E) Apical meristem

94. What function do lenticels, pneumatophores, and root hairs all have in common?

    (A) They are used to absorb nutrients in the form of nitrogen- and phosphorus-bearing compounds.
    (B) They are used to anchor plants into the ground.
    (C) They are used for gas exchange, often in plants whose roots are underwater.
    (D) They are involved in the maintenance of hydrostatic pressure at the roots for proper conduction of sugar within the phloem.
    (E) They are responsible for forming mycorrhizal associations with bacteria and fungi.

95. Restoration of the resting state in muscles begins when neural stimulation stops and calcium ions are transported back into the

    (A) Post-synaptic terminal of the nearby axon
    (B) Sarcoplasmic reticulum
    (C) Presynaptic axon terminal
    (D) Neuromuscular junction
    (E) T-tubules

96. Glucose is a simple sugar with the formula $C_6H_{12}O_6$. If two glucose molecules are joined into a molecule of fructose, the disaccharide fructose is formed with the molecular formula

    (A) $C_6H_{12}O_6$
    (B) $C_{12}H_{24}O_{12}$
    (C) $C_{12}H_{23}O_{11}$
    (D) $C_{12}H_{22}O_{11}$
    (E) $C12H_{23}O_{10}$

97. Early methods of food preservation included drying, salting, or sugar-curing food. The main reason for these methods is that they

    (A) prevented bacterial growth through creating poisoning bacteria with too much sugar or salt.
    (B) created a hypertonic environment in which osmotic potential would burst bacterial cells.
    (C) dissolved bacterial cell membranes resulting in cell lysis.
    (D) interfered with bacterial flagella, preventing the movement of bacteria on the food.
    (E) caused the formation of thickened bacterial cell walls that blocked diffusion of key substances into the bacterial cells.

98. Where in the tertiary structure of a water-soluble protein would you expect to find amino acids with hydrophobic R groups?

    (A) At both termini of the polypeptide chain
    (B) Pointing toward the outside on a β-chain
    (C) Folded into the interior of an α-helix
    (D) Covalently bonded to other R groups
    (E) On pleated sheets within α-helices

99. Which of the following might interfere most directly with the process of glycolysis?

    (A) A compound that reacts with NADH and oxidizes it to NAD^+
    (B) A substance that binds to oxygen and blocks it from acting as the terminal electron acceptor
    (C) A compound that inactivates pyruvate by binding to it
    (D) An agent that inhibits the formation of acetyl-CoA
    (E) A substance that closely mimics the structure of glucose but is nonmetabolic

100. Two plants, X and Y, are grown as potential food crops. Plant X is able to maintain a high rate of photosynthesis as oxygen level in the air around it increases from a low of 10% to a high of 50%, yet plant Y's rate of photosynthesis drops drastically under these circumstances. The best conclusion to draw from this data is that

    (A) plant X is a CAM plant.
    (B) plant Y is performing only the Calvin cycle in higher oxygen partial pressures.
    (C) plant X is a CAM plant and plant Y is a C_4 plant.
    (D) plant Y is performing only the light reactions of photosynthesis.
    (E) plant X is a C_4 plant, and plant Y is a C_3 plant.

101. Which of the following terms or phrases would not be associated directly with photosystem II in plants?

    (A) Photophosphorylation
    (B) The splitting of water
    (C) Harvesting light energy by chlorophyll
    (D) Oxygen released from water
    (E) chlorophyll a

CONTINUE ON THE NEXT PAGE

102. Which of the following statements concerning cellular respiration is correct?

    (A) Aerobic respiration most likely evolved before anaerobic respiration due to the complex proteins involved.
    (B) Fermentation proceeds in the absence of aerobic respiration and can work without enzymatic assistance.
    (C) Eight molecules of CO_2 are produced in one turn of the citric acid cycle.
    (D) Fatty acids can be β-oxidized into three-carbon units during respiration for entry into the Krebs cycle.
    (E) The Krebs cycle directly generates ATP via substrate-level phosphorylation only.

103. Substrates may bind into the active sites of enzymes by all the following EXCEPT

    (A) Hydrogen bonds
    (B) Peptide bonds
    (C) Polar covalent bonds
    (D) Dipole-dipole interactions
    (E) Hydrophobic interactions

104. Identical twins can develop in humans when

    (A) a single egg is fertilized by two different sperm.
    (B) two identical eggs are fertilized by two nonidentical sperm.
    (C) haploid eggs are fertilized by diploid sperm.
    (D) two different haploid eggs are fertilized by two different sperm.
    (E) one egg fertilized by one sperm splits into two zygotes during development.

105. The structures of hemoglobin, chlorophyll, and the cytochromes are similar in that they all contain

    (A) Iron
    (B) Porphyrin rings
    (C) Pyrimidine bases
    (D) Histone proteins
    (E) Phosphorous

106. The reaction that converts pyruvic acid into acetyl-CoA at the mitochondrial membrane is known as a(n)

    (A) Amination
    (B) Hydrolysis
    (C) Decarboxylation
    (D) Phosphorylation
    (E) Oxidation-reduction

107. Organisms in which of the following groups possess stinging cells on tentacles?

    (A) Cnidaria
    (B) Avis
    (C) Porifera
    (D) Platyhelminthes
    (E) Insecta

108. All of the following are examples that are evidence of the endosymbiont theory EXCEPT

    (A) mitochondria and chloroplasts are both affected by drugs that halt protein synthesis in prokaryotes.
    (B) mitochondrial inner membranes are similar in structure to bacterial cell membranes.
    (C) some modern-day bacteria can live outside eukaryotic cells.
    (D) fossils carbon-dated to over 500 mya show prokaryotes lived within eukaryotic cells.
    (E) both mitochondria and chloroplasts possess their own DNA and can reproduce independently of the cell they are within.

109. A chemiosmotic gradient is responsible for the active transport of glucose from the proximal tubule of nephrons into the bloodstream. This type of transport would also be found in

    (A) The tracheae of insects
    (B) Immature sclerenchyma cells
    (C) Islet cells of the pancreas
    (D) Sieve-tube members of plants
    (E) The electron transport chain of thylakoids

110. As a leading English biologist in the early 1900s, Lankester defined a species as "an assemblage of all variants that are potentially the offspring of the same parents." Later, a species was defined as "individuals connected by a blood relationship that form a single faunistic unit in an area...and these areas are separated from each other by gaps." This new definition changed the biological species concept by

    (A) defining species by their intrinsic properties and not by relation to other co-existing species.
    (B) defining species as populations or groups of populations rather than as types.
    (C) allowing the definition of a species to be applied to biological entities.
    (D) clarifying that all similar-looking organisms within a particular area were blood-related and capable of interbreeding.
    (E) explaining the existence of balanced polymorphisms within a particular population of organisms.

111. Which of the following scientists most likely said, "I am inclined to regard the separation of parental traits...as complete...I have never observed gradual transitions between the parental traits or a progressive approach toward one of them"?

    (A) Thomas Hunt Morgan
    (B) Gregor Mendel
    (C) Hugo de Vries
    (D) Francis Crick
    (E) E.O. Wilson

112. The highest primary productivity in kilojoules per year would be found in which of the following ecosystems?

    (A) Desert
    (B) Tundra
    (C) Taiga
    (D) Tropical rain forest
    (E) Temperate deciduous forest

113. Which of the following is true of both the wing of a bird and the flipper of a dolphin?

    (A) Both will continue to grow as long as the organism is alive.
    (B) Their bone structure is identical because of divergent evolution.
    (C) They have attachment sites for tendons and ligaments.
    (D) Both contain hollow bones with air sacs for extra oxygen retention.
    (E) Direct diffusion of oxygen into muscle tissue is accomplished via large arteries.

CONTINUE ON THE NEXT PAGE

114. Both sexual reproduction in plants and nondisjunction in animals can result in which of the following?

(A) Polyploidy
(B) Speciation
(C) Alternation of generations
(D) The formation of haploid offspring
(E) Pleiotropy

115. The water-soluble vitamin that acts as a co-enzyme in nucleic acid metabolism and can be found in vegetables and grains is

(A) Vitamin B_6
(B) Niacin
(C) Folic acid
(D) Vitamin A
(E) Ascorbic acid

116. Bombykol receptors on the surfaces of chemoreceptor cells in the male silkworm moth are used to sense the presence of a hormone called bombykol, which is released by female moths in search of a mate. This hormone would be considered a(n)

(A) Histone
(B) Transcription factor
(C) Accomodator
(D) Enhancer
(E) Pheromone

117. Prostaglandins are modified fatty acids derived from a molecule called arachidonic acid. They are released by cells into the extracellular fluid and act as local mediators to regulate blood vessel contraction and pain. In such a way, prostaglandins would be considered what kind of regulators?

(A) Endocrine
(B) Merocrine
(C) Paracrine
(D) Synaptic
(E) Exocrine

118. Individuals who are of blood group AB

(A) have circulating "anti-A" and "anti-B" antibodies.
(B) can donate blood to any other individual.
(C) will be Rh-negative.
(D) are considered "universal recipients" for blood transfusions.
(E) have the same haplotype.

119. A mutant bacterial cell line that was lacking a functional thymidine kinase gene was exposed to a preparation of DNA from normal bacterial cells. Under appropriate growth conditions, cells were allowed to grow and a colony of cells was isolated that could produce thymidine kinase. This is an example of

(A) Transduction
(B) Transformation
(C) Conjugation
(D) Transposition
(E) Phosphorylation

Questions 120-122

(A) Endosperm
(B) Gametophyte
(C) Protoderm
(D) Pericarp
(E) Cotyledon

120. Embryonic seed leaf that has a large surface area to absorb nutrients during seed germination

121. The thickened wall of the fruit, derived from the ovary

122. Multicellular haploid form of a plant that undergoes mitosis to produce haploid gametes

Questions 123–124

(A) Gene amplification
(B) Gene cloning
(C) Gene flow
(D) Gene pool
(E) Genetic drift

123. The loss or gain of alleles from a population due to immigration or emigration

124. The replication of select areas of DNA resulting in multiple copies of a gene

Questions 125–127

(A) Ciliary body
(B) Aqueous humor
(C) Organ of Corti
(D) Lateral line system
(E) Sensillae

125. Small hairs on the feet and mouthparts of insects used for taste reception

126. Contains mechanoreceptors that are used by fishes to detect movement

127. Contains the receptor cells of the mammalian ear and is located within the cochlea

Questions 128–130

(A) Erythrocytes
(B) Lymphocytes
(C) Platelets
(D) Pluripotent stem cells
(E) Leukocytes

128. A term that includes all five types of white blood cells, whose functions deal with fighting off infection

129. Blood cells packed with hemoglobin in order to carry oxygen around the body

130. B cells and T cells are the two main types of this group of white blood cells

Questions 131–132

(A) Gibberellins
(B) Auxins
(C) Cytokinins
(D) Ethylene gas
(E) Abscisic acid

131. Initiates and enhances the conditions under which a fruit ripens and ages

132. Responsible for phototropisms in plants

CONTINUE ON THE NEXT PAGE

Questions 133–135 refer to the following diagram showing changes in the membrane potential of a neuron as it carries an impulse.

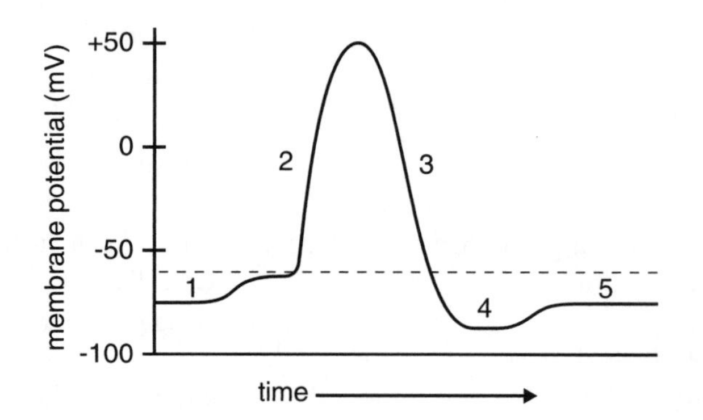

(A) 1
(B) 2
(C) 3
(D) 4
(E) 5

133. The phase of the action potential in which only Na^+ channels are open

134. The phase of the action potential representing the absolute refractory period

135. The part of the diagram representing the initial resting membrane potential

Questions 136–140 refer to the following diagram of organic molecules below.

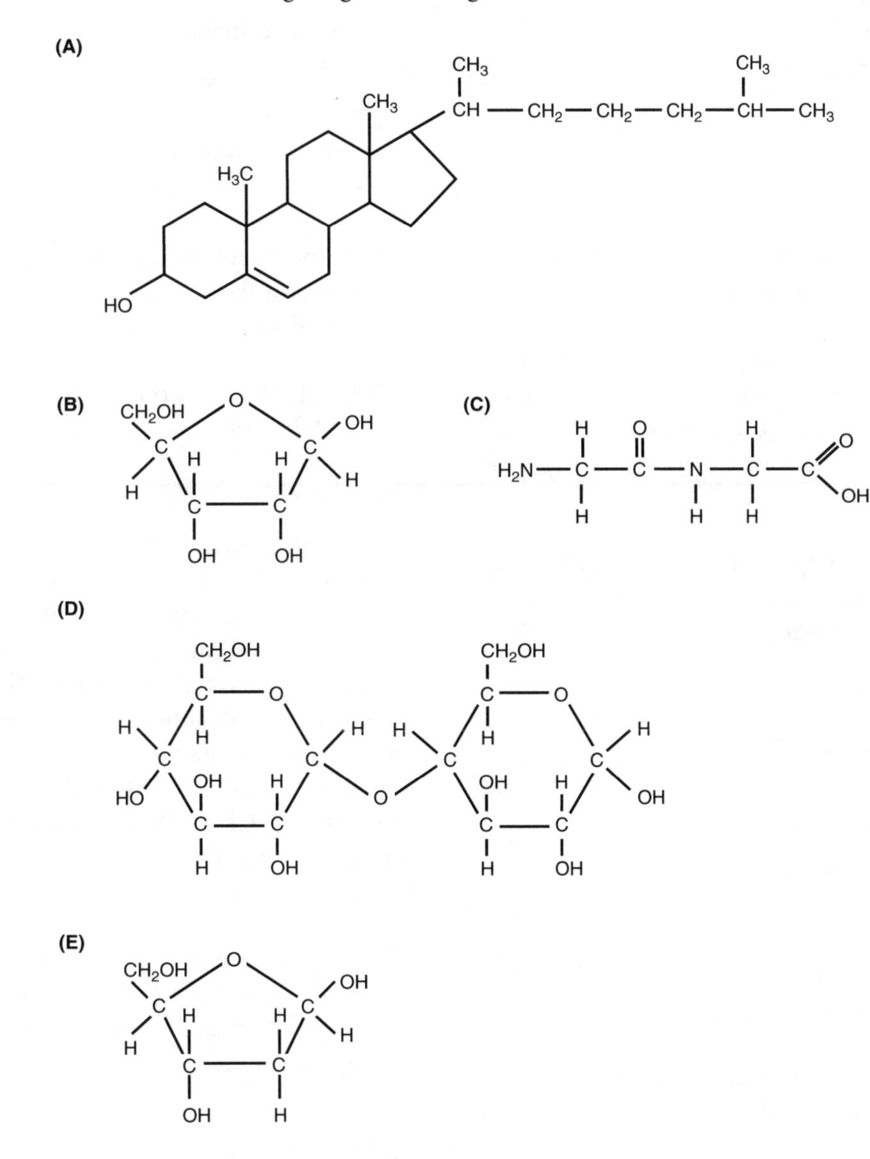

136. The amino acid glycine is a building block of this molecule.

137. Able to diffuse through the plasma membrane of cells for direct transcriptional control of genes

138. This type of sugar molecule makes up the backbone of single-stranded genetic material used in the process of translation.

139. Estrogen and progesterone are forms of this type of molecule.

140. Disaccharidases in the intestinal lumen will destroy this molecule.

CONTINUE ON THE NEXT PAGE

Questions 141–143

(A) Monophyletic
(B) Paraphyletic
(C) Polyphyletic
(D) Homology
(E) Analogy

141. A single ancestor gives rise to all species within a particular taxon.

142. Similar environmental pressures have led two unrelated cactuslike plants to evolve the same adaptation for water conservation, although one is found in Africa and the other is found in North America.

143. A restriction mapping analysis of DNA from related organisms shows highly conserved sequences of homeotic genes.

Questions 144–148

(A) Platyhelminthes
(B) Cnidaria
(C) Porifera
(D) Echinodermata
(E) Annelida

144. Organisms in this phylum have a brain and two or more nerve trunks.

145. Organisms in this phylum have a central nerve ring with radial nerves.

146. Possess a nerve net in place of a more complex nervous system.

147. A brain and central nerve cord containing segmented ganglia is the hallmark of this phylum.

148. These organisms show radial cleavage during early embryonic development, and the anus develops from the blastopore in embryos from this phylum.

Questions 149–150

(A) Exoskeleton
(B) Cartilage
(C) Hinge joint
(D) Ball and socket joint
(E) Endoskeleton

149. Responsible for the formation of "discs" that cushion the vertebral column in some chordates

150. Used for extension and flexion motions in vertebrates

Questions 151–152

The heterozygosity (H) of a population is the frequency of heterozygotes in the population. If genetic drift is operating in the absence of selection, a population will become completely homozygous over time. In fact, the rate at which heterozygosity declines is 1/(2N) each generation, where N represents the effective population size. In a population of naked mole rats, the effective population size is 200 despite the fact that it fluctuates widely over many generations from a high of 600 one year to a low of 80 another year.

151. Considering only the effective size of the mole rat population, the heterozygosity of the population decreases each generation by approximately

(A) 0.0%
(B) 0.25%
(C) 0.025%
(D) 0.5%
(E) 5%

152. After four generations maintaining a fairly small effective population size, the naked mole rat population becomes extremely inbred. Although its heterozygosity originally was x, after four generations it is $0.5x$. If the original mating pair of mole rats had possessed two different alleles at a particular gene locus, the probability that an individual in generation four would have two identical alleles at the same locus is

(A) $0.25 + x$
(B) $0.25x$
(C) 0.5
(D) 0
(E) 0.59

Questions 153–155

Differences in DNA base pair sequences that code for tusk length are measured in six species of elephants, A–E, in order to investigate the phylogenetic relationships between the six species. The DNA differences (in nucleotides) between the species are listed in the following table:

	A	B	C	D	E	F
A						
B	12					
C	29	29				
D	30	30	9			
E	22	22	21	22		
F	21	21	20	21	5	

153. Which of the following cladograms best illustrates the evolutionary history of these six species of elephant?

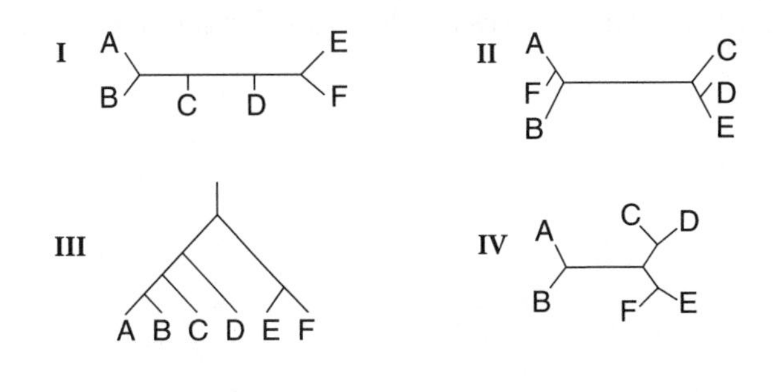

(A) I
(B) II
(C) III
(D) IV
(E) There is not enough information given to answer the question.

154. Which of the following species are most closely related according to the information presented in the phylogenetic chart above?

(A) A and B
(B) C and D
(C) D and E
(D) E and F
(E) A and E

155. The taxon made up solely of species C and D would be considered

(A) paraphyletic
(B) monophyletic
(C) polyphyletic
(D) paedomorphic
(E) aphyletic

CONTINUE ON THE NEXT PAGE

Questions 156--157

A systematist measures a particular trait in a group of South American birds and finds that the birds fall into three general classes: A, B, and C. The relationships between these birds is already known, and the researcher plots the trait classes on the evolutionary tree for the group:

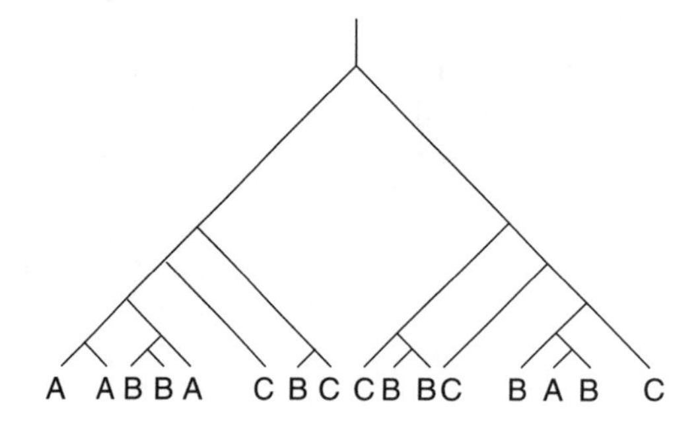

156. What is the most likely trait class for the common ancestor of the entire clade?

(A) A
(B) B
(C) C
(D) All trait classes evolved independently.
(E) It cannot be determined from the information given.

157. What is the best estimate for the number of times class A has evolved independently?

(A) 1
(B) 2
(C) 3
(D) 4
(E) It cannot be determined from the information given.

Questions 158–162

Total mRNA from five different cell lines was isolated and analyzed by Northern blotting as shown in figure A below. Expression levels of the cytokines IL-1α and Il-1β, as well as the expression of GADPH (glyceraldehyde-3-phosphate dehydrogenase), were examined. As seen in figure B, total cellular protein was then extracted from the five different tumor cell lines and run on a denaturing polyacrylamide gel. The protein was subsequently transferred to nitrocellulose by electroblotting and IL-1α protein levels were examined by Western analysis using a commercially available IL-1α polyclonal antibody.

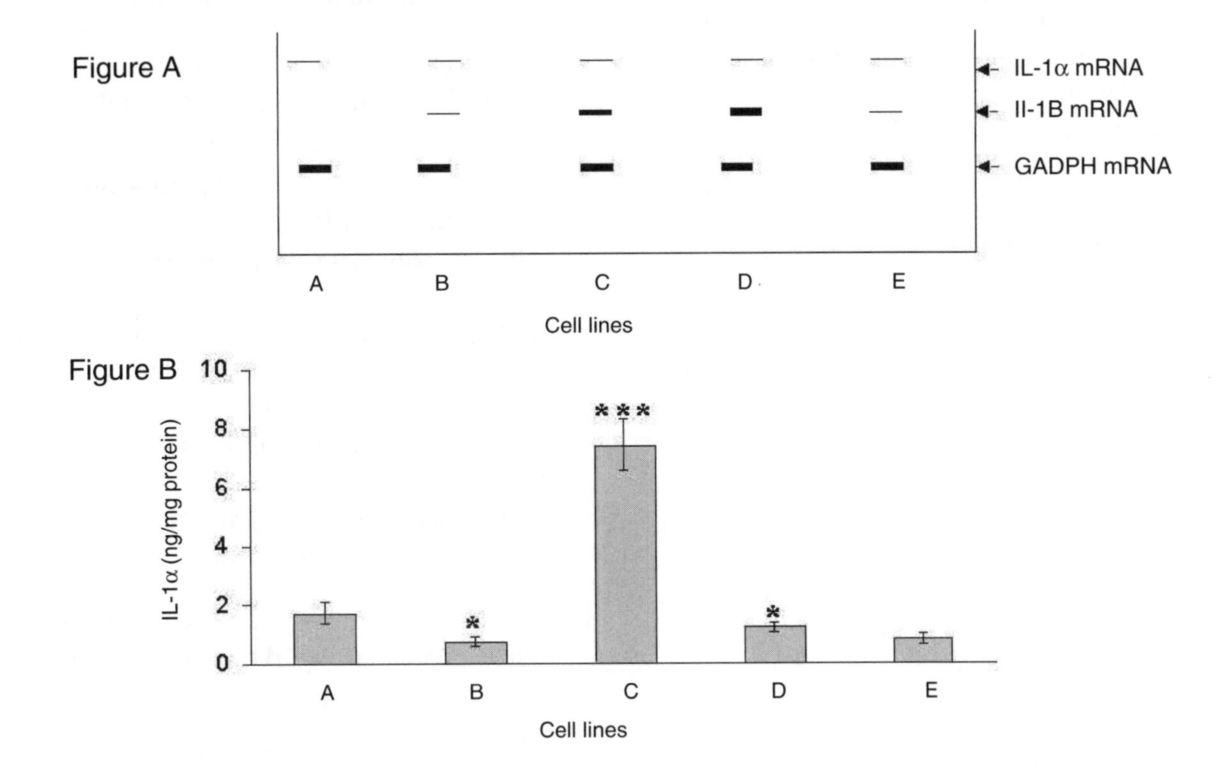

158. It has been proposed that the some of the properties of different tumor cell types might be associated with differential expression of cytokines such as IL-1α. Based on the above figures, at which level is the expression of IL-1α most likely regulated?

(A) At the level of DNA replication
(B) At the level of transcription
(C) At the level of translation
(D) IL-1α is most likely unregulated within the cell lines investigated.
(E) IL-1α regulation is dependent upon IL-1β formation.

159. High levels of IL-1α expression have been associated with rapid tumor growth. Based on this association, which of the following cell lines might be expected to exhibit the highest rate of tumor growth?

(A) A
(B) B
(C) C
(D) D
(E) E

CONTINUE ON THE NEXT PAGE

160. It has been postulated that inhibition of IL-1α expression may inhibit tumor growth. Which of the following therapies is most likely to be successful based on this hypothesis?

    (A) Introduction of IL-1α antisense cDNA into the tumor
    (B) Restriction enzyme digest of IL-1α genes
    (C) Introduction of IL-1α sense cDNA into the tumor
    (D) Upregulation of IL-1β secretion
    (E) Destruction of cells producing any IL-1α

161. The GADPH gene encodes glyceraldehyde-3-phosphate dehydrogenase, which catalyzes the reversible oxidative phosphorylation of glyceraldehyde-3-phosphate in the presence of inorganic phosphate and nicotinamide adenine dinucleotide (NAD^+). It is expressed constitutively in all cells and its expression has been shown many times to remain constant in the above cell lines under all conditions. The most likely explanation for the inclusion of GADPH mRNA in the Northern blot is

    (A) as a comparison of GADPH expression with that of IL-1α in these cell lines.
    (B) to verify that GADPH expression does not vary in these cell lines.
    (C) as a control to verify that equal amounts of mRNA have been added to each lane in the Northern blot.
    (D) to verify that mRNA has been successfully isolated.
    (E) to show the effect GADPH production has on IL-1α production.

162. Based on the figure above, at which level is the expression of IL-1β most likely regulated?

    (A) At the level of DNA replication
    (B) At the level of transcription
    (C) At the level of translation
    (D) IL-1β is most likely unregulated within the cell lines investigated
    (E) IL-1β regulation is dependent upon IL-1α formation

Questions 163–165

Human tumor growth was studied in mice with SCID, or severe combined immunodeficiency disorder. Each group of mice in the experiment was injected with a particular type of human cell line, and tumor responses were measured. One group was injected with HT-15 cells that overexpress the enzyme manganese superoxide dismutase; another group was injected with HT-15mCAT cells that overexpress both manganese superoxide dismutase and catalase; a third group was injected with CMV cells, human fibrosarcoma cells that act as controls; and, a final group was injected with CMV-mCAT cells that overexpress the catalase enzyme. The results are shown in the graph below.

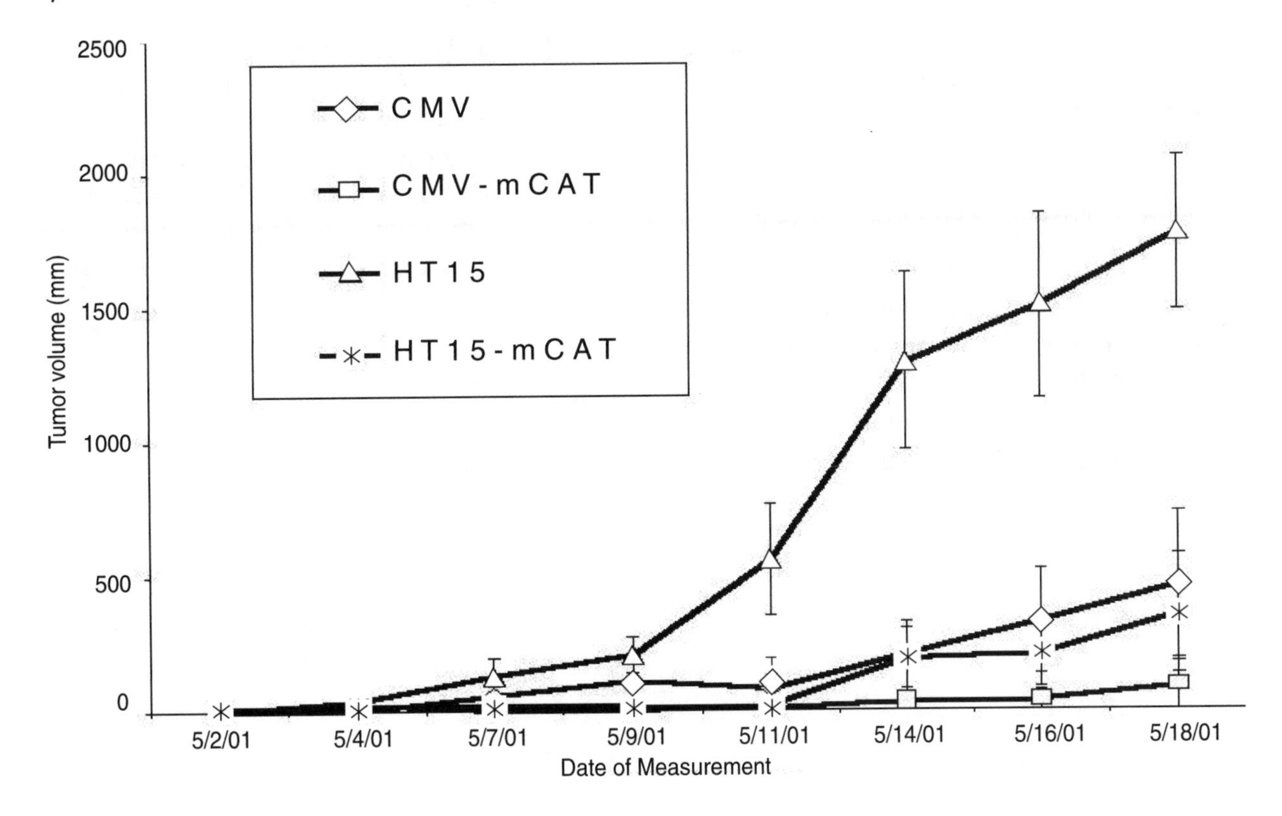

CONTINUE ON THE NEXT PAGE

163. Based on the figure above, the greatest tumor volume is exhibited by

 (A) HT15 cells.
 (B) HT15-mCAT cells.
 (C) CMV cells.
 (D) CMV-mCAT cells.
 (E) cells that overproduce catalase enzyme.

164. Superoxide dismutase converts superoxide produced in the mitochondria to hydrogen peroxide. Hydrogen peroxide is then converted to water and oxygen by catalase. Based on the figure above, which of the following possibilities is most likely?

 (A) Superoxide promotes tumor growth
 (B) Hydrogen peroxide inhibits tumor growth
 (C) Superoxide dismutase promotes tumor growth
 (D) Catalase promotes tumor growth
 (E) Superoxide dismutase inhibits tumor growth

165. The most likely reason that SCID mice, rather than wild-type mice, were used in this study is that

 (A) too many tumors would form in wild-type mice, which would make it difficult to quantitate tumor volume.
 (B) SCID mice are a more popular type of laboratory strain of mice than wild-type.
 (C) SCID mice lack a functional immune response; thus human CMV tumor cells are able to grow within them.
 (D) superoxide dismutase and catalase are not functional in SCID mice cells and can be substituted by human enzymes.
 (E) CMV cells are lethal to wild-type mice.

Questions 166–169

Studies of organ transplants to repair injuries during World War II established that transplant rejection was the result of an inflammatory reaction mediated by the recipient's immune system. The studies also showed that the rejection of a second transplant from the same donor, known as "second set" rejection, occurred much faster than rejection of the first transplant ("first set" rejection). The results of these experiments are shown in the table below.

Donor Strain	Recipient Strain	Prior Treatment	Rejection
A	B	None	Slow
A	B	Received previous transplant from Strain A donor	Rapid
C	B	Received previous transplant from Strain A donor	Slow

166. What form of rejection is anticipated after a strain C animal that has already been exposed to a graft from strain A receives a graft from a strain B mouse?

(A) Rapid second set rejection
(B) Slow second set rejection
(C) Rapid first set rejection
(D) Slow first set rejection
(E) No rejection at all

167. Which of the following experimental observations would demonstrate that humoral immunity plays a role in graft rejection?

(A) Strain B animals never exposed to foreign cells mount a first set rejection against a strain A graft after being injected with serum from strain B animals that had been previously exposed to strain A.
(B) Strain B animals never exposed to foreign cells mount a second set rejection against a strain A graft after being injected with lymphocytes from strain B animals that had been previously exposed to strain A.
(C) Strain B animals never exposed to foreign cells mount a first set rejection against a strain A graft after being injected with lymphocytes from strain B animals that had been previously exposed to strain A.
(D) Strain B animals never exposed to foreign cells mount a second set rejection against a strain A graft after being injected with serum from strain B animals that had been previously exposed to strain A.
(E) Strain B animals exposed to foreign cells from strain A mount a rapid second set rejection, although they have never before been exposed to Strain A cells.

CONTINUE ON THE NEXT PAGE

168. Before a specific immune response can be mounted against a foreign pathogen, the immune system launches a nonspecific defense that is active against a broad range of pathogens. Which of the following represents a nonspecific defense?

(A) T-cell mediated immunity
(B) B-cell mediated immunity
(C) Antibody-mediated immunity
(D) MHC-mediated hypersensitivity reactions
(E) Mucous secreted by cells lining the lungs

169. During maturation of T-cells in the thymus, large numbers of T-cells that are able to bind to an individual's own MHC molecules are killed off. Which of the following is most likely to occur is such cells are allowed to survive?

(A) T-cells will launch an immune response against the individual's own cells.
(B) T-cells will be less able to recognize and bind to foreign MHC molecules.
(C) More T-cells than B-cells will be produced by the individual's body.
(D) The individual will be less susceptible to bacterial infections.
(E) The individual will produce more antibodies during first set rejections.

Questions 170–174

In vertebrates, the ability to store sufficient quantities of energy-dense triglycerides in adipose tissue allows survival during periods of food deprivation. Two pure-breeding strains of mice were located, Ob1 and Ob2, both of which had defects in their weight regulatory system. Both strains of mice had markedly increased amounts of adipose tissue. When plasma from the mice was analyzed, a 16-kD protein called leptin was missing in the Ob1 mice. To determine the biological effects of leptin, both obese strains of mice and a wild-type strain received daily injections of leptin for 30 days. Food intake and body mass were measured for the entire 30-day period and then 30 days following for a total of 60 days of measurement.

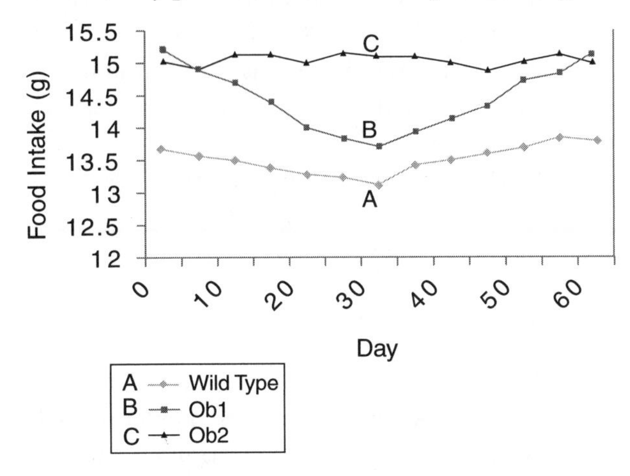

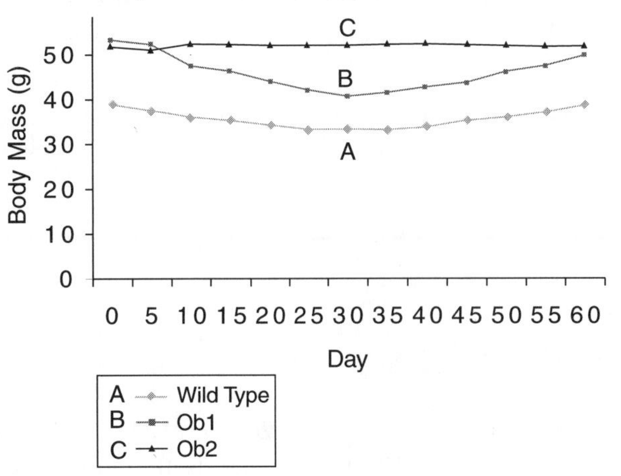

170. Obesity linked to mutations in the leptin gene is inherited as an autosomal recessive trait. What chance does an offspring have of being obese if it is produced by the mating of an Ob1 mouse and a homozygous dominant wild-type mouse?

(A) 0%
(B) 12.5%
(C) 25%
(D) 50%
(E) 100%

171. Which of the following is true of the Ob1 strain?

(A) It has a response to the leptin injections that is similar to that of the wild-type mice.
(B) It is likely to possess a mutation in the gene coding for leptin such that leptin protein is not produced.
(C) It is likely to be lacking leptin receptors on its cell surfaces.
(D) Before injections of leptin, the food intake of the Ob1 mice was much lower than during the 30 days following the start of leptin injections.
(E) Leptin acts on food intake in the Ob1 strain by increasing metabolic activity and blood insulin levels.

172. The obese phenotype in the Ob2 strain of mice can be reasonably explained by a defect in the:

(A) weight regulatory system that does not involve leptin.
(B) leptin genes resulting in a decreased production of leptin.
(C) genes that play a role in activating the transcription of leptin genes.
(D) genes coding for leptin receptors on the cell surfaces.
(E) islet cells of the pancreas so that insulin production is decreased.

CONTINUE ON THE NEXT PAGE

173. Leptin acts to prevent hypothalamic secretions of neuropeptide Y, a potent feeding stimulant. As a result, the levels of neuropeptide Y are most probably
    - (A) reduced in Ob2 mice but not reduced in Ob1 mice.
    - (B) reduced in both Ob1 and Ob2 mice.
    - (C) equal to levels in the wild-type mice.
    - (D) elevated in Ob2 mice but not in Ob1 mice.
    - (E) elevated in both Ob1 and Ob2 mice.

174. If leptin serves the same function in humans as it does in mice, which of the following groups of people is LEAST likely to lose weight due to daily administration of leptin?
    - (A) Individuals with normal body weight
    - (B) Overweight individuals with mutations in the leptin gene
    - (C) Overweight individuals with mutations in the leptin receptor genes
    - (D) Overweight individuals with mutations in genes that are not part of the leptin pathway
    - (E) Individuals with pancreatic oversecretion of insulin

Questions 175–178

Bacterial plasmids range in size from 1,000 to 200,000 base pairs of DNA, and they are used extensively for cloning purposes. Plasmids make useful cloning vectors because they have unique restriction enzyme sites for insertional cloning. In other words, foreign genes of interest, if cleaved from an organism's genome by a particular restriction enzyme, can be inserted into a bacterial plasmid that has been cleaved by the same enzyme. The plasmid drawn below has cutting sites for the following restriction enzymes: EcoR1, Sal1, and BamH1. The distance in base pairs (bp) between cutting sites is listed between the respective sites.

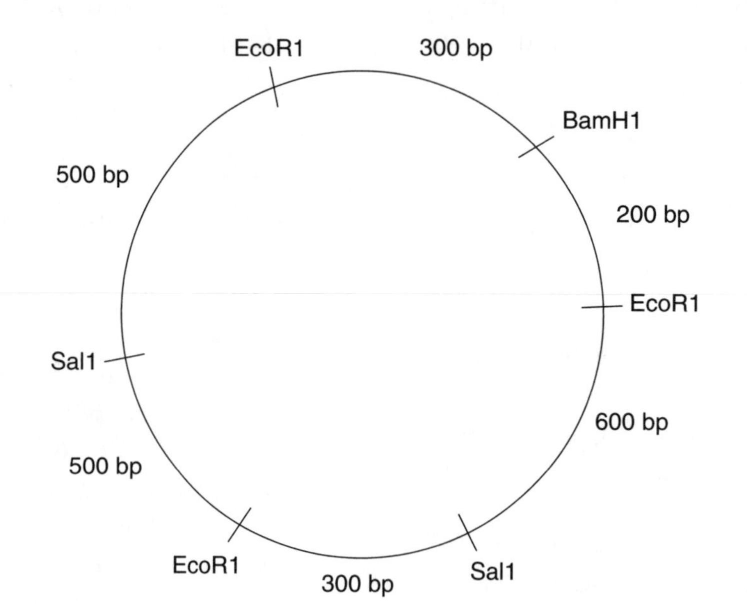

To answer questions 175-177, use your knowledge of biology and the diagrams below labeled (a) through (e).

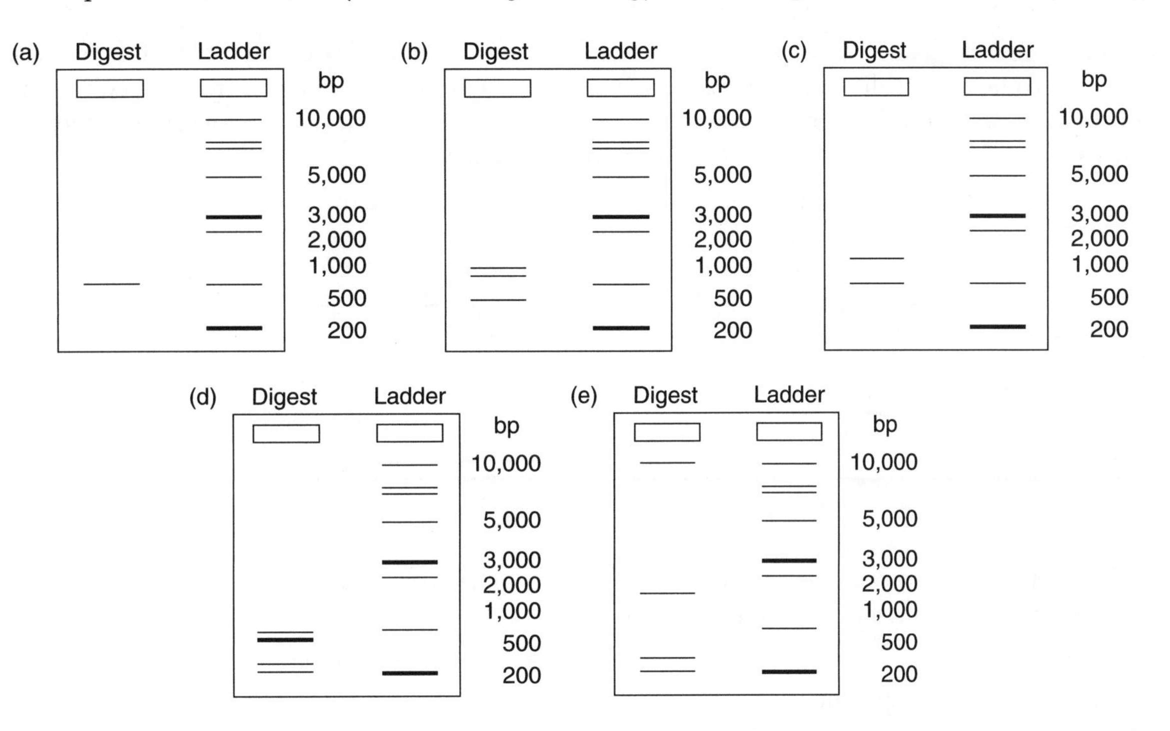

175. Which of the gel electrophoresis results pictured above would you expect after cutting the cloning plasmid with the restriction enzyme Sal1?

176. Which of the gel electrophoresis results pictured above would you expect after cutting the cloning plasmid with the restriction enzyme EcoR1?

177. Which of the gel electrophoresis results pictured above would you expect after cutting the cloning plasmid with all three restriction enzymes — EcoR1, Sal1, and BamH1—all at the same time?

178. Important features of a useful cloning vector include all of the following EXCEPT

(A) the ability to replicate within host cells.
(B) some sort of genetic marker to be able to select for host cells that contain the vector.
(C) unique restriction enzyme sites for insertional cloning.
(D) a minimum amount of nonsense, or junk, DNA.
(E) reporter genes that code for production of a cellular protein normally produced in large quantities.

CONTINUE ON THE NEXT PAGE

Questions 179–182

A particular homeotic gene in *Drosophila* is essential for proper embryonic development. However, it is not active in embryonic cells at all times during growth, and at certain times the gene is transcribed into mRNA without being further translated into protein. The Western and Northern blots below show expression of this homeotic gene at various points during development in a fruitfly.

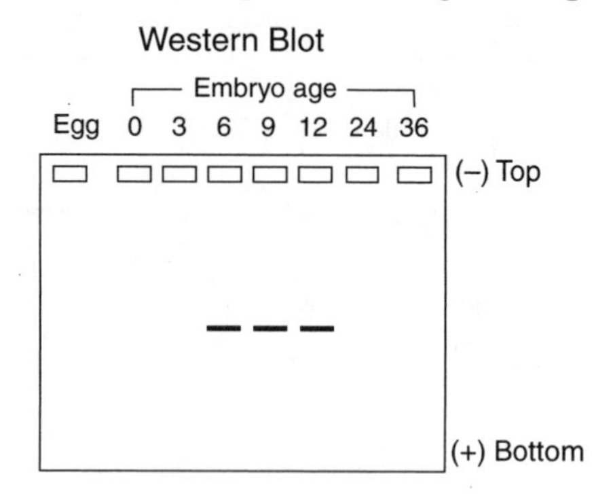

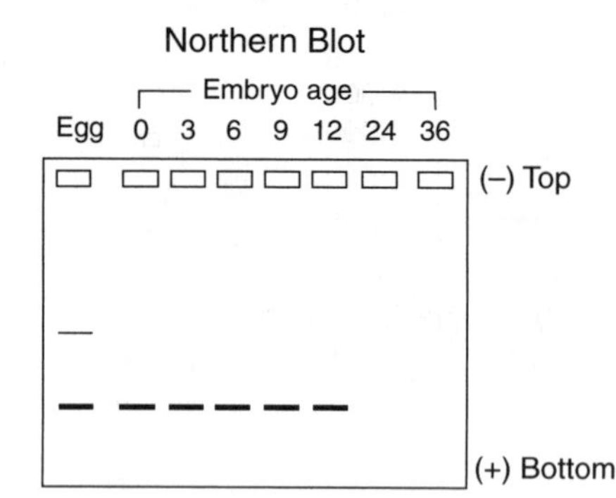

179. The results of the assays completed above suggest that transcription of the homeotic gene takes place only

    (A) early on in development.
    (B) in the egg cell.
    (C) after six hours of development.
    (D) after 24 hrs of development.
    (E) up through 12 hours of development.

180. The proteins produced by homeotic genes such as the one described above produce proteins known as homeodomain proteins, whose function is to bind to and regulate certain genes responsible for development and differentiation. These homeodomain proteins, then, could be considered

    (A) Enhancers
    (B) Promoters
    (C) Transcription factors
    (D) Transposons
    (E) Major histocompatibility proteins

181. The homeodomain protein produced by the *Drosophila* gene described above is most likely regulated by what mechanism during very early development?

    (A) Transcriptional
    (B) Alternative splicing
    (C) Translational
    (D) Spatial regulation
    (E) Localized gene expression

182. After 24 hours of *Drosophila* development has elapsed, which of the following statements can be said to be true?

    (A) The protein product of the homeotic gene is being secreted from the cell as fast as it is being made, and thus cannot be measured in appreciable quantities by the Western blot.
    (B) The homeotic gene has been inactivated.
    (C) The homeotic gene is not actively being transcribed but will become active again during the larval stage, where it will code for the development of adult fly structures.
    (D) The gene has been cleaved out of the fly genome in most cells due to selective gene loss as embryonic development progresses.
    (E) The gene is being transcribed but is not being translated.

Questions 183–185

A helix-loop-helix transcription factor Y (TFY) is examined for its DNA binding ability and its expression in different cell lines. The specific DNA binding sequence has been determined and is used as a probe in an Electromobility Shift (EMS) assay. EMS assays, also known as gel shift assays, show DNA binding ability. Normally, short DNA sequences will run from one end of a gel to the other, yet when used as probes, these DNA sequences can bind to proteins that slow them down and prevent them from traveling through the gel.

Specific antibodies are also present for use in Western blot analysis. In this technique, antibodies bind to proteins that are run out on a gel to indicate the presence of a particular protein at a specific cellular location. The EMS and Western blot results for TFY activity in muscle cells and nerve cells are shown below.

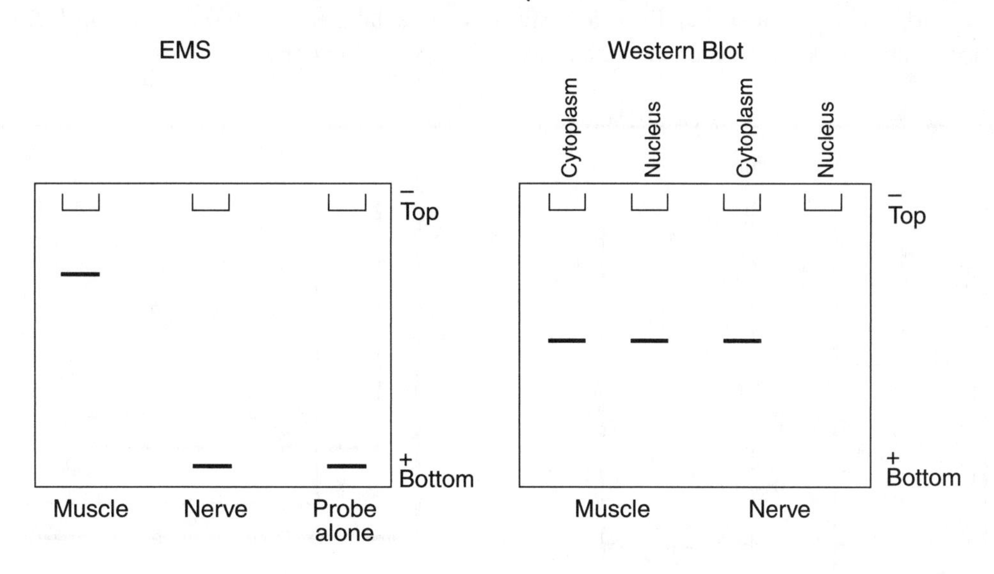

183. The EMS results suggest that
    - (A) TFY is expressed only in muscle cells.
    - (B) TFY binds DNA in muscle cells, but not in nerve cells.
    - (C) TFY binds DNA in nerve cells, but not in muscle cells.
    - (D) TFY is expressed only in nerve cells.
    - (E) TFY is expressed both in nerve and muscle cells.

184. According to the results of the Western blot,
    - (A) TFY is expressed only in muscle cells.
    - (B) TFY binds DNA in muscle cells, but not in nerve cells.
    - (C) TFY binds DNA in nerve cells, but not in muscle cells.
    - (D) TFY is expressed only in nerve cells.
    - (E) TFY is expressed both in nerve and muscle cells.

185. The data from the experiment seem to suggest that the most likely mechanism for the regulation of TFY is
    - (A) transcription.
    - (B) translocation to the nucleus.
    - (C) translocation to the cytoplasm.
    - (D) phosphorylation.
    - (E) ADP-ribosylation.

CONTINUE ON THE NEXT PAGE

Questions 186–187

The four small plots on the left show population growth for two species of freshwater diatoms, *Asterionella formosa* and *Synedra ulna*, in relation to the amount of two critical resources: phosphate and silicate. The estimated population sizes over time are indicated by dotted lines, while the amount of the particular resource is indicated by a solid line showing depletion over time. Each of the four small plots shows the results of these populations growing in isolation.

The larger figure to the right shows the supply rate of resources as points A–D, with each axis representing the amount of a particular resource. The unlabeled dark lines (ZGNIs) show the critical thresholds for each species below which they do not have enough of a particular resource to survive. The leftmost ZGNI is for *Asterionella* and the rightmost ZGNI is for *Synedra*. The labeled, dotted arrows indicate the resource consumption vector for each of the two species as their populations grow.

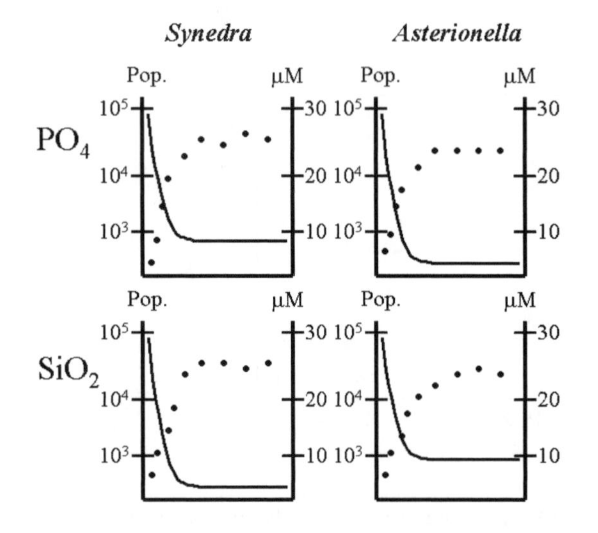

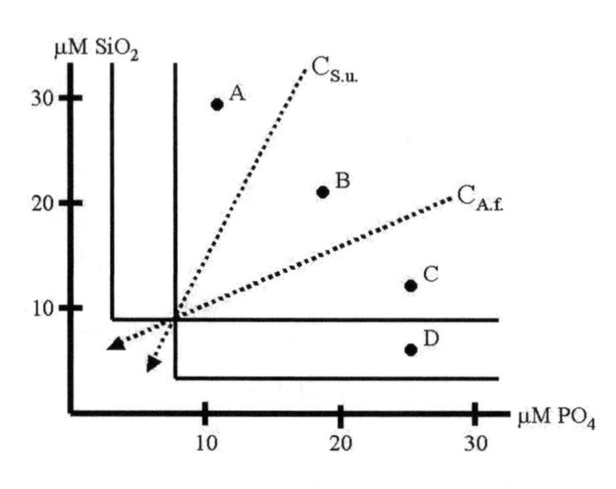

186. Which point, A–D, in the figure to the right represents the resource supply point at which both species of diatom can coexist?
   (A) Point A
   (B) Point B
   (C) Point C
   (D) Point D
   (E) Coexistance is not possible for these competing species.

187. If a habitat has phosphate and silica concentrations that would place it at point C on the figure to the right, and if small founding populations of *Synedra* and *Asterionella* were introduced, what would happen to the size of the *Synedra* population as time went on?
   (A) Initial increase to an asymptote.
   (B) Initial increase, followed by a decline to an asymptote.
   (C) Initial increase, followed by a decline to extinction.
   (D) Initial decrease, followed by an increase when the *Asterionella* population goes extinct.
   (E) Initial decrease to an asymptote.

Questions 188–189

The figure below presents a 15 base-pair homologous sequence of DNA analyzed across four species and used to generate the indicated phylogeny of these species: A, B, C, and D. The branches in the phylogenetic tree on the left are determined based on multiple nucleotide differences in this 15 bp sequence as shown to the right.

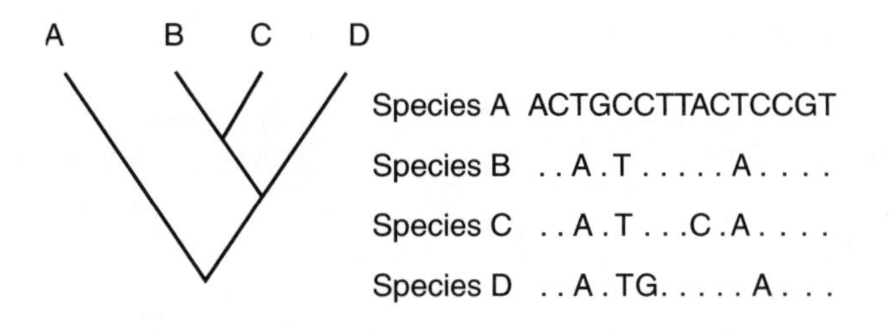

188. Position 11 of the 15 bp sequence (counting from left to right) can be used to define which of the following for species B and C?

    (A) synapomorphy
    (B) homoplasy
    (C) phenotype
    (D) symplesiomorphy
    (E) autapomorphy

189. If we consider species B and D as a group unto their own, it would be a ______ group?

    (A) biphyletic
    (B) cladistic
    (C) monophyletic
    (D) paraphyletic
    (E) polyphyletic

Questions 190–193

The study of population settlement on newly formed islands involves not only an assessment of the distance from the islands to the mainland nearby, but also research into the factors involving the movement of species from the mainland to the islands. Immigration of organisms from mainland populations establishes island populations, yet the extinction of species on these islands is possible as they struggle to take hold. The plot below shows the immigration and extinction rate curves, λ and μ, for two recently formed volcanic islands, A and B, located near a large mainland source.

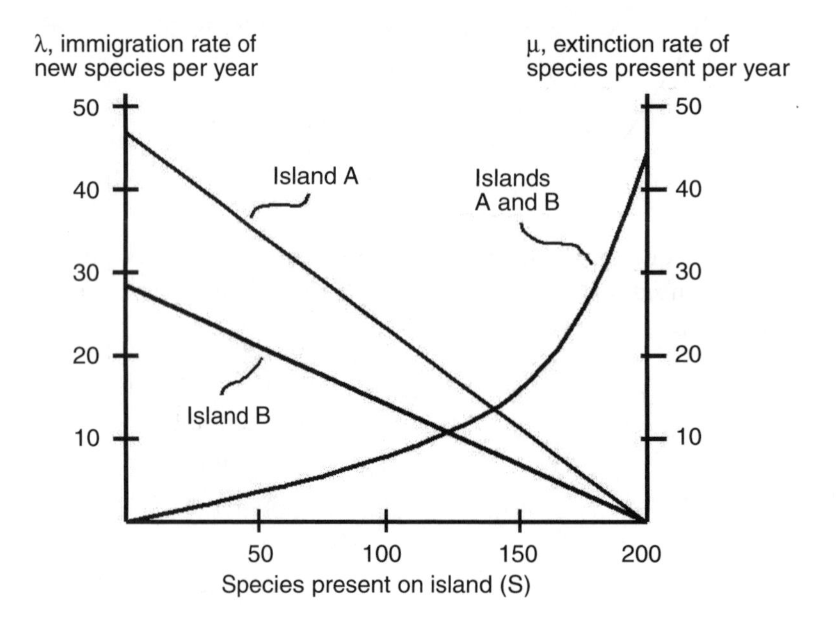

190. Which of the following must not be true given the immigration curves for island A and island B above?

    (A) Island B is farther from the mainland than island A.
    (B) Island B is the older of the two islands.
    (C) Island A experiences a higher number of immigrants from the mainland.
    (D) Both islands are the same size and elongate, but for island A the longest axis is oriented perpendicular to the mainland, whereas for island B the longest axis is oriented towards the mainland.
    (E) Islands A and B are very dissimilar in their diversity and quality of habitats.

191. An analysis of these curves reveals which of the following pieces of information?

    (A) Some species colonizing these islands are better colonizers than others.
    (B) The dispersal of these organisms is due to wind rather than to water currents.
    (C) Island A is larger in size than island B.
    (D) As species accumulate on the islands, their individual extinction probabilities increase.
    (E) Islands A and B are very dissimilar in their diversity and quality of habitats.

192. When island A reaches equilibrium, which of the following quantitative statements is correct?

(A) The rate of species turnover on island A is 10 species per year.
(B) The rate of species turnover on island A is 14 species per year.
(C) The rate of species turnover on island B is 10 species per year.
(D) The equilibrium number of species on island A is approximately 125 species.
(E) The equilibrium number of species on island A is approximately 200 species.

193. As organisms spread out from the mainland to settle on islands A and B, it is possible that new species may evolve on these islands from a single, mainland species. This evolutionary phenomenon is known as

(A) adaptive radiation
(B) hybrid inviability
(C) sexual isolation
(D) sympatric speciation
(E) panmictic mating

Questions 194-197

Several subspecies of *Drosophila* were measured in three locations on the Pacific coast of North America for the presence of allelic variants of the *abc* gene. The measured frequencies and their standard deviations for two alleles of the *abc* gene, ^{H}abc and ^{C}abc, are shown in the graph below.

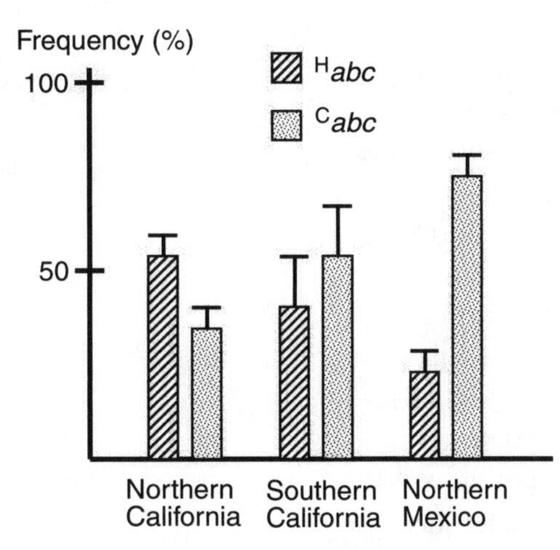

194. Which of the following can be inferred from the data presented above?

(A) There are only two *abc* alleles present in these subspecies of *Drosophila.*
(B) Behavioral factors determine the fitness of individuals with ^{H}abc or ^{C}abc.
(C) Individuals with the ^{C}abc allele are more fit overall when averaged across the entire geographic range of the species.
(D) Fewer specimens were collected in southern California than in other locales.
(E) Genetic drift accounts for the differences in the allele frequencies in the different locales.

CONTINUE ON THE NEXT PAGE

195. The ^{H}abc or ^{C}abc alleles follow a clear trend in frequency across the geographic range of Drosophila on the Pacific coast. Assuming that the trend in frequency seen is statistically significant, this data is suggestive of

(A) changes in allele frequency due to inbreeding.
(B) a response to an ecological cline.
(C) presence of more competitors in Mexico.
(D) higher use of insecticides in California.
(E) a high mutation rate at the abc locus.

196. Hybrid flies created by the sexual union of flies from the southern California population and the northern Mexico population could be expected to

(A) resemble flies from northern Mexico because of the large differences in the frequencies of ^{H}abc and ^{C}abc alleles between the two populations.
(B) exhibit a balanced polymorphism between the expression of ^{H}abc and ^{C}abc alleles.
(C) resemble flies from southern California more because the difference in frequencies between the ^{H}abc and ^{C}abc alleles in the southern California population is far less than that in the northern Mexico population.
(D) will likely exhibit ^{H}abc and ^{C}abc allele frequencies intermediate between Californian and Mexican populations.
(E) may show an increase in the frequency of the ^{C}abc allele without a decrease in the ^{H}abc allele.

197. If hybrids between the northern California subspecies and the southern California subspecies were not as fit as offspring born within the northern and southern populations, which of the following would be true?

(A) The lower fitness of the hybrids would act as a barrier to the flow of alleles between northern and southern populations.
(B) Average frequencies of ^{H}abc or ^{C}abc alleles would be lower in the hybrid zone than in either of the parent populations.
(C) The hybrid zone would generally be a wide area in between the two parent populations.
(D) Hybrids would become reproductively isolated, leading to the formation of a new *Drosophila* subspecies.
(E) The frequencies of ^{H}abc and ^{C}abc alleles will always be an exact average between the frequencies of those alleles in the parent populations.

Questions 198-200

A study of the symbiotic interactions among a group of species is represented in the diagram shown below. Positive interactions are represented by arrows from the organism acting as benefactor to the organism that is benefiting. Negative interactions are represented by a line and terminal circle, with the circle closest to the species that is adversely affected by the relationship.

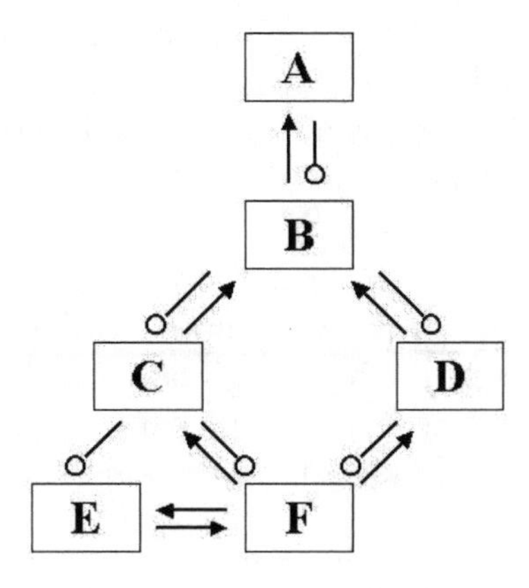

198. The nature of the interaction between the species labeled C and D is best described as

    (A) predation.
    (B) exploitative competition.
    (C) interference competition.
    (D) mutualism.
    (E) amensalism.

199. If species B goes extinct, which of the following species would benefit?

    (A) Species A, B and D only.
    (B) Species C and F only.
    (C) Species C and D only.
    (D) Species C, D and E only.
    (E) Species A, C and D only.

200. The ecological interactions that involve species E can best be described as

    (A) mutualism and amensalism.
    (B) commensalism and predation.
    (C) amensalism and commensalism.
    (D) competition and predation.
    (E) predation and mutualism.

END OF TEST

ANSWERS AND EXPLANATIONS

1. (B)

The citric acid cycle, otherwise known as the Krebs cycle, takes place within the matrix (or inner fluid) of the mitochondria. All other answer choices are correctly placed within their proper areas of occurrence.

2. (D)

Although enzymes may decrease the activation energy needed to start a reaction, the overall free energy change between products and reactants does not change. In other words, products and reactants maintain the same potential energy difference in uncatalyzed reactions as they do in catalyzed ones. Be careful of (E) since this statement is the opposite of what enzymes do.

3. (E)

The molecule pictured here is a sequence of three amino acids linked together by peptide bonds. When trying to figure out what molecule this might be, the giveaway is the series of N-C-C bonds that occur. Amino acid chains all have this sequence with R groups hanging off the middle C of each amino acid. Saccharides are sugars, while glycerides are lipids, so those answer choices should be eliminated.

4. (C)

The bond shown here is a disulfide bridge, formed between cysteines that are across from each other in a protein's tertiary structure. The bonds help anchor and hold the 3-D tertiary structure together. Keep in mind that the primary structure of a protein is simply the amino acid sequence, while the secondary structure is made up of beta-pleated sheets and alpha helices caused by hydrogen-bonding between nearby amino acid side chains. Covalent bonds, such as disulfide bridges, make up tertiary structure only.

5. (B)

Chromosomes separate to opposite poles of the cell during anaphase of mitosis due to the rapid shortening of kinetochore microtubules, those parts of the spindle that are attached to the centromere regions of the chromosomes. Taxol interferes with the breakdown of these microtubules, thereby halting the separation of the chromosomes and stopping mitosis in its tracks, perfect for halting the spread of a tumor. No new transcription is required at this point (C and D are wrong) and spindle formation occurs earlier on in mitosis.

6. (E)

Barr bodies are inactivated X chromosomes commonly found in cells of female mammals. One of the two X chromosomes in each cell within the organism condenses to the side of the nucleus and cannot be used for transcription. The X chromosome of each pair that inactivates is apparently random, and results in some cells expressing traits coded for by alleles found on one X chromosome, while other cells express traits coded for by alleles on the other X. This leads certain patches of calico tabby fur to express orange coloration while other patches remain black.

7. (B)

Epistasis occurs when one gene alters the expression of another gene that is independently inherited. In this instance, the brown fur color allele is most likely affected by the presence of recessive alleles on a different chromosome that code for the expression of "no color." Don't confuse epistasis with epigenesis, the phenomenon where genes are expressed differently depending upon which parent they are inherited from.

8. (E)

If you were to draw out the chromosome and map the genes to this piece of DNA, you would see that they are in the following order (shown with map units between them): A—5—W—3—E —12—G. Thus, genes A and G are furthest from each other and most likely to split apart during a crossing over event in meiosis. A and G have the highest frequency of recombination. Mapping genes to a chromosome is like a puzzle—you have to try various positions of the genes until they fit together linearly in such a way that their distances from each other all work out according to the information given.

9. (C)

The structure pictured, showing nine microtubule triplets, can be found in basal bodies, which form the base of eukaryotic cilia and flagella. Within the cilia and flagella, however, are nine microtubule doublets surrounding a central pair of microtubules. Centrioles have the same structure as basal bodies, yet are often found as pairs oriented at 90-degree angles to one another within the cytoplasm.

10. (C)

Proteins can be activated in cells via phosphorylation. Kinases are enzymes that phosphorylate other enzymes or proteins. Yet, being proteins themselves, the kinases must generally be phosphorylated in order to work. So in an unphosphorylated state, the kinase described in the question stem would be inactive. Calmodulin is the transducer (messenger) of intracellular calcium signals here.

11. (A)

The key here is to recognize that the question describes two protein products that are both coded for by the same gene. In general, these differences come from the fact that various exons within a gene can be spliced together in several ways to allow for the production of many different, but related, proteins from a single stretch of DNA. In this case, the proteins coded for by *double-sex* are different because of alternative RNA splicing after transcription.

12. (E)

The light reactions of photosynthesis produce electrons, which get picked up by $NADP^+$, as well as ATP due to a hydrogen ion gradient that is formed within the thylakoid membrane. The $NADP^+$ gets oxidized to NADPH as it gains electrons, and it is used along with the ATP to power the Calvin cycle, where sugars are produced. Don't be tempted by choice (B), since oxygen is a waste product of the light reactions, and is not used in the Calvin cycle.

13. (C)

The key function of fermentation is to regenerate NAD^+. The reason this is so essential is that as glucose gets broken down within the cytoplasm of a cell, electrons fly out of the broken bonds. These electrons are normally caught by NAD^+ and ferried into the mitochondria and ETC. Yet, fermentation occurs in the absence of the mitochondria and ETC. Thus, in order to keep a ready supply of NAD^+ around in order to catch electrons coming off the broken glucose molecules, the NADH must be re-oxidized into NAD^+ by the process of fermentation. This allows the breakdown of glucose to continue and the subsequent production of small amounts of ATP via glycolysis.

14. (B)

The main problem with trying to insert a eukaryotic gene into a bacterium is that bacteria have no introns, and thus have no machinery to remove them. Choice (A) is incorrect here for that reason. Because of this, the most successful transformations can be achieved using cDNA, or complementary DNA, which is made off an mRNA template of the gene of interest. The mRNA has already been spliced to remove introns, so using that mRNA to create a complementary DNA molecule lets you use the "pure" gene without exons in your experiment. Although the use of reporter genes and antibiotic resistance may be important after the transformation occurs (in order to select transformed colonies), reporter genes are not needed for the success of the experiment (choice C) nor can one select for antibiotic resistance *before*hand.

15. (D)

All of the choices listed describe post-transcriptional modifications, those that take place *in the nucleus* after transcription, except choice (D), the attachment of mRNA to polyribosomes. Polyribosomes are groups of multiple ribosomes that can attach to mRNA to read the codons and produce multiple polypeptide chains at the same time. This takes place outside the nucleus during translation.

16. (B)

RFLPs are restriction fragment length polymorphisms. If two individuals' DNA is cut up using the same restriction enzyme, there will be differences in the lengths of the fragments created because the individuals have slight differences in nucleotide sequences within their DNA. These restriction fragments can be compared using a gel to separate the fragments by size. The banding patterns on the gels for each individual can be analyzed to determine where genetic differences may lie. RFLPs may be caused by single nucleotide differences, as in choice (C); however, DNA fingerprinting is not based on the analysis of single nucleotide differences between two individuals, but rather much larger sections of DNA.

17. (E)

Gene linkage leads to phenotypes in the offspring that are similar to the parental phenotypes. In fact, the more closely linked the genes are on the same chromosome, the less likely it is that recombinant (nonparental) phenotypes will occur due to crossing-over. Do not confuse gene linkage with X-linkage, which is simply when genes are located on the X-chromosome.

18. (D)

The LDL receptor proteins cannot have an ER-retention signal, which as the name implies causes proteins to remain in the ER after synthesis. In fact, proteins with ER-retention signals will be shipped to the Golgi first, along with all other proteins made on the ER membrane and then returned to the ER. All answer choices except (D) seem reasonable to conclude about this protein receptor. It is anchored into the cell membrane, suggesting it has regions of polar and nonpolar amino acids; its production would be increased during times of increased lipoprotein in the blood; and, it matures in the Golgi before being shipped to the cell membrane. Choice (E) also makes sense as a possible mutation.

19. (C)

This is a diagram of a mitochondrion. It is here that the reactions of cellular respiration take place. Pyruvate, produced from the breakdown of glucose in glycolysis, enters the mitochondrial matrix along with oxygen and NADH, the Krebs cycle takes place (as well as electron transport and oxidative phosphorylation), and ATP is produced. In fact, the letter X represents ATP, whose production is the most essential aspect of the entire cellular respiration process.

20. (C)

Irreversible inhibitors are molecules that fit into an enzyme's active site and bind covalently to the enzyme structure, thereby permanently disabling that enzyme. Allosteric effectors and non-competitive inhibitors act as sites *other* than the active site.

21. (A)

Being invertebrates and not chordates, rotifers do not have a notochord or dorsal nerve cord. In addition, being pseudocoelomate means that their internal body cavity is only partially covered with mesoderm. The question stem states that rotifers have a complete digestive system, so that is equivalent to their having a mouth and an anus. Make sure to know the basic vocabulary of animal phyla for the test!

22. (D)

Of all the answer choices, the only one that does not deal with bacterial infection of other organisms' cells is choice (D). While the ability of bacterial cells to exchange genetic information through pili make help antibiotic-resistance spread, all the other choices are more direct causes of increased virulence within a given organism. Look out for answer choices such as (D) that do not fit in with the pattern given by all the other choices.

23. (A)

Neither NADH nor NAD^+ serve as electron acceptors within the electron transport chain of eukaryotes. However, oxygen does. Recall that oxygen is the terminal electron acceptor in the ETC of most organisms, allowing water to be formed from hydrogen and electrons coming off the ETC. Without the presence of oxygen, ATP production would rapidly stop.

24. (C)

RNA is a single-stranded molecule that does not hybridize to itself readily. However, DNA does. If a DNA probe contains the genes for the rRNA and can hybridize to it, it will be the best molecular probe to search the bacteria for the presence of certain RNA sequences. The double-stranded DNA in choice (E) cannot hybridize to anything in its double-stranded state.

25. (C)

The word *lymphatic* should tip you off here that SALT is associated with the immune system. Although the lymphatic system is generally part of the circulatory system because it removes excess fluids from the tissues and returns them to the general circulation, it is a repository for immune system cells. B and T lymphocytes fill the lymphatic system, particularly the lymph nodes. In SALT, immune system cells known as Langerhans cells are phagocytic guardians that engulf and destroy pathogens that enter through wounds.

26. (A)

E. coli is a bacterial species, and bacteria have no introns within their genes. This is in contrast to eukaryotes, which not only have introns but also possess multiple origins of replication (rather than *E. coli*'s single origin), and linear chromosomes (rather than the single, circular one of bacteria). Choice (C) must also be eliminated since RNA genes must be found on DNA if they are to be transcribed at all.

27. (B)

Neurons send their signals in an all-or-none fashion, meaning that once they initiate an action potential, the internal voltage of the neuron climbs rapidly to a certain level no matter the scale of the initial stimulus. In other words, most neurons reach an internal voltage of +50 mV for each action potential no matter how intense the stimulus is. The reason that you can feel one stimulus as being more intense than another has to do with the number of different neurons involved or with the frequency of firing, not the voltage-change of any individual neuron. All other statements regarding nerve cell potentials are correct.

28. (B)

Neurotransmitters that are inhibitory would not be expected to open sodium channels, because the opening of sodium channels rapidly increases the flow of positive ions into the axon, resulting in depolarization. Again, if this depolarization is great enough, the axon's membrane will pass a threshold point and an action potential will be initiated. Inhibitory neurotransmitters will, however, open chloride ion or potassium ion channels, both of which will hyperpolarize the inside of the axon relative to the outside. Chloride channels will cause the inflow of negative chlorine ions, making the inside of the axon more negative and, thus, harder to depolarize; and, the opening of potassium channels will cause the outflow of positive potassium ions, again causing the inside of the axon to become more negative and harder to depolarize.

29. (D)

The molecule pictured is urea, a nitrogenous waste product formed as amino acids and proteins are broken down by cells. Urea picks up free amino groups (—NH_2) and links them via a carbon atom so that they can be filtered out by the kidney and eliminated in the urine. Remember that amino acids all have a carboxyl group (—COOH) as well and urea does not.

30. (C)

The only mutation listed that could change one amino acid into another amino acid without altering the rest of the protein's primary sequence is a point mutation, changing one base out of the three that make up the codon coding for cyteine. A nonsense mutation, choice (D), would result in a change that would cause a premature termination of the protein rather than replace one amino acid with another. Frameshift mutations, including single base-pair deletions would shift the entire reading frame of the mRNA strand from the point of mutation onward, thus changing many amino acids in the primary sequence.

31. (E)

The bone marrow is the birthplace for blood cells, both red and white. It is in the bone marrow that pluripotent stem cells reside, which can turn into a variety of different blood cells. Anemia is the term given to the condition caused by having too few red blood cells, and hence less oxygen available to the tissues of the body. Both the spleen and liver are involved in the removal of old, worn-out red blood cells, and the thymus is an organ in the neck where a specific white blood cell, the T-lymphocyte, matures. If the bone marrow is damaged by radiation, anemia is the most likely result due to a rapid decrease in red blood cell production.

32. (A)

The toxin blocks sodium channels; therefore, depolarization, or the initial rise in positive charge within the neuron, is blocked. The opening of sodium channels is what causes a rush of positive sodium ions into the nerve cell when a stimulus arrives and it is this initial rush of sodium that causes the action potential to start. The phase after depolarization, when potassium channels open to restore the resting potential of –70mV, is called repolarization.

33. (A)

You can remember parasympathetic actions in the body by recalling "rest and digest." In contrast, sympathetic actions can be remembered by "fight or flight." Rest and digest actions include increased digestive system function (acid secretion and gastric motility), as well as slower heartrate, decreased blood pressure, and constricted pupils. With less parasympathetic activity after vagotomy, you would experience decreased gastric motility, decreased HCl secretion, dilation of the eyes and increased blood pressure; but, you would not experience decreased heartrate because the increase in sympathetic activity would speed up the heart.

34. (B)

The transfer of electrons in the respiratory chain results in the transport of H^+ ions from the mitochondrial matrix to the intermembrane space. When an alternative energy source, such as fatty acids, is introduced it can enter the respiratory chain by transferring its electrons to unbiquinone. Ubiquinone can then proceed to transfer the electrons to other intermediates of the respiratory chain, causing the movement of hydrogen ions across the inner mitochondrial membrane. Choice (C) is incorrect because taking electrons from succinate does not reduce NADH. However, succinate's electron transfer will have an effect on ATP synthesis—it will increase it—thus, choices D and E are incorrect.

35. (B)

Maintaining the mitochondria in a solution at pH 8 until saturation will result in a slightly basic environment throughout the organelles. When you see basic solution, you need to think of a decrease in H^+ ions. When the mitochondria are suddenly moved to a solution of pH 4, the medium outside the inner membrane will immediately become acidic while the matrix will still be basic. This pH gradient across the inner mitochondrial membrane will drive the synthesis of ATP in the presence of ADP and P_i.

36. (D)

Recall that a sarcomere is a unit of layered actin and myosin filaments, and within a given muscle fiber there are repeating sarcomere units that traverse the length of the muscle cells. The shortening of the sarcomeres is regulated by a nerve impulse outside the fiber, resulting in calcium ion release from the sarcoplasmic reticulum, a specialized ER found in muscle cells. Calcium allows a conformational change to take place within the sarcomere such that tropomyosin is pulled off of the binding site (troponin) for the myosin heads on the actin filaments. With the myosin binding sites uncovered, myosin can attach to actin and slide past it, shrinking the length of the sarcomere.

37. (B)

If proteins meant for the lysosomes are, in fact, found outside of the cell, that means these proteins have been incorrectly targeted after translation. Proteins are "labeled" for transport by both the ER and the Golgi, where various enzymes cleave and shape the newly formed protein and add sugar chains to these proteins. If certain sugar chains don't get added, though, the proteins will not be targeted properly—e.g., to the cell membrane, out of the cell, to the lysosomes, etc. Choice (B) is the most likely reason that the enzymes are not targeted to the lysosomes. The defect with the ribosomal enzymes in Choice (A) would lead to far more widespread problems, as perhaps every enzyme made (not only the lysosomal ones) would be defective. Choice (C) is incorrect because lysosomes do not bind to the cell membrane. That is a job for vesicles. Choice (D) would leave the enzymes in the nucleus rather than outside the cell, and Choice (E) makes no sense—proteins are not lipids (steroids).

38. (D)

Passive immunity results when antibodies are transferred from one individual to another. Often this occurs when a pregnant woman passes on her own antibodies to her fetus through the placenta, or to her baby via breast milk. Since antibodies are proteins, they are eventually degraded, and passive immunity may last only for a few days or weeks. All choices other than (D) are examples of active immunity, whereby one's immune system, after being exposed to a particular pathogen, mounts a response and creates memory B and T lymphocytes that can be activated very quickly in the event of reexposure.

39. (D)

Compared to inhaled air, the pulmonary capillaries have a higher partial pressure of CO_2 and a lower partial pressure of O_2. Thus, in the alveoli, gas exchange occurs when CO_2 flows down its concentration gradient from the capillaries to the alveoli, and O_2 flows down its gradient from the alveoli into the capillaries. The presence of surfactant lowers the surface tension of the alveoli and keeps the membranes moist to facilitate gas exchange. Therefore, statements I and III contribute to gas exchange—choice (D).

40. (E)

Keep in mind for this question that pregastrula cells have not yet differentiated. In other words, the blastula cells in chordates have the potential to turn on or off any of their genes as gastrulation occurs and endo-, meso-, and ectoderm begins to form. Removal or movement of nondifferentiated cells, even cells that have yet to become organizers, cannot cause any effect because these cells have not yet differentiated. Selective gene loss would not occur in pregastrula cells and, if it did, it would result in certain cells already taking steps to develop along certain lines. The removal or rearrangement of these cells, then, could result in drastic changes in the embryo. In addition, selective gene loss generally does not occur in chordates.

41. (C)

Anchoring junctions are those that are found between cells subjected to fair amounts of stress, either from shearing forces or contacting forces. They connect two cells via a series of linker proteins often involving actin filaments.

42. (C)

Be careful of the trap answer here, choice (D). Insects do not transport oxygen via their hemolymph, a blood-like fluid that bathes all of their cells via an open circulatory system. Rather, insects have a complex tracheal tube system, where openings called spiracles in the insect underbelly feed oxygen into a series of branching tubes that reach every cell of the insect.

43. (E)

Lichens are symbiotic unions of a fungus and an alga. The fungus supplies water and other absorbed nutrients, while the alga supplies various products of photosynthesis. They live in many unpolluted areas, on rocks, bark, and on the ground.

44. (D)

All of the statements are true of restriction enzymes except that they can be used for nuclear transfer in cloning. In nuclear transfer, you may remember, an entire nucleus is transferred from the cell of one organism into an enucleated egg for the purpose of creating a genetically identical offspring. Restriction enzymes would certainly not be helpful for this purpose, as they rip apart DNA.

45. (A)

Gametangia are the separate hyphae, one male and one female, that contain several haploid nuclei each and reach toward each other to fuse, a process called plasmogamy. The fusion of these hyphae gives rise to dikaryotic hyphae, or hyphae that have two nuclei, which only later fuse to create a fully diploid cell. Mycorrhizae are mutualistic associations between fungi or plants and symbiotic bacteria that help fix nitrogen and other substances for use in the fungus or plant. Haustoria are parasitic hyphae that contain nutrient-absorbing tips to penetrate the host's cells.

46. (E)

The only major difference between marsupials and placentals is that marsupials' young are born (leave the uterus) at a much earlier stage than the young of placentals. Marsupials finish out their development within a pouch filled with mammary glands for nourishment. Placentals also have mammary glands (although not in a pouch). All mammals have a layer of insulating hair and only the monotremes (such as the duck-billed platypus) lay eggs.

47. (C)

This is a completely deductive question that requires you to know almost nothing about the specifics of human evolution. If fossils of *erectus* were found in Asia and dated to the same time period in which *sapiens* lived, this one finding proves that *sapiens* did not evolve *from* erectus (at least in Asia)—otherwise, there would be no *erectus* fossils to find there at that time. The fact that there have been numerous *Homo* or hominid species in the past would neither strengthen nor weaken the multiregionalist hypothesis, nor would Leakey's or Dart's findings on Australopithecines directly address the *erectus/sapiens* issue.

48. (D)

Recall that the mother is type-A, but has children who do not have type-A blood. Thus, she must be able to pass onto her kids an allele other than an I^A. Because her daughter is type-O, this second allele of the mother's must be a recessive i. In addition, because the mother is Rh-positive, but must also be able to pass on an Rh-negative allele to her son, the mother must have the Rh genotype of Rr. This excludes every answer other than choice (D).

49. (C)

The diatoms are algae known for their box-like silica shells. Accumulations of these shells on the seafloor can be collected and used for many commercial purposes. You should have eliminated choice (B), zygomycetes, because the suffix *-mycetes* should signal to you that that is a fungus or fungus-related.

50. (B)

Although all of the phyla listed as answer choices are bilaterally symmetric (except for Porifera, the sponges), only one has organisms classified as deuterostomes—Chordata. You should remember for the GRE that deuterostome embryos have indeterminate cleavage. Each cell produced in the early zygote by mitotic cleavage retains the ability to become a complete embryo on its own. There is no selective gene loss in chordates. In contrast, the other phyla mentioned in the answer choices (again, with the exception of Porifera) are either protostomes or are not coelomate organisms (and, thus, are not classified as proto- or deuterostomes).

51. (B)

Organisms may adapt and respond to their environment during their lifetime, yet these changes are not heritable (able to be passed down to offspring). Choice (B) is much more Lamarckian than Darwinian in nature, and it should be picked as the correct answer here. All other statements are Darwinian in content.

52. (D)

A community is defined as a group of different species living together in the same habitat. The word community, as differentiated from the word ecosystem, does not include the non-living, or abiotic, factors within the habitat of these populations.

53. (A)

If one plant dies while the other is strengthened because of this separation, it is likely that one plant was preying on the other as a parasite. Only after removal does the host species grow taller, but the parasite will die off because of lack of nutrition. It is not clear that the separation would affect plants in a commensalistic relationship, and those in a competitive relationship would both be strengthened by the separation.

54. (B)

The molecule pictured is a phospholipid, found in all membranes that make up cells and their organelles. Because secretory and Golgi vesicles derive from the membranes of organelles, they too are made of phospholipids. The only structure listed that is not made of lipid, but rather of the amino sugar chitin, is the chitinous cell wall of fungi.

55. (A)

Given that diarrhea is an excess of water in the stool, usually caused by improper water and salt absorption in the intestines, the most probable mechanism of Rotavirus action is that it damages intestinal epithelial cells so that they can no longer absorb proper amounts of water and salt. Choice (C) can be eliminated since an increase in absorption of water and salt could lead to constipation but not to diarrhea; and, choice (D) can be eliminated since the kidneys are not part of the digestive system, and that is where diarrhea arises. (E) makes little sense, since viruses would be too small to absorb enough, if any, food particles to cause any great effect on the stool. Finally, eliminate (B) because viruses have no protein making machinery so that the only proteins that would be secreted would have to be produced by intestinal host cells in the first place. While plausible, this is not the best answer to consider here.

56. (C)

A tumor virus genome can acquire one or more genes from the cells of the host it infects. These genes can undergo conversions within the virus as mutations occur, changing proto-oncogenes (normal cellular genes controlling cell growth) into oncogenes, cancerous versions of the normal genes. *Ras* genes exist in non-cancerous cells, yet mutant *ras* genes can lead to uncontrolled cell division.

57. (E)

Along with progesterone, estrogen is able to exert a negative feedback on the brain, blocking the secretion of both LH and FSH. However, *high* concentrations of estrogen, released prior to ovulation, causes a sudden surge in LH secretion by the anterior pituitary that results in ovulation. In addition, the corpus luteum is maintained by high LH concentration and is responsible for its own destruction because it secretes estrogen and progesterone that feedback inhibit the pituitary. As soon as estrogen and progesterone levels rise, approximately 10–12 days after the corpus luteum forms, LH concentration drops and the corpus luteum breaks apart.

58. (C)

The major difference between endo- and ectotherms comes not from the temperature at which either type of organism maintains its body, but rather from the source of body heat that each type of organism uses. Ectotherms derive most of their body heat from the surrounding environment—surrounding temperature is termed the "ambient" temperature. However, endotherms can generate body heat or cool themselves off using metabolic reactions to keep their body temperature fairly constant.

59. (B)

All of these statements concerning asexual reproduction are correct, except that asexual reproduction is best in favorable, stable environments, ones that don't change rapidly. The reason for this is that asexual reproduction, in contrast to its sexual counterpart, results in the formation of identical offspring. Although asexual organisms can often produce many more offspring in a single reproductive event than sexual organisms, these asexually produced young do not usually have the genetic variation caused by meiosis and crossing-over to be able to survive a rapidly changing environment or times of environmental stress.

60. (E)

Bark is made up of phloem and periderm. The periderm is a tissue built from a cylinder of meristem called cork cambium, which can be found in the outer parts of the stem. Xylem, made up of tracheids and vessel elements, is found internally. The reason that trees will die if you cut off their bark is that they lose the means by which to ferry sugars down from the leaves to the roots, since their phloem tissue has been destroyed.

61. (B)

Phage lambda must enter a bacterial host to replicate. The phage has two methods of reproduction: the lytic cycle and the lysogenic cycle. In the lytic cycle, the phage attaches to a host bacterial cell and injects its DNA into the bacterium. The virus utilizes the nucleotides, enzymes, and ribosomes of the host bacterium to replicate, and it organizes replicated viral DNA as well as viral proteins to build new phages. The viral particles will kill the host bacterium as they burst out of its cell membrane. In a lysogenic cycle, the phage attaches to a host bacterial cell and injects its DNA into the cell, after which the viral DNA gets integrated into the bacterial genome, remaining there and copying itself over many generations as the bacterial cell replicates.

62. (C)

In irreversible inhibition, a molecule other than the substrate covalently bonds to the active site of the enzyme, preventing substrate molecules from accessing the active site. Although competitive inhibition, where chemicals compete with substrate molecules for a spot within the active site, may slow down an enzymatic reaction, irreversible inhibition may stop a reaction altogether and permanently. Noncompetitive inhibition takes place when a chemical decreases the affinity of an enzyme for a substrate through binding to a location on the enzyme other than the active site, inducing a conformational change in the active site without ever entering that area.

63. (A)

Organisms at the top of the food chain have the least biomass, while organisms at the bottom have the greatest biomass. In this food chain, grass is at the bottom and has the greatest biomass.

64. (B)

Bryophytes (mosses) are avascular plants that alternate between haploid and diploid generations and have flagellated sperm. Make sure to know the basic differences between plant and animal phyla!

65. (E)

Dicots, not monocots, have parts of their flowers (petals) in multiples of four or five; monocots, however, have petals in multiples of three. All other characteristics in the answer choices are traits of monocots.

66. (D)

Members of the phyllum Mollusca (snails, for example) undergo torsion during development and possess a mantle and anus above their heads. You could also have taken a stab at the answer by recognizing that the word mantle should be associated with creatures such as molluscs.

67. (C)

According to the heterotroph hypothesis, the earliest forms of life were probably unicellular organisms that used organic molecules as their sources of food.

68. (C)

Punctuated equilibrium postulates that most evolutionary change happens quickly in small isolated populations that break off from larger groups, so this eliminates (B). Between bursts of change, phenotypes in a population hover around some mean value, not changing very much. Even though change is rapid in punctuated equilibrium, it relies on variation in the population, not on the sudden production of novel features. In the fossil record, two different phenotypes would mark one of these punctuation events and would, therefore, not be arbitrary.

69. (A)

Under neutral conditions, it should take $4*N_e$ generations for an allele to become fixed in a population where N_e is the effective population size. Thus, these 500 alleles all shared a common ancestor 40,000 generations ago.

70. (C)

The initial frequencies are irrelevant as long as each allele can mutate into the other. The important thing to note is that 80% (0.004/(0.004+ 0.001)) of the mutations form the allele *a*, while only 20% (0.001/(0.004+ 0.001) form *A*. Thus, it is these frequencies (.80 and .20) that will be the equilibrium frequencies, although it may take the population several (many?) generations to reach this equilibrium.

71. (E)

Carrying an allele has nothing to do with dominance or recessiveness. There is no selection (as stated, the population is in Hardy-Weinberg equilibrium), and the question is not asking about phenotypes. Thus, the only way for males in the F_1 not to carry allele *A* is if this is an X-linked allele and the F_1 males got allele *B* from their mothers. In addition, males carrying allele *A* would pass it only to females (assuming it was an X-linked allele). The results described in the question stem come from tracking the transmission of an X-linked gene. Thus, choice (E) is best here.

72. (C)

Organisms in the phylum Porifera lack true tissues. These multicellular animals have cells that are not specialized (i.e., developed into muscle or nerve, or connective tissues). All other animals possess specialized cells and tissues.

73. (C)

Two genes that are close together on a chromosome will have a low frequency of separating from one another during crossover events. The frequency that they do split up and end up in different gametes from each other is called the recombination frequency. The closer two genes are, the lower the frequency. Here, genes P and Q will split up 35% of the time, giving them a "chromosomal distance" of 35 "map units." Genes R and Q will split up 20% of the time, giving them a chromosomal distance of 20 map units. The distances for the other genes can be read off the chart. Once you know all the distances, you can work out the relative positions of the genes to one another much like a puzzle. How can you fit together P, Q, R, and T in a linear pattern that matches the units in the chart? The answer is choice (C).

74. (D)

Only choice (D) here is a Lamarckian explanation of why a vestigial organ would disappear. All other explanations are Darwinian in nature and can be considered valid reasons that a nonfunctional organ may be reduced in size to the point of disappearing. In (E), a negative genetic correlation between two traits means that the development of one trait is inversely proportional to the development of another. As selection favors sense organs other than eyes in cave-dwellers, eyes may disappear if the development of eyes were negatively-correlated with the development of other sensory appendages.

75. (B)

Speciation caused by polyploidy typically takes place within the parent population geographically. In other words, a new species of plant may crop up sympatrically within a parent population and be unable to interbreed because of differences in chromosome number. This is considered postzygotic isolation because backcross offspring produce offspring that have too many or too few chromosomes and are usually infertile. Think about this type of isolation that occurs after (post) zygotes have formed, isolating the new zygotes from the original, parent population.

76. (C)

In an iteroparous life history, organisms save their energy for later reproductive efforts, delaying their maturity and investing energy early on in their own growth and self-maintenance. This type of reproductive strategy is found in species that live in environments where adult mortality is low and where the costs of reproduction are high, such that the organisms need to be large and strong before reproducing yet the reproductive effort is so great that they will be unlikely to repeat it (think of salmon spawning as adults only once at the end of their lives). In contrast, semelparous organisms reproduce once early on in life and then usually die. This kind of strategy is found where juvenile survival is high but the chances of making it to adulthood are low. It generally allows for extremely rapid population growth because the cycling of generations takes place much faster than in iteroparous species.

77. (A)

Polygamy is a mating system that involves one male and many females. The most fit males are able to mate with a multitude of females, thereby passing their genes to a maximum number of offspring.

78. (C)

Paedomorphies are features present in adults that are typical of features in the juvenile stage of that organism's ancestor. Paedomorphic features often arise because of a delay in growth of a particular organ, so that it remains underdeveloped in the adult. An example occurs in some salamander species whose adults retain the gills that are present only in juvenile salamanders of related species.

79. (A)

Because males can pass to males, this allows us to rule out that the trait is sex-linked. Fathers will only pass Y-chromosomes to their sons. Because parents who do not show the disease can have kids who do show the disease, this allows us to rule out that the trait is dominant. Choice (A) is correct here.

80. (A)

Vasopressin, otherwise known as ADH or antidiuretic hormone, regulates extracellular fluid volume and blood pressure by affecting how the kidneys handle water. Specifically, ADH increases the permeability of cells lining the collecting duct to water, but does not work by increasing salt absorption.

81. (E)

All three statements are correct concerning linkage disequilibrium, an important concept related to speciation, sexual selection, and molecular evolution. Think of gene linkage when you are asked about linkage disequilibrium. Mechanisms that maintain linkage disequilibrium keep linked genes linked. Selective mating, low recombination, and mutations that are not separated from linked genes through recombination are all reasons for two genes to remain linked.

82. (A)

Turner syndrome, otherwise known as monosomy X, occurs when a fertilized egg possesses only one X chromosome because of a nondisjunction event during male or female meiosis. Klinefelter syndrome occurs in males with genotype 47, XXY (they possess an extra X chromosome in every cell). Down syndrome is also known as trisomy-21, where individuals have three copies of chromosome 21 in every cell; and, cri-du-chat syndrome occurs from a deletion of part of the short arm of chromosome 5. It was given its name because a crying infant affected by this syndrome sounds like a mewing cat.

83. (A)

All of the statements in the answer choices are correct, except for choice (A). Both gram-positive and gram-negative bacteria can possess capsules, which are layers of sugar or protein external to the bacterial cell wall. Capsules allow for bacteria to be much more pathogenic (disease-causing) and to evade destruction by the immune system.

84. (C)

Retroviruses transcribe their RNA genome into DNA after they infect a host cell. They do this by carrying the enzyme reverse transcriptase into the host cell with them, so that they are able to turn ssRNA into dsDNA. This DNA can then integrate into the host cell genome or can be used for immediate transcription of viral genes.

85. (E)

An agglutination test would be best to determine which blood type the person had. Antibodies to each of the major blood cell surface proteins (A and B) could be used. If blood type A were mixed with antibodies to blood type A, one would observe under the microscope a rapid agglutination, or clumping, reaction in which the red blood cells would be physically clumped together by antibodies binding them to each other. This reaction can be fatal in the body if too much blood of the wrong blood type is given during a transfusion.

86. (A)

The only type of RNA that does not leave the nucleus is the primary transcript of DNA, which is called heterogeneous nuclear RNA, or hnRNA. This type of RNA is formed and processed wholly within the nucleus; its noncoding regions are excised and the resultant mRNA strand exits the nucleus through a nuclear pore.

87. (D)

Recall that a sex-linked (or X-linked) allele is carried on the X-chromosome and can be passed from fathers to daughters only, and from mothers to both sons and daughters. An early-acting lethal disease will kill all males in infancy, since males can never carry sex-linked alleles (they have only one X chromosome). Because of this, there will be no males of reproductive age to pass along the allele to daughters, so the disease can be passed on only by female carriers, who will pass the allele to 50% of their sons and daughters. A female can never be homozygous, since she could never have received an allele from her father (males don't survive past infancy with the disease).

88. (D)

All of the statements are correct with the exception of choice (D). Estrogen is released within the ovaries, but acts anywhere in the body that estrogen receptors exist. This includes acting on cells in the hypothalamus and pituitary in order to regulate the secretion of GnRH, LH, and FSH.

89. (D)

When fibers of a muscle are exposed to very frequent stimuli, the muscle cannot fully relax. The contractions begin to combine, becoming stronger and more prolonged. This is known as frequency summation. If the stimuli become so frequent that the muscle cannot relax, the contractions become continuous, which is known as tetanus.

90. (E)

All of the answer choices should stand out as examples of why imperfections in the fossil record prevent paleontologists from readily identifying evolutionary patterns, except choice (E). If fossils show rapid transition from one phenotype to another over a short period, that may be indicative of punctuated equilibrium as a pattern for this species.

91. (D)

Evolution generally does not proceed by creating novel features de novo. Natural selection and phenotypic change relies on the development of new features from adaptations of pre-existing ones. In the case of the reptile-to-mammal transition, a secondary articulation (attachment) developed between the lower jaw and skull in early mammals that allowed one of the bones previously involved in this connection (the quadrate) to "move" backward toward the ear and become one of the ear bones. In certain mammalian ancestors (notably, the advanced cynodonts), the skulls possess both a fully developed mammalian jaw attachment as well as a much weaker reptilian jaw connection. The duplication in the articulation between jawbone and skull allowed for the co-opting of the quadrate as an ear bone. You'll never have to know details such as this for the GRE; however, you should understand the concept that evolution works by using existing structures to form new features, and the only way this is possible, usually, is for existing structures to be duplicated in some manner so that one structure can act as it always has (i.e., to hinge the jaw) and the copied structure can change over time to fit into a new role (i.e., ear bone). It is this concept that is tested here. This also exists in molecular biology, as gene duplications allow for the formation of entirely new genes coding for new proteins while the original genes continue to code for the proteins they always had coded for.

92. (A)

Only an exoskeleton, usually made of the amino sugar chitin, is a characteristic of arthropods, out of the three traits listed here. Although arthropods are bilaterally symmetric, they are not acoelomate, meaning without an internal body cavity. Arthropods, in fact, possess a fully developed coelom and are considered coelomate organisms. Only echinoderms have a water-vascular system, not to be confused with the open circulatory system that arthropods have.

93. (C)

Secondary growth results in the outward growth of a tree (in width, not height). During secondary growth, secondary xylem and phloem are built from secondary (vascular) cambium, and these tissues expand outward creating the familiar tree rings that can be studied to tell the age of a tree. The initial xylem and phloem of a tree is built from the embryonic tissue called procambium, which builds a cylinder of vascular tissue called the stele in young trees. The pericycle is the outer region of this stele.

94. (C)

Lenticels, common in woody stems, are an avenue of gas exchange for very active tissues within the bark. Both the vascular cambium and the cork cambium are extremely active during growth and oxygen is needed for proper phloem transport as well. Pneumatophores, otherwise known as air roots, are present in trees such as mangroves, whose roots must use these extensions to exchange air above the water that submerges the base of the roots. Root hairs themselves exchange oxygen and carbon dioxide with small pockets of air within the soil, which is why you can kill a plant by overwatering it, thereby filling in these air pockets with water.

95. (B)

Calcium ions are stored in muscle cells within the sarcoplasmic reticulum (SR), and they are pulled back into the SR after muscle contraction for use in the next contraction. All other answer choices deal with structures that are involved in the neuronal stimulation of the muscle fiber (using acetylcholine), not in the contraction itself.

96. (D)

Don't forget that molecules such as glucose combine with each other via dehydration synthesis, a process that extrudes one molecule of water for each bond formed. In this case, two molecules of attached glucose would have one molecule of water removed as they bonded, leaving the final formula $C_{12}H_{22}O_{11}$ rather than $C_{12}H_{24}O_{12}$.

97. (B)

Early methods of food preservation created such a solute-rich, or hypertonic, environment on the preserved food that small organisms such as bacteria could not grow there. The high extracellular solute content would rive water out of their cells, causing them to shrivel and die due to an increased osmotic potential from outside to inside.

98. (C)

As you have already learned, the most common structure that allows a protein to span the hydrophobic regions of a membrane is an alpha-helix. Hydrophobic residues in a globular protein will fold into the protein's interior.

99. (E)

The enzymes that metabolize glucose in glycolysis would be stopped in their tracks by a compound that is chemically like glucose yet metabolically inactive. Compounds listed in choices (B), (C), and (D) would interfere in later parts of the glucose metabolic process.

100. (E)

Plant X must be a C4 plant. The enzyme used by most plants (C3 plants) to capture carbon dioxide for the Krebs cycle is called rubisco. Yet rubisco tends to accept oxygen rather than carbon dioxide when oxygen concentrations are high. This process, known as photorespiration, consumes oxygen and generates no ATP. However, C4 plants, known for living in hot and dry climates, use the enzyme PEP carboxylase, which has a much higher affinity for carbon dioxide than rubisco. This allows these plants to continue accepting carbon dioxide, and hence photosynthesizing, long after oxygen concentrations rise around them.

101. (A)

Light hitting the photosystems in the thylakoid membranes does provide the energy to produce ATP (a process called photophosphorylation, since the ATP is regenerated from ADP + P_i in the presence of light). Yet, this only indirectly involves the actual photosystem. The photosystems themselves are directly involved with process such as the splitting of water to fill electron holes, receiving the light energy from the sun, and releasing oxygen. In addition, chlorophyll a is present in both photosystems. Thus, the best answer here is that photophosphorylation is not directly associated with photosystem II.

102. (E)

It is true that the Krebs cycle generates ATP by substrate-level phosphorylation only, meaning that ADP is phosphorylated into ATP without the use of oxygen as happens in the ETC. All other statements here are incorrect regarding cellular respiration: anaerobic respiration likely preceded aerobic due to its relative simplicity; no biosynthetic reactions, including fermentation, can occur with the help of enzymes; two molecules of CO_2 are produced in one turn of the TCA cycle; and, fatty acids are oxidized into 2-carbon derivatives for entry into the Krebs cycle.

103. (B)

Substrates will not bind into active sites via peptide bonds. The amino acids that make up the active site are already bonded to one another via peptide bonds and there are no more places on their structure for another bond of that type to form. The bonds that do form occur between R groups sticking out into the active site and are of an ionic (hydrophobic, hydrophilic) nature or a covalent nature that does not involve amino and carboxyl groups (i.e., a peptide bond).

104. (E)

Identical twins arise when the developing egg splits into two embryos. Usually, this occurs at a very early stage, perhaps the 8-cell stage or earlier. In this case, the two embryos have exactly the same genetic content since they have both arisen from a single fertilization event. In contrast, fraternal twins (which can be of opposite sexes unlike identical twins) arise when two different sperm fertilize two different eggs, and the resulting embryos are no more genetically identical than are two siblings born at different times.

105. (B)

A porphyrin is any of a class of heterocyclic compounds, often found in biological pigments, that have four fused rings and nitrogen atoms. When their derivatives are combined with proteins and metal ions, the resulting compounds include the hemoproteins (e.g., hemoglobin, cytochromes, and chlorophyll). Although you may have been unfamiliar with this biochemical structure, you should have been able to eliminate (C) pyrimidines, and (D) histones, very quickly since these structures are associated with nucleic acids.

Phosphorus is not found in proteins, and you know hemoglobin is a protein, so (E) is also wrong. Choice (A) could also have been eliminated if you knew that the main atom holding the rings of chlorophyll together is magnesium, not iron (as it is in the case of hemoglobin).

106. (C)

Whether or not you remember that the enzyme used to convert pyruvate into acetyl-CoA at the mitochondrial membrane is called pyruvate *decarboxylase*, you should recall that this is the first reaction in cellular respiration that produces carbon dioxide as a waste product. More is produced later in the Krebs cycle. Reactions that remove carbon dioxide from a substrate are generally known as decarboxylation reactions. Aminations, choice (A), add amino groups ($—NH_2$); hydrolysis splits up complex biomolecules by the addition of water; and, phosphorylation adds phosphorus to molecules

107. (A)

The cnidarians, characterized by the jellyfish (and corals), possess stinging cells known as nematocysts on their tentacles. These cells discharge powerful compounds that can incapacitate marine organisms and even kill humans on occasion. If you missed this question, you may wish to refamiliarize yourself with the chart of animal phyla and their characteristics in Section III.

108. (C)

The fact that some modern-day bacteria live outside eukaryotic cells tell nothing of the origin of eukaryotic organelles such as the mitochondria and chloroplasts. All other statements listed accurately describe examples that support the endosymbiont theory, which proposes that many eukaryotic organelles were once their own free-living prokaryotic species.

109. (D)

The sieve-tube members are cells that make up the phloem of plants. These cells load the sucrose made in the leaves into the phloem using a chemiosmotic gradient and active transport similar to that used in the kidneys. However, the main difference in plants is that this active transport is via a sucrose-H^+ symport pump, while the kidneys use a Na^+-glucose symport pump.

110. (B)

Most biologists agree that the words "reproductively isolated" are a key aspect of the definition of a species: a group of interbreeding organisms that are reproductively isolated from other populations around them. Yet biologists still disagree about how to properly define a species. Given the two quotations in the question stem, it should be clear that one biologist (Lankester) refers to species as a type of organism, while the other sees them as reproductively isolated units and as separated populations.

111. (B)

This statement was said by Gregor Mendel, famed father of modern genetics, in referring to the finding that led to the establishment of a law we now know as the "law of independent assortment." The law states that alleles for different traits are passed on from parent offspring independently of one another. Because of Morgan's later work, we know that this law is only partially correct and does not hold for linked genes that are found nearby on the same chromosome.

112. (D)

The highest primary productivity, or production of food from light, would occur in the tropical rainforest because of its year-round growing season, ample light, and huge plant biomass. No other biomes listed could achieve the same productivity due to shorter growing seasons or lower plant biomass.

113. (C)

Both wings and flippers must have attachment sites for tendons, which connect muscle to bone, and ligaments, which connect bone to bone. The reason for this is that these are structures that move a great deal and must have a complex muscle and bone foundation. Flippers and wings do not continue to grow indefinitely, and stop growing by adulthood (choice A). Also, only birds have hollow bones filled with air sacs for lightness during flight, so (D) is out, as is (E), because oxygen is diffused via capillary walls, not through arteries or veins.

114. (A)

Both sexual reproduction in plants and meiotic nondisjunction in animals can result in polyploidy, offspring that possess *more than two* full sets of chromosomes (diploid). Many plant species have "accidents' during cell division that result in polyploidy, and it is common for two different species of plant to create a polyploid hybrid (allopolyploidy) that can propagate itself asexually. Nondisjunction in animals (e.g., humans) can result in conditions such as Down syndrome (trisomy-21) or Edward syndrome (trisomy-18).

115. (C)

You learned back in the chapter on enzyme inhibition that folic acid is needed for nucleic acid biosynthesis. This vitamin is found in vegetables and whole grains and is also used for a variety of metabolic reactions involving amino acids. Vitamin B_6 is found in meats and vegetables and is needed for amino acid metabolism, while niacin is part of the coenzymes NAD^+ and $NADP^+$. Vitamin A is essential for healthy vision, and ascorbic acid (vitamin C) is used in collagen synthesis and as an antioxidant to combat damage due to oxygen radicals formed in certain cellular reactions.

116. (E)

Pheromones are sex attractants, usually hormones that are released by female or male organisms and act as communication signals between animals rather than within a single animal's body like regular hormones. Histones, transcription factors, and enhancers can be eliminated here because they all involve the nucleus and the question stem states that cell surface receptors bind to the bombykol molecules.

117. (C)

Paracrine signaling results when a secreting cell affects nearby target cells by the production and secretion of various compounds that act only locally. Here, the question stem strongly suggests that prostaglandin secretion is done in a paracrine, or local, manner. Recall that endocrine is secretion into the bloodstream, and exocrine is secretion into ducts that lead out of the body (e.g., sweat glands, digestive system, etc.).

118. (D)

Those with both A and B proteins on their red blood cell surfaces can be given blood from individuals that are A-type, B-type, or O-type without having an agglutination reaction typical of when incorrect blood types are mixed. The reason for this is that their bodies are already "sensitized" to A and B proteins, and their immune systems do not reject any blood type. Thus, they can be considered "universal recipients" for blood transfusions.

119. (B)

DNA transformation occurs when a cell picks up "naked" DNA from the outside without requiring any contact with other cells. However, when two bacterial cell cultures are mixed, such as the question stem describes, not only will some cells burst and die as a normal consequence of life, but also cells may exchange plasmids (although plasmids are generally poorly transformed). In either case, DNA has moved from a donor to a recipient, which is a more general definition of transformation. Transduction, choice (A) occurs when a *virus* infects another cell and inserts genetic material into that cell, so that choice is incorrect here.

120. (E)

In angiosperms, embryos within the seed form one (monocot) or two (dicot) seed leaves that absorb endosperm and start the growth of the new plant. The single cotyledon of a monocot is also known as the scutellum.

121. (D)

The pericarp is the thickened wall that surrounds the fruit of an angiosperm. The fruit begins to develop only after certain chemical changes that cause the ovary of the fruit to grow rapidly. As the fruit grows and matures, much of the rest of the flower that surrounds it deteriorates, eventually causing the fruit to fall to the ground.

122. (B)

The gametophyte is the form of the plant that produces gametes via mitosis, in contrast to the spore-producing sporophyte. These haploid gametes will fertilize each other to develop into a single sporophyte organism.

123. (C)

Gene flow is defined as the movement of genes into or out of a population, and it generally occurs due to the movement of organisms into or out of a population—immigration or emigration. Contrast this term with "gene pool," the sum of all the genes within a population at a particular time.

124. (A)

Gene amplification is the selective copying of certain DNA regions, resulting in the creation of multiple copies of particular genes. Many organisms perform gene amplifications during certain times of development in order to increase the amount of a particular protein that is made. Proteins controlling embryonic development, for example, may be secreted much more heavily during early development due to temporary gene amplifications.

125. (E)

Sensillae are the small hairs on the feet and mouthparts of insects. They are used primarily for taste reception and are analogous to our taste buds. Each hair contains chemoreceptor cells that extend dendrites out to the tip of the hair. Each chemoreceptor cell is sensitive to a particular taste varying from sugars and salts to proteins and fats.

126. (D)

The lateral line system is series of pores filled with mechanoreceptors along the sides of fishes, much like the inner ear of humans, that allow the fishes to detect movements nearby (including sound waves) and alert them to the direction and intensity of particular stimuli.

127. (C)

The organ of Corti is the hearing organ of the vertebrate ear. It is located within the cochlea of the inner ear and contains sensitive hair cells that are complexed with nerve cells and convey precise information about sound waves entering the ear. Each group of hair cells within the organ of Corti are sensitive within a particular range of sound frequencies.

128. (E)

The collective name for all types of white blood cells is leukocytes. The lymphocytes are specialized B and T cells, white blood cells involved in antigen recognition and binding as well as secretion of antibodies (in the case of B cells).

129. (A)

The erythrocytes are the red blood cells, each packed with up to 1,000,000 molecules of hemoglobin protein. Every hemoglobin molecule can carry up to four oxygen molecules around the body. Try to remember the prefix *erythro-* as meaning "red," which can help with other vocabulary: erythropoetin is a hormone released by the kidney to upregulate production of erythrocytes.

130. (B)

Lymphocytes are specialized B and T cells, a specific arm of the immune system that can bind to infected host cells and kill them, or secrete antibody proteins that bind to foreign cells and toxins in order to mark them for destruction.

131. (D)

Ethylene gas is responsible for fruit ripening. If anyone has ever told you to place unripened fruit in a paper bag until it gets more ripe, this technique works because it exposes the developing fruit to its own ethylene gas, causing it to ripen much faster even after it has been picked off the plant.

132. (B)

Auxins are the plant hormones responsible for phototropisms, the movements of plants caused by the direction of light. Auxins are secreted predominantly on the dark sides of plants, within the cells not exposed to direct light. Auxin secretion causes cells on the unexposed side of the plant to elongate and grow more rapidly, allowing the plant to bed toward the light source.

133. (A)

Sodium channels are open during depolarization, the period of time at the start of an action potential where the inside of the neuron becomes rapidly more positive. The action potential can cause a change in membrane voltage from approximately –70 mV (millivolts) at resting to approximately +50 mV or more as sodium channels open up due to a particular stimulus. These sodium channels rapidly close as potassium ones open and bring the membrane potential back down toward resting.

134. (C)

The absolute refractory period is the time during which sodium channels in a particular location on the axon membrane will not open, no matter how great the stimulus. It is a brief period lasting only milliseconds, but it forces the action potential to travel in one direction only—down the axon toward the axon terminal. During the refractory period, the membrane potential in areas of closed sodium channels is often much less than that at resting.

135. (A)

The resting membrane potential of most neurons is approximately –70 mV, depicted here on the graph by regions 1 and 5. Be careful of the word "initial" in the question stem. Number 5 also represents a time period during which the neuron is at the resting potential, yet region number 1 is the initial resting potential.

136. (C)

Glycine is an amino acid, and two are attached here, recognizable by the C-N peptide bond and the presence of amino (—NH_2) and carboxyl (—COOH) groups at opposite termini of the molecule. The molecule pictured is, in fact, a dipeptide, two amino acids attached together.

137. (A)

Steroid hormones are able to diffuse directly through the lipid membrane of cells in order to exert transcriptional control over DNA in the nucleus. Steroids can be recognized by their four fused-ring structure, and the molecule drawn is cholesterol, a steroid although not a steroid hormone.

138. (B)

Single-stranded genetic material used in the process of translation is RNA, so look for ribose sugar among the answer choices. Recall that ribose sugar, unlike deoxyribose, has —OH groups hanging off both the 2' and 3' carbons in its ring structure.

139. (A)

Estrogen and progesterone are both steroid hormones and share the four fused-ring structure common to steroids. Although this is not the case with all steroids, progesterone and estrogen happen to have *steroid*-sounding letters within their names, which should help you remember that they are indeed steroids.

140. (D)

Molecule D here is a disaccharide and will be destroyed by disaccharidases within the small intestine. The bond formed between sugars such as these is called a glycosidic bond, and it occurs in both an alpha form and a beta form that can determine the overall structure of long chains of attached sugars.

141. (A)

The term monophyletic means that all members of a particular evolutionary group have arisen from a single ancestor. This ancestor will have given rise to no other organisms in other taxa. The term is useful in cladistics, as certain groups of organisms can be placed together in an evolutionary (phylogenetic) tree based on shared characteristics.

142. (E)

Analogous structures are those that serve a similar purpose in two evolutionarily unrelated species. Examples could include the wings of an insect and the wings of a bat, or similarly placed spines in unrelated plant species in order to repel predators. It is thought that nature has a limited number of solutions to survival problems faced by organisms, and that those organisms in similar environments (even if they are entirely unrelated) will often evolve through natural selection the same types of coping mechanisms and structures.

143. (D)

In contrast to analogous structures, homologous ones derive from a shared common ancestor. While homologous structures have a similar underlying foundation (e.g., the bone structure of a bird's wing and a human's hand), they may be used in completely different ways. Highly conserved sequences of DNA in developmental genes, meaning sequences that are nearly identical when compared nucleotide by nucleotide, are indicative of homology.

144. (A)

The flatworms and other platyhelminthes are known for their brain and nerve trunks that split off from the brain to travel down the length of their bodies.

145. (D)

Although Echinodermata is a complex group developmentally (sharing deuterostome status only with the chordates), they are radially symmetric in adult stages. Think about sea stars (starfish) or sea urchins, which shoot spikes or arms out from a central cavity. In such a way, they possess a central nerve ring (analogous to a central nerve cord in chordates) with radial nerves.

146. (B)

When you see the term "nerve net," think Hydra or jellyfish, the phylum Cnidaria. These simple creatures possess no central nervous system such as a brain or central nerve cord.

147. (E)

The annelids possess both a brain and a central nerve cord containing ganglia. The word *segmented* should be a tip-off here.

148. (D)

Deuterostomes, which show both radial cleavage and the development of the anus from the blastopore, include both chordates and echinoderms. Because chordates is not a choice here, phylum Echinodermata must be the answer.

149. (B)

The notochord is the tissue in chordates around which the bony vertebral column develops. Yet, pieces of the notochord remain between these bony vertebrae to cushion them so that they do not rub against each other and get damaged. Although the vertebrae are bone, the discs remain as cartilage and never turn into bone.

150. (C)

Hinge joints such as the knee and elbow allow for flexion (bending inward) and extension (bending outward). As the name implies, hinge joints act as hinges (like on a door) to allow movement in two directions.

151. (B)

The answer can be found by calculating 1/(2N) for the effective population size described in the question stem: 200. 1/(2*200) = 1/400, or 0.25%.

152. (C)

Every individual mole rat will be heterozygous or homozygous. Since the original parents had different alleles (hetero), and since heterozygosity has decreased by 1/2, then 1/2 of the population must be homozygotes and have identical alleles at this locus.

153. (D)

The best way to choose an answer here is to figure out what the tree looks like from the distances in the table. First, group species that have the least distance between them. You'll notice that there seem to be two classes of distances: 20 and larger, and 12 and smaller. And notice that the 12 and smaller ones only group 2 species each: E and F, C and D, and A and B. So these will be next to each other on a cladogram. The only tree that correctly depicts relationships based on the distances in the chart is (D), cladogram IV.

154. (D)

Scientists can measure the genetic relationships between species by analyzing protein sequences or DNA sequences, and it can be assumed that the elephants most closely related evolutionarily by tusk length differences are ones with the fewest nucleotide differences between them. Here, species E and F have a distance of 5 between them.

155. (A)

Paraphyletic taxons exclude other species that share a common ancestor. Because all elephant species in the chart can be assumed to share a common ancestor, especially if you picked the correct cladograms in question 153, a taxon with just C and D would be exclusive and paraphyletic.

156. (C)

This may be easy for some to "eyeball," meaning that trait class C should jump out as being the original trait class. Because all major branch points of the clade would need to have a C class ancestor in order for the clade to evolve as it has, this is the most likely choice. Alternatively, one could go to each common ancestor and assign a trait class by having each descendant clade "vote" with its own state. If the two descendants of a common ancestor, for example, are A and B, the situation is ambiguous. But if you go up to the next common ancestor and you have A/B (ambiguous) and B, then A gets 1/2 vote and B gets 3/2 votes and the situation gets resolved. State C emerges as the most likely candidate.

157. (B)

The best estimate would be the one in which class A evolved independently the fewest number of times. Keep in mind that however many times you have A evolve impacts how many times the others evolve. So one might say that A evolved only once at the top of the tree and that was it, but that would force B and C to evolve many times. From the previous question, you have already seen that C was the probable trait value of the ancestor, so A must have originated once on the right and once on the left. Any other combination of originations and changes to other states would involve more changes and therefore would be far more complex.

158. (C)

In Panel A it is shown that IL-1α mRNA levels remain constant in all cell lines (unlike the levels of IL-1β, which vary among the cell lines.) However, as depicted in Panel B the protein levels vary greatly among the different cell lines. This suggests that IL-1α regulation is occurring at the level of the protein, not the mRNA—in other words, at the level of translation rather than transcription. Choice (A) is incorrect because neither Panel A nor Panel B refer to DNA replication. Choice (D) is incorrect, because Panel B clearly shows regulation in various cell lines, and choice (E) is wrong since there is no evidence that IL-1α has any interaction with IL-1β.

159. (C)

In Panel B, cell line C clearly exhibits the highest expression of IL-1α. Thus, if the rate of tumor growth is positively associated with IL-1α growth, cell line C would exhibit the most rapid growth.

160. (A)

The theory behind antisense therapy is that the antisense cDNA will bind to the sense (also known as the coding strand) mRNA of the targeted gene thus creating a cDNA-mRNA doubled-stranded heteroduplex. Such a structure is not an appropriate template for translation and should prevent the mRNA from being translated into protein. (C) is incorrect because the expression of the "sense" cDNA might have one of two effects: first, its expression could result in the production of more IL-1α. Second, if it "sense" cDNA were to bind to the antisense mRNA, which doesn't code for the protein, levels of IL-1α would not be affected. Choice (E) is also out since the destruction of all cells producing this cytokine (assuming one could accomplish this unlikely feat), which is most likely needed in certain concentrations in particular areas, could cause drastic problems in cell-to-cell communications.

161. (C)

In the question stem, it is stated that the levels of the GADPH mRNA remain constant under all conditions and that GADPH is expressed constitutively, meaning all the time. Given this, it is unlikely that the researchers were interested in the role of GADPH in cancer or as a comparison with IL-1α. It is also stated in the question that the levels of GADPH have been examined many times, so we can exclude (B). Genes such as GADPH, which exhibit constitutive expression, are routinely used in Northern analysis as controls for loading. Since its expression remains constant, the researcher can verify that differences in the expression of other genes, such as IL-1β, are not due to differences in total mRNA added in different lanes.

162. (B)

Unlike IL-1α, IL-1β exhibits great variety in expression in the different cell lines. For example, it is not detected at all in cell line A, but it exhibits very high expression in cell line D. GADPH levels do not vary, so we can be assured that the differences in expression from cell line to cell line do indeed reflect the levels of the IL-1β mRNA. Since mRNA is produced during transcription, we can assume that there is cellular control occurring at this level. Though there may be other levels of regulation occurring, we are not given any additional data to support this, so we must choose B.

163. (A)

This question requires only that you can correctly interpret the graph. Tumor volume is shown on the Y-axis, and the greatest tumor volume is exhibited by the line containing the triangles, which we can see in the figure legend is the HT-15 cell line.

164. (C)

The highest tumor growth is exhibited by the HT-15 cell line, which overexpresses superoxide dismutase. We are told in the question stem that superoxide dismutase converts superoxide to hydrogen peroxide. Thus, we can conclude that the greatest tumor growth is due to the removal of superoxide by superoxide dismutase, or the presence of hydrogen peroxide.

165. (C)

It is indicated in the passage that SCID mice lack B and T-cells, so they are immunodeficient. It is also indicated that the cells injected into the mice were of human origin. Taken together, the answer that is most likely is C. Wildtype mice possessing a normal immune system would reject the human tumors and no tumors would grow. The figure gives no indication about the popularity of any strain of laboratory mice, nor can you make assumptions about tumor formation in wildtype (healthy) mice, except to say that CMV will grow without rejection in the SCID mice.

166. (D)

This question is analogous to the scenario represented in the last row of table 1. In both cases, a particular strain is sensitized to another strain, but is then given a graft from a third strain. Since that strain has never been exposed to this third strain, it will mount a slow, first set rejection to the graft. It can only mount a second set rejection if this is its second time being exposed to the same kind of tissue.

167. (D)

Substances present in body fluids mediate humoral immunity. These substances include serum proteins called antibodies that can be transferred in a cell-free serum injection. In order to demonstrate that humoral immunity plays a role in transplant rejection, one must show that the transfer of serum from a sensitized animal into one never before exposed will transfer immunity against a particular graft. Choice D clearly supports the humoral model.

168. (E)

This question draws on your knowledge of the difference between the specific and nonspecific defenses of the body. Nonspecific defenses include physical and chemical barriers, the inflammatory response, and widely released chemical such as cytokines. Physical barriers include the intact skin and mucous membranes. These barriers are aided by various microbe-catching fluids such as saliva and mucous, choice (E).

169. (A)

T-cells mediate rejection of foreign grafts by recognizing foreign-MHC molecules and then lyse the infected cells on which these molecules are displayed. If T-cells bind to self-MHC molecules, this can result in lysis of the individual's own cells. In fact, failure to eliminate self-reactive lymphocytes is believed to be the basis for many autoimmune diseases.

170. (A)

This mating represents a cross between a homozygous recessive strain of Ob1 mice and a homozygous dominant wild-type strain. The cross between LL and ll individuals will result in all Ll offspring, thus no obese mice can be expected in the progeny.

171. (B)

Ob1 mice respond nicely to administration of leptin. Their food intake rapidly drops over the 30 day course of the drug. This is suggestive of the fact that they are not producing leptin themselves, yet their cells retain the ability to respond to it when it is in their bloodstream. (B) is the best answer here. Leptin exerts its effects by binding to receptors in the hypothalamus and activating a series of reactions that decreases appetite, metabolic rate, and blood insulin level.

172. (D)

Both figures in the passage show that the Ob2 strain of mice did not respond to daily administrations of leptin. This could be explained of the Ob2 strain had a defect in the gene coding for the leptin receptor. If not, the Ob2 mice would likely respond to increased levels of leptin in their bloodstream much as the Ob1 strain did.

173. (E)

In order for levels of neuropeptide Y to be reduced, mice must produce functional leptin and be capable of responding to it. Since the Ob1 strain is defective in the former and the Ob2 strain is defective in the latter, both strains will have elevated levels of neuropeptide Y.

174. (C)

Again, leptin promoted weight loss in rodents without functional leptin proteins, but not in rodents without functional leptin receptors. All others should (and do, according to the data in the figures) respond to leptin administration by losing weight.

175. (C)

After cutting with Sal1, you would expect two bands of DNA to appear on the gel, one approximately 1600 bp in length and the other approximately 800 bp in length. Notice where the cutting sites are for this enzyme. If Sal1 cuts the plasmid, it cuts in two places: leaving one segment of DNA that is 1600 bp long (moving clockwise from 8 o'clock, add the distance between Sal1 and EcoR1, plus the distance from EcoR1 to BamH1, plus the distance from BamH1 to the next EcoR1 site, and finally the distance to the other Sal1 site); and, leaving another segment of 800 bp.

176. (B)

By the same method as above, digestion with EcoR1 at its three cutting sites, should leave you with three distinct bands, one 1000 bp long, another 900 bp long, and the last 500 bp long.

177. (D)

Cutting with all three enzymes will leave you with four small bands: 600 bp, 500 bp, 300 bp, and 200 bp. Although there will be two segments that are 500 bp long, the DNA will collect at the same place within the gel because both segments are the same size even though they do not have the same sequence necessarily.

178. (E)

It is clearly important to be able to find the cells which have taken up the recombinant plasmid or other vector so that you can work with these cells directly. However, reporter genes must code for proteins not normally made by the transformed cells. Otherwise these reporter genes would not be "reporting" the presence of a cloning vector within cloned cells. Reporter genes are best off producing proteins such as fluorescent markers, that will show up easily and not be diluted by normal protein products.

179. (E)

A Northern blot is an assay for mRNA, not for protein. Recall that the information presented told you that the gene was sometimes transcribed without being translated. According to the Northern blot, this Drosophila gene is transcribed from the egg stage all the way through 12 hours of development.

180. (C)

Transcription factors are regulatory proteins that bind to DNA and stimulate transcription of certain genes. As homeotic (you may also see them called homeobox) genes are transcribed, homeodomain proteins are produced that can act as transcription factors to turn on or off genes involved in cell differentiation and morphogenesis. Enhancers are upstream genetic elements that can act to "enhance," or activate, the transcription of a particular gene, yet enhancers are DNA regions themselves, not proteins. Promoters, too, are DNA sequences that are placed immediately adjacent to a structural gene and provide a place of attachment for RNA polymerase. MHC, or major histocompatibility proteins, are cell-surface proteins and do not interact directly with DNA as transcription factors do.

181. (C)

During very early development, the homeotic gene is being transcribed but is not being translated. The Western blot shows this as an absence of bands from the egg stage up through the three-day stage, while the Northern blot shows bands of mRNA formation during that same period of time. This suggests strongly that the cell's control over the expression of the protein is taking place just prior to translation, and has little or nothing to do with transcriptional control or alternative gene splicing. Although many homeotic genes are active only in certain regions of the organism (e.g., in the limb buds or in mesodermal tissue), there is no evidence in this problem to suggest either that the expression of this homeotic gene is spatially regulated or that it is expressed in a localized (cell- or tissue-specific) fashion. Therefore, choice (C) is best here.

182. (B)

Very simply here, the gene has become inactivated. According to the data in both the Western and the Northern assays, neither is the gene being transcribed nor the mRNA from the gene being translated. In many cases, homeotic genes are active only for short periods during development, although sometimes they may also regulate the development of adult structures much later on. However, there is no evidence to suggest that this homeotic gene will be turned on later to do that, so choice (C) must be eliminated. Choice (E) is incorrect because neither the Western blot nor the Northern blot shows bands of mRNA or protein expression past 24 hours of development.

183. (B)

When the probe is alone, it runs fully through the gel, as can be seen by the gel band at the far bottom. In addition, the probe runs through the entire gel when mixed with nerve cell proteins. However, in muscle cells, a protein is present that binds to the TFY binding site on the piece of DNA used as a probe. This makes the probe-plus-protein combination too large to make it fully through the gel, as shown by a mid-gel band in the muscle cell lane.

184. (E)

Western blots probe for the presence of a protein product. If the product is expressed in a particular location, a protein-specific antibody will detect it. In this case, TFY protein is found in both muscle and nerve cells as seen by the presence of bands in both the muscle and the nerve cell lanes on the gel.

185. (B)

TFY is detected in both muscle cell cytoplasm and nuclei (see bands on the Western blot), whereas it is detected only in the cytoplasm of nerve cells. Therefore, it is most likely that a factor in muscle cells allows TFY to translocate to the nucleus where it can bind DNA and induce transcription, yet this factor does not exist in nerve cells as shown by the lack of a band in the EMS assay.

186. (B)

Since the leftmost ZGNI is the one for *Asterionella* and the right ZGNI is that for *Synedra* and since each species preferentially consumes the resource it is more limited by (indicated by the resource consumption vectors), coexistance is possible for regions of the plot between the consumption vectors and above both ZNGIs. In fact, this question tests your graph-reading skills more than it tests your ecological knowledge, since point B on the graph represents the mid-point between the consumption vectors of both species, and therefore represents the best habitat for both to coexist.

187. (A)

According to the graphs on the right, *Synedra* can live at lower molar concentrations of silicate than the *Asterionella* (notice how the silicate graph on the right for *Synedra* dips to a lower minimum [2-4] when compared to the silicate minimum for *Asterionella* [10 mM] Therefore, as *Synedra* uses up the phosphate and silicate resources, it will be growing exponentially and the system will cross the *Asterionella* ZNGI first, causing the *Asterionella* population to decrease and die out. Yet, *Synedra* thrives until the resources available decrease to the *Synedra* ZNGI, where the population stabilizes. Another way to look at this question in the absence of ecological knowledge is that both populations are shown to increase exponentially and then plateau at an asymptote in all four graphs to the right. Thus, the expectation should be that any population introduced into the habitat would do the same thing. This leaves choice (A) the only logical choice here.

188. (A)

A synapomorphy is a derived character state shared by two or more groups of organisms (taxa) that most likely evolved in their common ancestor. Here, position 11 is a thymine (T) in species A and D, yet an adenine (A) in species B and C. The fact that A and D share the same nucleotide at this position, but B and C share a derived nucleotide at the same position, qualifies as a synapomorphy. Other vocabulary you might see on the exam includes: homoplasy—the possession of two or more species of a similar character state not

derived from a common ancestor (think: convergent evolution); symplesiomorphy—a shared ancestral characteristic. If all taxa here (A through D) had the same nucleotide in position 11, this might be considered a symplesiomorphy, yet the word is usually used for full traits (e.g., possession of hair in the ancestor of kangaroos and whales). In contrast, an autapomorphy is a unique characteristic of a group of organisms in a cladogram, one that no other organisms in that phylogenetic tree share.

189. (D)

A group that is defined by the descendants of a common ancestor but omits some descendants of that common ancestor is called a paraphyletic group. In contrast, a monophyletic group consists of all species derived from a certain common ancestor, and a polyphyletic group consists of members that do not al share the same common ancestor.

190. (E)

Island B is further from the mainland, because it receives fewer organisms than Island A—a lower immigration rate — thus, choice (A) is incorrect because it is true. The age of islands is unlikely to have an effect on the immigration rate, so choice (B) should be ruled out as well. Island B's immigration curve is much lower than Island A's, so choice (C) is incorrect because it is true. In choice (D), elongate islands oriented as stated would lead us to predict island A to have the higher immigration rate, an incorrect answer because it is likely true. In choice (E), the fact that the islands are very dissimilar in their diversity and quality of habitats may alter the immigration rates as shown, but would also be expected to result in different extinction rates, not seen here.

191. (D)

Differences in the colonizing ability of different species are reflected by a nonlinear immigration curve, because if certain species are better colonizers than others, all species that arrive on the island are not as likely to be successful there. Thus, immigration curves do not rise indefinitely in linear fashion, as seen here, so choice (A) is incorrect. No data is present as to the nature of dispersal so choice (B) must be eliminated. Island A may have a higher immigration rate than island B because of larger size or it may be closer to the mainland, but the curves do not tell which is the case, so choice (C) is incorrect. The presence of competition and, therefore, increased extinction probability increases with greater numbers of species present. This is reflected by a nonlinear extinction curve as shown, so choice (D) is correct. If the islands were so dissimilar, as mentioned above their extinction curves would likely differ, which they do not; thus, choice (E) is incorrect as well.

192. (B)

At equilibrium, the intersection of the immigration curve and extinction curve for each island reveals the number of species present on the horizontal axis and the species turnover on the vertical axis (for each island). Only choices (B) and (C) appear correct at first glance. However, choice (C) is incorrect because when island A is at equilibrium, there is no guarantee that island B is as well, and the question stem asks only about island A.

193. (A)

The evolution of new species from a common ancestor as the populations spread out into new niches is known as adaptive radiation. This phenomenon commonly occurs on newly formed volcanic islands that arise near a mainland.

194. (D)

The process of elimination may be the best strategy to use for this question. The northern California frequencies sum to less than 100%, which indicates that other alleles of *abc* are present but not indicated in the graph. This effectively eliminates choice (A). No data on behavior is given at all, eliminating choice (B). Frequencies of alleles in populations can be due to natural selection, genetic drift, or both; yet, to determine which one is responsible requires more data than a single frequency measurement, ruling out choices (C) and (E). Fewer specimens collected would lead to larger standard errors in the frequency estimates, as the southern California data shows, so choice (D) is correct.

195. (B)

This question asks for what the data is suggestive of, a looser constraint than the previous question. All of the factors given may be consistent with the pattern observed; however, the north to south pattern correlating with the frequencies suggests a geographic explanation and makes choice (B) the best answer. Because temperature gets warmer as one travels south, and perhaps drier into Mexico, variations in populations are likely to parallel this environmental gradient—the definition of an ecological cline. The lack of other information weakens the arguments for choices (C) and (D), which cannot explain a difference in frequency across all three subspecies of *Drosophila*. In addition, it is not clear that choices (A) or (E) could explain the consistent pattern seen in the graph.

196. (D)

There is not enough information in the data to know whether or not intermediates between California and Mexico will exhibit higher ^{H}abc or ^{C}abc allele frequencies than simply an average between the two populations. In most hybrid zones, all forms of intermediate organisms can be found. While the discrepancy between the ^{H}abc or ^{C}abc alleles does increase markedly in the Mexico population as compared to the southern California one (typical of a cline), it is not clear that hybrids would have any other allele frequencies than an intermediate frequency between the two populations, given the small amount of information in the question. Thus, answer choice (D) is best here.

197. (A)

A lower fitness of hybrids would act as a barrier to the flow of alleles between the two populations because mating between organisms of the two groups of *Drosophila* would not lead to successful offspring, which would allow genes to effectively flow from north to south or vice-versa. In addition, if there is specific selection against either the ^{H}abc or ^{C}abc allele, any alleles linked to these alleles will also be affected, which should make sense—alleles linked to other alleles being selected against will themselves be selected against. Hybrid zones with organisms of relatively low fitness are not generally wide due to the genetic incompatibility between subspecies, so choice (C) can be eliminated, and it is unlikely that low fitness hybrids could be stable enough to form their own species (Choice D). As explained in the answer to question 196, almost all forms of intermediate organism may be found within a hybrid zone, so choices (B) and (E) should be eliminated as well.

198. (B)

Species C and D do not interact directly and the only indirect interaction listed in the answer choices is exploitative competition, choice (B). Both species deprive each other of the benefit of resource F, a species that both C and D benefit from (most likely because they depend on F for food). Just as commensalism means an interaction between species where one benefits and the other species is unaffected, amensalism is an interaction in which one species is harmed and the other species is unaffected. Answer choice (E) should be eliminated, then, as species C and D clearly have no direct relationship.

199. (C)

Absence of B is bad for A (loss of prey), good for C (loss of predator), good for D (loss of predator), indirectly bad for E (more C individuals will adversely affect E), and indirectly bad for F (more C and D individuals will prey on F). Thus, choice (C) is the only possible answer here.

200. (A)

Species E is involved in an amensal (one way negative) interaction with species C and a mutual (two way beneficial) interaction with species F. Remember that amensalism means that one species is harmed while another is unaffected. An example of amensalism is the crab that unearths many burrowing animals in its quest for clams and mollusks to eat. While the crabs do not eat these worms and other burrowers, the burrowers must expend a great deal of energy to rebury themselves after being excavated. Hence, they suffer while the crabs benefit. This interaction can also be considered interference competition.